An Introduction to the Finite Element Method for Differential Equations

An Introduction to the Finite Element Method for Differential Equations

M. Asadzadeh

WILEY

This edition first published 2021.

Registered Office
John Wiley & Sons, Inc., 111 River Street, Hoboken, NJ 07030, USA

Editorial Office
111 River Street, Hoboken, NJ 07030, USA

For details of our global editorial offices, customer services, and more information about Wiley products visit us at www.wiley.com.

Wiley also publishes its books in a variety of electronic formats and by print-on-demand. Some content that appears in standard print versions of this book may not be available in other formats.

Library of Congress Cataloging-in-Publication Data

Names: Asadzadeh, M., author.
Title: An introduction to the finite element method for differential equations / M. Asadzadeh.
Description: Hoboken, NJ : Wiley, [2020] | Includes bibliographical references and index.
Identifiers: LCCN 2020008313 (print) | LCCN 2020008314 (ebook) | ISBN 9781119671640 (cloth) | ISBN 9781119671671 (adobe pdf) | ISBN 9781119671664 (epub)
Subjects: LCSH: Finite element method. | Differential equations.
Classification: LCC TA347.F5 A83 2020 (print) | LCC TA347.F5 (ebook) | DDC 515/.35–dc23
LC record available at https://lccn.loc.gov/2020008313
LC ebook record available at https://lccn.loc.gov/2020008314

Cover Design: Wiley
Cover Image: Courtesy of: Mohammad Asadzadeh and Larisa Beilina

Set in 9.5/12.5pt STIXTwoText by SPi Global, Chennai, India

Contents

Preface

This book is an introduction to finite element methods (FEMs) used in the numerical solution of differential equations based on the piecewise polynomial approximations of the solutions. The presented material is accessible for upper undergraduates and starting graduate students in natural science and engineering. We mention three books for further and deeper study of FEM for differential equations.

Brenner, S.C. and Scott, L.R. The Mathematical Theory of Finite Element Methods. Springer, ed 3, 2017.

Ern, A. and Guermond, J.-L. Theory and Practice of Finite Elements. Springer, 2004.

Larsson, S. and Thomée, V. Partial Differential Equations with Numerical Methods. Springer, 2003.

The material is presented in three main theme.

(I) *Basic theory*: Chapters 1 and 2, contains introduction (and in some extend derivation) of basic ordinary and partial differential equations, their classifications, well-posedness (as the proof of Reisz and Lax–Milgram Theorems) and formulation of the corresponding initial- and initial-boundary value problems. The concept of fundamental solutions using Green's functions approaches are discussed and necessary mathematical tools and environment are introduced in some details.

(II) *One-space dimensional problems*: Chapters 3–7 concern the polynomial approximations, polynomial interpolation, quadrature rules (numerical integration), iterative numerical methods to solve linear system of equations and finite element procedure for the one-space dimensional boundary value problems (BVPs), initial value problems (IVPs), and initial boundary value problems (IBVPs).

(III) *Problems in higher (≥ 2) dimensional cases.* This part is the matter of Chapters 8–10 and is devoted to the generalization/extension of the results of Part II to higher dimensions. The proofs in higher dimensions do not, in general, require any additional ideas. More specifically, we have introduced the higher-dimensional interpolation procedure and study the stability and convergence aspects of certain finite element approximation for, higher-dimensional, Poisson, heat, wave, and convection–diffusion equations. The convergence analysis are given both in the “a priori” (exact solution dependent) and “a posteriori” (computed solution/residual dependent) settings.

Whole or selected parts of the book is suitable as course material and for particular purposes. As outlined in table below, I have used some minor parts of Chapters 1 and 2 and the whole Chapters 3, and 5–7, together with a theory and a computer assignment, for a 7.5 credit points for undergraduates at Chalmers University of Technology. The whole book: Chapters 1–3 (a less pronounced cover of Chapters 5–7) together with Chapters 8–10 has been the course material for upper undergraduates and some graduates at Chalmers for more than two decades (since 1995). This part is associated with two, somewhat involved, theory and computer assignments and generates the same credit points 7.5, for a somewhat higher-level audience. There is a, 5 credit points, intensive course on Chapters 2–3 and 8–10 for some graduates. (According to Bologna unifying system now a full-time semester should correspond to 30 credit points.) Here, the first combination is accessible for the students with basic knowledge of calculus of single- and several-variables, linear algebra and some Fourier analysis. The extended combination requires more of mathematical tools (in Chapter 2) and can be of interest for beginning graduates in applied math and engineering disciplines.

To conclude, the theory combined with approximation techniques and computer projects can give a better understanding of this useful tool (FEM) to solve differential equations. Finally, there are some easily implementable Matlab codes presented at the end of the book that are useful for freshmen in finite elements to test and check the theory through implementations.

Suggestions for possible course syllabus (what I have had)

Chapters/Sections	1 Semester	7–8 wk	Credits
1.1–1.4, 2.5, 2.6, **3, 5–7**	3 h/wk	6 h/wk	**5**
1.1–1.4, 2.5, 2.6, **3, 5–10**	4 h/wk	8 h/wk	**7**
except 9.2, 9.3, 10.2.3–10.2.4, 10.5.4–10.5.6			
Whole material (includes Chapter **4**)	6 h/wk	10h[a]/wk	**10**

a) An intensive course for graduates in applied math/engineering.

All above configurations are associated with home and computer assignments. Examples of some assignments are given in Appendix C.

Acknowledgments

Many colleagues have been involved in the design, presentation, and correction of the material in this book. I wish to thank Niklas Eriksson and Bengt Svensson who have read the entire manuscript and made many valuable suggestions. Niklas has contributed to a better presentation of the text as well as to simplifications and corrections of many key estimates that has substantially improved the quality of the book. Bengt has made all *xfig* figures. The final version is further polished by John Bondestam Malmberg and Tobias Gebäck who, in particular, have many useful input in the Matlab codes. Many others have been involved in teaching and/or assisting me in teaching whole or parts of the content of the book as well as helping in design of computer assignments. They have supplied invaluable feedback and raised the quality of this work. Here are some (far from all): Fredrik Benzon, Christoffer Cromvik, Tommy Gustafsson, Kristin Kirchner, Fredrik Lindgren, Anders Logg, Fardin Saedpanah, Christoffer Standar, Maximilian Thaller. I owe all of them and the un-named supporting colleagues my most sincere gratitude.

Göteborg, Sweden
August 2019

Mohammad Asadzadeh

1

Introduction

There are two ways of spreading light:
to be the candle or the mirror that reflects it.

Edith Wharton

This book presents an introduction to the Galerkin finite element method (FEM) as a powerful and general tool for approximating solution of differential equations. Our objective is twofold.

i) To present the main ordinary and partial differential equations (ODEs and PDEs) modeling different phenomena in science and engineering and introduce mathematical tools and environments for their analytic and numerical studies.
ii) To construct some common FEMs for approximate solutions of differential equations and analyze their well-posedness (existence, uniqueness, and stability of such approximate solutions) as well as the accuracy of the approximation.

In its final step, a finite element procedure yields a linear system of equations (LSEs) where the unknowns are the approximate values of the solution at certain nodes. Then, an approximate solution is constructed by adapting piecewise polynomials of certain degree to these, approximate, nodal values.

The entries of the coefficient matrix and the right-hand side of FEM's final LSEs consist of integrals which, e.g. for complex geometries or less smooth, and/or more complex, data, are not always easily computable. Therefore, numerical integration and quadrature rules are introduced to approximate such integrals. Furthermore, iteration procedures are included in order to efficiently compute the numerical solutions of such obtained LSEs.

An Introduction to the Finite Element Method for Differential Equations, First Edition.
Mohammad Asadzadeh.

Interpolation techniques are presented for both accurate polynomial approximations and also to derive basic a priori and a posteriori error estimates necessary to determine qualitative properties of the approximate solutions. That is to show how the approximate solution, in some adequate measuring environment, e.g. a certain norm, approaches the exact solution as the number of nodes, hence, the number of unknowns increases. For convenience, the frequently used classical inequalities, such as the *Cauchy–Schwarz'* and Poincare, likewise the *inverse and trace estimates*, that are of vital importance in error analysis and stability estimates, are introduced. In the theoretical abstraction, we demonstrate the fundamental solution approach based on Green's functions and prove the Riesz (Lax–Milgram) theorem which is essential in proving the existence of a unique solution for a minimization problem that in turn is equivalent both to a variational formulation as well as a corresponding boundary value problem (BVP).

Galerkin's method for solving a general differential equation is based on seeking an approximate solution, which is

1. Easy to differentiate and integrate
2. Spanned by a set of "nearly orthogonal" base functions in a finite-dimensional vector space.
3. Satisfies *Galerkin orthogonality* relation.
   Roughly speaking, this means *a closeness relation* in the sense that:
   I). In a priori case, the difference between the exact and approximate solution is orthogonal to the finite dimensional vector space of the approximate solution.
   II). In a posteriori case, the *residual* of the approximate solution (=the difference between the left- and right-hand side of an expression obtained from the differential equation where exact solution is replaced by the approximate solution) is orthogonal to the finite dimensional vector space of the approximate solution.

1.1 Preliminaries

In this section, we give a brief introduction to some key concepts in differential equations. A standard classification and some general properties are presented in Trinities below.

- A *differential equation* is a relation between an unknown function u and its derivatives.
- If the differentiation in the equation is with respect to only one variable, e.g. $x \in \mathbb{R}$, $t \in \mathbb{R}^+$ (or x_1 in $(x_1, x_2, \ldots, x_n) \in \mathbb{R}^n$), then the equation is called an *ordinary differential equation* (ODE).

Example 1.1 As a simple example of an ODE, we mention the population dynamics model

$$\frac{du}{dt}(t) - \lambda u(t) = f(t), \quad t > 0. \tag{1.1.1}$$

If $f(t) \equiv 0$, then the equation is called *homogeneous*; otherwise, it is called *inhomogeneous*. For $\lambda > 0$, the homogeneous equation $\frac{du}{dt}(t) - \lambda u(t) = 0$ has an exponentially growing analytic solution given by $u(t) = u(0)e^{\lambda t}$, where $u(0)$ is the initial population. On the contrary, $\lambda < 0$ yields a population that vanishes (dies out) with time.

- The *order* of a differential equation is the order of the highest derivative of the function u that appears in the equation.
- If the function u depends on more than one variable, and the differential equation possesses derivatives with respect to at least two variables, then the differential equation is called a *partial differential equation* (PDE), e.g.

  $$u_t(x,t) - u_{xx}(x,t) = 0,$$

  is a *homogeneous* PDE of the second order, whereas for $f \neq 0$, the equations

  $$u_{xx}(x,y) + u_{yy}(x,y) = f(x,y),$$

  and

  $$u_t(x,y,t) - u_{xx}(x,y,t) - u_{yy}(x,y,t) + u_x(x,y,t) + u_y(x,y,t) = f(x,y,t),$$

  are *nonhomogeneous* PDEs of the second order.
- A solution to a differential equation is a function; (e.g. $u(x)$, $u(x,t)$, $u(x,y)$, or $u(x,y,t)$ above), which satisfies the corresponding differential equation.
- In general, the solution of a differential equation cannot be expressed in terms of elementary functions, and numerical methods are the only way to solve the differential equations through constructing *approximate solutions*. Then, the main questions are
  To what extent does the approximate solution preserve the physical properties of the exact solution, or satisfies a corresponding, discrete, version of the differential equation (*consistency*)?
  How sensitive is the solution to the change of the data (*stability*)?
  How close is the approximate solution to the exact solution (*convergence*)?
  Which are the adequate environments to measure this *closeness*?
- These are some of the questions that we want to deal within this text when approximating with the FEMs.
- A linear ODE of order n has the general form:

  $$L(t,D)u = u^{(n)}(t) + \sum_{k=0}^{n-1} a_k(t)u^{(k)}(t) = b(t),$$

where $D = d/dt$ denotes the derivative, with respect to t, and $u^{(k)} := D^k u$, with $D^k := \frac{d^k}{dt^k}$, $1 \leq k \leq n$ (the k-th order derivative). The corresponding *linear differential operator* is denoted by

$$L(t, D) = \frac{d^n}{dt^n} + a_{n-1}(t)\frac{d^{n-1}}{dt^{n-1}} + \cdots + a_1(t)\frac{d}{dt} + a_0(t).$$

1.2 Trinities for Second-Order PDEs

Problems modeled by PDEs of the second order can be classified using, the so-called, trinities. Below we introduce basic ingredients of this concept. For detailed study see, e.g. [68].

The usual three operators in PDEs of second order in $\mathbb{R}^n$.

$$\text{Laplace operator } \Delta_n := \frac{\partial^2}{\partial x_1^2} + \frac{\partial^2}{\partial x_2^2} + \cdots + \frac{\partial^2}{\partial x_n^2}, \tag{1.2.1}$$

$$\text{Diffusion operator } \frac{\partial}{\partial t} - \Delta_n, \tag{1.2.2}$$

$$\text{d'Alembert operator } \Box := \frac{\partial^2}{\partial t^2} - \Delta_n, \tag{1.2.3}$$

where we have the space variable $\mathbf{x} := (x_1, x_2, x_3, \ldots, x_n) \in \mathbb{R}^n$, the time variable $t \in \mathbb{R}^+$, and $\partial^2/\partial x_i^2$ denotes the second partial derivative with respect to x_i, $1 \leq i \leq n$. We also define a first-order operator, namely the *gradient operator* ∇_n which is the vector valued operator

$$\nabla_n := \left(\frac{\partial}{\partial x_1}, \frac{\partial}{\partial x_2}, \ldots, \frac{\partial}{\partial x_n}\right).$$

Often, the dimension is obvious from the context and therefore, usually, the subindex n is suppressed and the operators Δ_n and ∇_n are simply denoted by Δ (or by ∇^2) and ∇, respectively.

Classifying general second-order PDEs in two dimensions.

I) **The constant coefficients case**

A second-order PDE in two dimensions, with constant coefficients, can be written in its general form as

$$Au_{xx}(x,y) + 2Bu_{xy}(x,y) + Cu_{yy}(x,y) + Du_x(x,y) + Eu_y(x,y) + Fu(x,y) + G = 0.$$

We introduce the *discriminant* $d = AC - B^2$: a quantity that specifies the role of the coefficients of the second-order terms, in determining the equation type in the sense that the equation is

Elliptic: if $d > 0$, *Parabolic*: if $d = 0$, and *Hyperbolic*: if $d < 0$.

Example 1.2 Below are the classes of the most common, second-order differential equations (in any dimension) and examples of their forms in 2*d*:

Potential equation	Heat equation	Wave equation
$\Delta u = 0$	$\dfrac{\partial u}{\partial t} - \Delta u = 0$	$\dfrac{\partial^2 u}{\partial t^2} - \Delta u = 0$
$u_{xx} + u_{yy} = 0$	$u_t - u_{xx} = 0$	$u_{tt} - u_{xx} = 0$
$A = C = 1, B = 0$	$B = C = 0, A = -1$	$A = -1, B = 0, C = 1$
$d = 1$ (elliptic)	$d = 0$ (parabolic)	$d = -1$ (hyperbolic).

II) **The case of variable coefficients**
In the variable coefficients case, one can only have a local classification.

Example 1.3 Consider the Tricomi equation of gas dynamics:

$$Lu(x, y) = yu_{xx} + u_{yy}.$$

Here, the coefficient y is not a constant, and we have $A = y,\quad B = 0$, and $C = 1$. Hence, $d = AC - B^2 = y$ and consequently, the domain of ellipticity is $y > 0$, and so on, see Figure 1.1.

The usual three types of problems in differential equations.

1. Initial value problems (IVPs)
The simplest differential equation is $u'(x) = f(x)$ for $a < x \leq b$. But for any such u, also $(u(x) + c)' = f(x)$ for any constant c. To determine a unique solution, a specification of the initial value $u(a) = u_0$ is generally required. For example for $f(x) = 2x,\quad 0 < x \leq 1$, we have $u'(x) = 2x$ and the general solution is $u(x) = x^2 + c$. With an *initial value* of $u(0) = 0$, we get $u(0) = 0^2 + c = c = 0$. Hence,

Figure 1.1 Tricomi equation: an example of a variable coefficient classification.

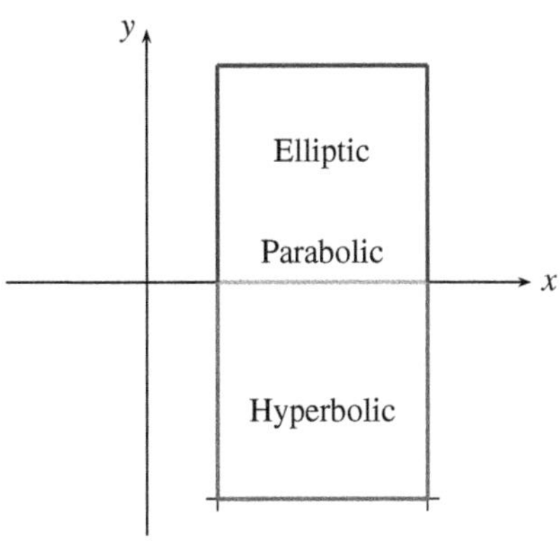

the unique solution to this IVP is $u(x) = x^2$. Likewise, for a time-dependent differential equation of *second order* (two time derivatives), the initial values for $t = 0$, i.e. $u(x, 0)$ and $u_t(x, 0)$, are generally required. For a PDE such as the heat equation, the initial value can be a *function* of the space variable.

Example 1.4 The wave equation, on the real line, augmented with the given initial data:

$$\begin{cases} u_{tt} - u_{xx} = 0, & -\infty < x < \infty, \quad t > 0, \\ u(x,0) = f(x), \quad u_t(x,0) = g(x), & -\infty < x < \infty, \quad t = 0. \end{cases}$$

Remark 1.1 Note that, here, for the unbounded spatial domain $(-\infty, \infty)$, it is required that $u(x, t) \to 0$ (or $\to u_\infty$ =constant) as $|x| \to \infty$. This corresponds to two boundary conditions.

2. Boundary value problems (BVP)

Example 1.5 (A BVP in $\mathbb{R}$)
Consider the one-dimensional *stationary heat equation*:

$$-(a(x)u'(x))' = f(x), \quad \text{for} \quad 0 < x < 1.$$

In order to determine a solution $u(x)$ uniquely (see Remark 1.2), the differential equation is complemented by boundary conditions imposed at the boundary points $x = 0$ and $x = 1$; for example $u(0) = u_0$ and $u(1) = u_1$, where u_0 and u_1 are given real numbers.

Example 1.6 (A BVP in $\mathbb{R}^n$).
The Laplace equation in $\mathbb{R}^n$, $\mathbf{x} = (x_1, x_2, \dots, x_n)$,

$$\begin{cases} \Delta_n u = \dfrac{\partial^2 u}{\partial x_1^2} + \dfrac{\partial^2 u}{\partial x_2^2} + \dots + \dfrac{\partial^2 u}{\partial x_n^2} = 0, & \mathbf{x} \in \Omega \subset \mathbb{R}^n, \\ u(\mathbf{x}) = f(\mathbf{x}), & \mathbf{x} \in \partial\Omega \ \text{(boundary of } \Omega). \end{cases}$$

Remark 1.2 In general, in order to obtain a unique solution for a (partial) differential equation, one needs to supply physically adequate boundary data. For instance for the potential problem $u_{xx} + u_{yy} = f$, stated in a rectangular domain $(x, y) \in \Omega := (0, 1) \times (0, 1)$, to determine a unique solution we need to give two boundary conditions in the x-direction, i.e. the numerical values for $u(0,y)$ and $u(1,y)$, and another two in the y-direction: the numerical values of $u(x,0)$ and $u(x,1)$. Whereas to determine a unique solution for the wave equation $u_{tt} - u_{xx} = 0$, it is necessary to supply two initial conditions in the time variable t, and two boundary conditions in the space variable x. We observe that in order to obtain a unique solution, we need to supply the same number of boundary (initial) conditions as the order of the differential equation in each spatial (time) variable. The general rule is that one should supply as many conditions as the highest order of the derivative in each variable. See also Remark 1.1, in the case of unbounded spatial domain.

3. Eigenvalue problems (EVPs)
Let $\mathbf{A}$ be a given square, say $n \times n$ matrix. The relation $\mathbf{Av} = \lambda\mathbf{v}, \quad \mathbf{v} \neq 0, \quad (\mathbf{v} \in \mathbb{R}^n)$, is a linear equation system, where λ is an *eigenvalue* and $\mathbf{v}$ is an *eigenvector*. In the Example 1.7 below, we introduce the corresponding terminology for differential equations.

Example 1.7 A differential equation describing a time-independent *vibrating string* is given by

$$-u''(x) = \lambda u(x), \qquad u(0) = u(\pi) = 0,$$

where λ is an eigenvalue and $u(x)$ is an eigenfunction. $u(0) = 0$ and $u(\pi) = 0$ are boundary values.

To derive this equation, we consider the differential equation for a *time-dependent vibrating string*, with small displacement $u(x,t)$, which is fixed at the end points, viz.

$$\begin{cases} u_{tt}(x,t) - u_{xx}(x,t) = 0, & 0 < x < \pi, \quad t > 0, \\ u(0,t) = u(\pi,t) = 0, \quad t > 0, & u(x,0) = f(x), \; u_t(x,0) = g(x). \end{cases}$$

Then, using the separation of variables technique, this equation is split into two EVPs: Insert $u(x,t) = X(x)T(t) \neq 0$ (a nontrivial solution) into the differential equation to get

$$u_{tt}(x,t) - u_{xx}(x,t) = X(x)T''(t) - X''(x)T(t) = 0. \tag{1.2.4}$$

Dividing (1.2.4) by $X(x)T(t) \neq 0$ separates the variables so that we have

$$\frac{T''(t)}{T(t)} = \frac{X''(x)}{X(x)} = \lambda \quad \text{(a constant independent of } x \text{ and } t\text{)}. \tag{1.2.5}$$

Consequently we get two ODEs (two EVPs):

$$X''(x) = \lambda X(x), \qquad \text{and} \qquad T''(t) = \lambda T(t). \tag{1.2.6}$$

The usual three types of boundary conditions.

1. Dirichlet boundary condition
   Here, the solution is known at the boundary of the domain, as
   $$u(\mathbf{x}, t) = f(\mathbf{x}), \quad \text{for } \mathbf{x} = (x_1, x_2, \dots, x_n) \in \partial\Omega, \quad t > 0.$$
   This is the case, for example, describing a fixed temperature at the boundary.
2. Neumann boundary condition
   In this case, the derivative of the solution in certain direction is given:
   $$\frac{\partial u}{\partial \mathbf{n}} = \mathbf{n} \cdot \mathbf{grad}(u) = \mathbf{n} \cdot \nabla u = f(\mathbf{x}), \quad \mathbf{x} \in \partial\Omega,$$
   where $\mathbf{n} = \mathbf{n}(\mathbf{x})$ is, e.g. the *outward unit normal* to $\partial\Omega$ at $\mathbf{x} \in \partial\Omega$, and
   $$\mathbf{grad}(u) = \nabla u = \left(\frac{\partial u}{\partial x_1}, \frac{\partial u}{\partial x_2}, \dots, \frac{\partial u}{\partial x_n}\right).$$
3. Robin boundary condition (a combination of 1 and 2),
   $$\frac{\partial u}{\partial \mathbf{n}} + k \cdot u(\mathbf{x}, t) = f(\mathbf{x}), \quad k > 0, \quad \mathbf{x} = (x_1, x_2, \dots, x_n) \in \partial\Omega.$$
   In a homogeneous case, i.e. for $f(\mathbf{x}) \equiv 0$, Robin condition means that the heat flux $\frac{\partial u}{\partial \mathbf{n}}$ through the boundary is proportional to the temperature (in fact the temperature difference between the inside and outside) at the boundary.

Example 1.8 In two dimensions, $\mathbf{n} = (n_x, n_y)$ with $|\mathbf{n}| = \sqrt{n_x^2 + n_y^2} = 1$ and hence, for $u = u(x, y)$, we have $\mathbf{n} \cdot \nabla u = n_x u_x + n_y u_y$, see Figure 1.2.

Example 1.9 Let $u(x, y) = x^2 + y^2$. We determine the normal derivative of u in (the assumed normal) direction $\mathbf{v} = (1, 1)$. The gradient of u is the vector valued function $\nabla u = 2x \cdot \mathbf{e}_x + 2y \cdot \mathbf{e}_y$, where $\mathbf{e}_x = (1, 0)$ and $\mathbf{e}_y = (0, 1)$ are the

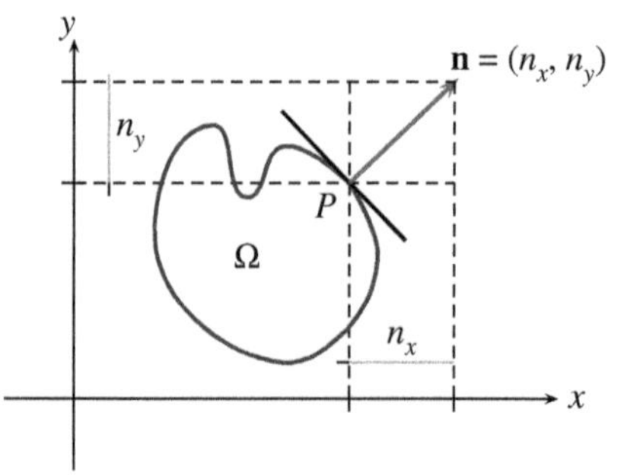

Figure 1.2 Outward unit normal **n** at a point $P \in \partial\Omega$.

unit orthonormal basis in $\mathbb{R}^2$: $\mathbf{e}_x \cdot \mathbf{e}_x = \mathbf{e}_y \cdot \mathbf{e}_y = 1$ and $\mathbf{e}_x \cdot \mathbf{e}_y = \mathbf{e}_y \cdot \mathbf{e}_x = 0$. Note that $\mathbf{v} = \mathbf{e}_x + \mathbf{e}_y = (1, 1)$ is not a unit vector. The normalized $\hat{\mathbf{v}}$ is obtained as $\hat{\mathbf{v}} = \mathbf{v}/|\mathbf{v}|$, i.e.

$$\hat{\mathbf{v}} = \frac{\mathbf{e}_x + \mathbf{e}_y}{|\mathbf{e}_x + \mathbf{e}_y|} = \frac{(1,1)}{\sqrt{1^2 + 1^2}} = \frac{(1,1)}{\sqrt{2}}.$$

Thus, $\hat{\mathbf{v}} \cdot \nabla u(x,y) = \frac{\mathbf{e}_x+\mathbf{e}_y}{|\mathbf{e}_x+\mathbf{e}_y|} \cdot (2x \cdot \mathbf{e}_x + 2y \cdot \mathbf{e}_y)$, which gives

$$\hat{\mathbf{v}} \cdot \nabla u(x,y) = \frac{(1,1)}{\sqrt{2}} \cdot [2x(1,0) + 2y(0,1)] = \frac{(1,1)}{\sqrt{2}} \cdot (2x, 2y) = \frac{2x + 2y}{\sqrt{2}}.$$

The usual three criteria to deal with

1) In theory

   A given differential equation problem is called *well-posed* if the following three conditions hold true:

   I1. Existence: there exists at least one solution u.

   I2. Uniqueness: we have either one solution or no solutions at all.

   I3. Stability: the solution depends continuously on the data.

   Note: A property that concerns behavior, such as growth or decay, or perturbations of a solution as time increases is generally called a stability property.

2) In applications

   II1. Construction: exact design of the solution.

   II2. Regularity: how smooth is the found solution.

   II3. Approximation: when an exact construction is not possible.

Three general approaches for solving differential equations

1. Separation of variables method: The separation of variables technique reduces the (PDEs) to simpler EVPs (ODEs). This method is known as *Fourier method, or solution by eigenfunction expansion*.
2. Variational formulation method: Variational formulation or the multiplier method is a strategy for extracting information by multiplying a differential equation by suitable test functions and then integrating (e.g. like Fourier or Laplace transforms). This is also referred to as *The Energy Method*. A discrete version of this is the subject of our study.
3. Green's function method: Fundamental solutions, or solution of integral equations, which we have briefly addressed in the chapter of mathematical tools is the subject of an advanced PDE course.

1.3 PDEs in $\mathbb{R}^n$, Further Classifications

In this section, we extend the overture of the Sections 1.1 and 1.2 to higher dimensions and give definitions for linearity, nonlinearity, and superposition concepts.

We recall the common notation $\mathbb{R}^n$ for the real Euclidean spaces of dimension n with the elements $\mathbf{x} = (x_1, x_2, \dots, x_n) \in \mathbb{R}^n$. In most of the applications, n will be 1, 2, 3, or 4 and the variables x_i, $i = 1, 2, 3$ denote coordinates in space dimensions, whereas x_4 represents the time variable. In this case, we usually replace (x_1, x_2, x_3, x_4) by the most common notation: (x, y, z, t). Further, we shall use the subscript notation for the partial derivatives, viz.

$$u_{x_i} = \frac{\partial u}{\partial x_i}, \quad \dot{u} = u_t = \frac{\partial u}{\partial t}, \quad u_{xy} = \frac{\partial^2 u}{\partial x \partial y}, \quad u_{xx} = \frac{\partial^2 u}{\partial x^2}, \quad \ddot{u} = \frac{\partial^2 u}{\partial t^2}, \quad \text{etc.}$$

A more general notation for partial derivatives of a sufficiently smooth function u (see Definition 1.1 below) is given by

$$\frac{\partial^{|\alpha|} u}{\partial x_1^{\alpha_1} \dots \partial x_n^{\alpha_n}} := \frac{\partial^{\alpha_1}}{\partial x_1^{\alpha_1}} \frac{\partial^{\alpha_2}}{\partial x_2^{\alpha_2}} \cdots \frac{\partial^{\alpha_n}}{\partial x_n^{\alpha_n}} u,$$

where $\frac{\partial^{\alpha_i}}{\partial x_i^{\alpha_i}}$, $1 \le i \le n$, denotes the partial derivative of order α_i with respect to the variable x_i, and $\alpha = (\alpha_1, \alpha_2, \dots, \alpha_n)$ is a multi-index of integers $\alpha_i \ge 0$ with $|\alpha| = \alpha_1 + \cdots + \alpha_n$.

Definition 1.1 A function f of one real variable is said to be of class $C^{(k)}$ on an open interval I if its derivatives $f', \dots, f^{(k)}$ exist and are continuous on I. A function f of n real variables is said to be of class $C^{(k)}$ on a set $S \subset \mathbb{R}^n$ if all of its partial derivatives of order $\le k$, i.e. $\partial^{|\alpha|} f / (\partial x_1^{\alpha_1} \dots \partial x_n^{\alpha_n})$ with the multi-index $\alpha = (\alpha_1, \alpha_2, \dots, \alpha_n)$ and $|\alpha| \le k$, exist and are continuous on S.

As we mentioned, a key defining property of a PDE is that there are derivatives with respect to more than one independent variable and a PDE is a relation between an unknown function $u(x_1, \dots, x_n)$ and its partial derivatives:

$$F(x_1, \dots, x_n, u, u_{x_1}, u_{x_2}, \dots, u_{x_1 x_1}, \dots, \partial^{|\alpha|} u / \partial x_1^{\alpha_1} \dots \partial x_l^{\alpha_l}, \dots) = 0. \tag{1.3.1}$$

Example 1.10 The one-space dimensional, homogeneous, heat, and wave equations (here $x_1 = t$) are among the simplest PDEs:

$$u_t - u_{xx} = 0, \quad \text{and} \quad u_{tt} - u_{xx} = 0.$$

Other examples are

$$u_t + u_x - u_{xx} = 0, \quad \text{convection–diffusion equation}$$

$$yu_{xx} + u_{yy} = 0, \quad \text{Tricomi equation.}$$

The most general PDE of first order in two independent variables, x and y is of the form

$$F(x,y,u(x,y),u_x(x,y),u_y(x,y)) =: F(x,y,u,u_x,u_y) = 0. \tag{1.3.2}$$

Likewise, the most general PDE of second order in two independent variables can be written as

$$F(x,y,u,u_x,u_y,u_{xx},u_{xy},u_{yy}) = 0. \tag{1.3.3}$$

As stated in Remark 1.2, when Eqs. (1.3.1)–(1.3.3) are considered in bounded domains $\Omega \subset \mathbb{R}^2$, in order to obtain a *unique solution* one should supply *boundary conditions*: conditions at the boundary of the domain Ω, denoted, e.g. by $B(u) = f$ or $B(u) = 0$ (as well as conditions for $t = 0$, initial conditions; denoted, e.g. by $I(u) = g$ or $I(u) = 0$; in the time-dependent cases), as in the theory of ODEs. B and I are expressions of u and its partial derivatives, stated on the whole or a part of the boundary of Ω (or, in case of I, for $t = 0$), and are associated with the underlying PDE. Below we shall discuss the choice of relevant initial and boundary conditions for a PDE.

A *solution* of a PDE of type (1.3.1)–(1.3.3) is a function u that identically satisfies the corresponding PDE, and the associated initial and boundary conditions, in some region of the variables $x_1, x_2, \dots, x_n$, or x, y (and t). Note that a solution of an equation of order k has to be k times differentiable. A function in $C^{(k)}$ that satisfies a PDE of order k is called a *classical (or strong) solution* of the PDE. We sometimes also have to deal with solutions that are not classical. Such solutions are called *weak solutions*. In this note, in the variational formulation for FEMs, we actually deal with weak solutions. For a more thorough discussion on weak solutions, see Chapter 2 or any textbook in distribution theory.

Definition 1.2 *Hadamard's criteria; compare with the three criteria in theory*
A problem consisting of a PDE associated with boundary and/or initial conditions is called well-posed if it fulfills the following three criteria:

1. **Existence** The problem has a solution.
2. **Uniqueness** There is no more than one solution.
3. **Stability** A small change in the equation or in the side (initial and/or boundary) conditions gives rise to a small change in the solution.

If one or more of the conditions abovementioned does not hold, then we say that the problem is *ill-posed*. The fundamental theoretical question of PDEs is whether the problem consisting of the equation and its associated side conditions is well-posed. However, in certain engineering applications, we might encounter problems that are ill-posed. In practice, such problems are unsolvable. Therefore, when we face an ill-posed problem, the first step should be to modify it appropriately in order to render it well-posed.

Definition 1.3 An equation is called linear if in (1.3.1), F is a linear function of the unknown function u and its derivatives.

Thus, for example, the equation $e^{x^2y}u_x + x^7u_y + \cos(x^2+y^2)u = y^3$ is a linear equation, while $u_x^2 + u_y^2 = 1$ is a nonlinear equation. The nonlinear equations are often further classified into subclasses according to the type of their nonlinearity. Generally, the nonlinearity is more pronounced when it appears in higher-order derivatives. For example, the following equations are both nonlinear

$$u_{xx} + u_{yy} = u^3 + u. \tag{1.3.4}$$

$$u_{xx} + u_{yy} = |\nabla u|^2 u. \tag{1.3.5}$$

Here $|\nabla u|$ denotes the norm of the gradient of u. While (1.3.5) is nonlinear, it is still linear as a function of the highest-order derivative (here u_{xx} and u_{yy}). Such a nonlinearity is called *quasilinear*. On the other hand, in (1.3.4), the nonlinearity is only in the unknown solution u. Such equations are called *semilinear*.

1.4 Differential Operators, Superposition

Differential and integral operators are examples of mappings between function classes as $C^{(k)}(\Omega)$ where $\Omega \subseteq \mathbb{R}^n$, $n \geq 1$. We denote by $\mathcal{L}[u]$ the operation of a mapping (operator) $\mathcal{L}$ on a function u.

Definition 1.4 An operator $\mathcal{L}$ that satisfies

$$\mathcal{L}[\beta_1 u_1 + \beta_2 u_2] = \beta_1 \mathcal{L}[u_1] + \beta_2 \mathcal{L}[u_2], \quad \forall \beta_1, \beta_2 \in \mathbb{R}, \tag{1.4.1}$$

where u_1 and u_2 are functions, is called a linear operator. We may generalize (1.4.1) as

$$\mathcal{L}[\beta_1 u_1 + \cdots + \beta_k u_k] = \beta_1 \mathcal{L}[u_1] + \cdots + \beta_k \mathcal{L}[u_k], \quad \forall \beta_1, \ldots, \beta_k \in \mathbb{R}, \tag{1.4.2}$$

i.e. $\mathcal{L}$ maps any linear combination of u_j's to corresponding linear combination of $\mathcal{L}[u_j]$'s.

For instance the *integral operator* $\mathcal{L}[f] = \int_a^b f(x)\,dx$ defined on the space of continuous functions on $[a,b]$ defines a linear operator from $C[a,b]$ into $\mathbb{R}$, which satisfies both (1.4.1) and (1.4.2).

A *linear partial differential operator* $\mathcal{L}$ that transforms a function u of the variables $\mathbf{x} = (x_1, x_2, \ldots, x_n)$ into another function $\mathcal{L}u$ is given by

$$\mathcal{L}[u] = a(\mathbf{x})u + \sum_{i=1}^{n} b_i(\mathbf{x})\frac{\partial u}{\partial x_i} + \sum_{i,j=1}^{n} c_{ij}(\mathbf{x})\frac{\partial^2 u}{\partial x_i \partial x_j} + \cdots \tag{1.4.3}$$

where u represents any function in, say $C^{(\ell)}$, and the dots at the end indicate higher-order derivatives, but the sums contain only *finitely* many terms.

The term *linear* in the phrase *linear partial differential operator* refers to the following fundamental property: if $\mathcal{L}$ is given by (1.4.3) and u_j, $1 \leq j \leq k$, are any set of functions possessing the requisite derivatives, and β_j, $1 \leq j \leq k$ are any constants, then relation (1.4.2) is fulfilled. This is an immediate consequence of the fact that (1.4.1) and (1.4.2) are valid for $\mathcal{L}$ replaced with the derivative of any admissible order. A linear differential equation defines a linear differential operator: the equation can be expressed as $\mathcal{L}[u] = f$, where $\mathcal{L}$ is a linear operator and f is a given function. The differential equation of the form $\mathcal{L}[u] = 0$ is called a *homogeneous equation*. For example, define the operator $\mathcal{L} = \partial^2/\partial x^2 - \partial^2/\partial y^2$. Then

$$\mathcal{L}[u] = u_{xx} - u_{yy} = 0,$$

is a homogeneous equation, while the equation

$$\mathcal{L}[u] = u_{xx} - u_{yy} = x,$$

is an example of an *inhomogeneous equation*. In a similar way, we may define another type of constraint for the PDEs that appears in many applications: the boundary conditions. In this regard, the linear boundary conditions are defined as operators B satisfying

$$B(\beta_1 u_1 + \beta_2 u_2) = \beta_1 B(u_1) + \beta_2 B(u_2), \quad \forall \beta_1, \beta_2 \in \mathbb{R}, \tag{1.4.4}$$

at the boundary of a given domain Ω.

Note that Laplace, heat, and wave equations are linear. Likewise, all the important boundary conditions (Dirichlet, Neumann, Robin) are linear.

The Superposition Principle. An important property of the linear operators is that if the functions u_j, $1 \leq j \leq k$, satisfy the linear differential equations $\mathcal{L}[u_j] = f_j$, and the *linear boundary* conditions $B(u_j) = g_j$ for $j = 1, 2, \ldots, k$, then any linear combination $v := \sum_{i=1}^{\ell} \beta_i u_i$, $\ell \leq k$, satisfies the equation $\mathcal{L}[v] = \sum_{i=1}^{\ell} \beta_i f_i$ as well as the boundary condition $B(v) = \sum_{i=1}^{\ell} \beta_i g_i$. In particular, if each of the functions u_i, $1 \leq i \leq \ell$, satisfies the homogeneous equation $\mathcal{L}[u] = 0$ and the homogeneous boundary condition $B(u) = 0$, then every linear combination of them satisfies that equation and boundary condition too. This property is called the *superposition principle*. It allows to construct complex solutions through combining simple solutions: suppose we want to determine all solutions of a differential equation associated with a boundary condition, viz.

$$\mathcal{L}[u] = f, \qquad B(u) = g. \tag{1.4.5}$$

We consider the corresponding, *simpler homogeneous problem*:

$$\mathcal{L}[u] = 0, \qquad B(u) = 0. \tag{1.4.6}$$

Now, it suffices to find just one solution, say v of the original problem (1.4.5). Then, for any solution u of (1.4.5), $w = u - v$ satisfies (1.4.6), since $\mathcal{L}[w] = \mathcal{L}[u] - \mathcal{L}[v] = f - f = 0$ and $B(w) = B(u) - B(v) = g - g = 0$. Hence, we obtain a general solution of (1.4.5) by adding the general (homogeneous) solution w of (1.4.6) to any particular solution of (1.4.5).

Following the same idea, one may apply superposition to split a problem involving several inhomogeneous terms into simpler ones each with a single inhomogeneous term. For instance, we may split (1.4.5) as

$$\begin{aligned} \mathcal{L}[u_1] &= f, \qquad B(u_1) = 0, \\ \mathcal{L}[u_2] &= 0, \qquad B(u_2) = g, \end{aligned}$$

and then take $u = u_1 + u_2$.

The most important application of the superposition principle is in the homogeneous case: linear homogeneous differential equations satisfying homogeneous boundary conditions (which we repeat from above).

The Superposition principle for the homogeneous case. If the functions u_j, $1 \leq j \leq k$, satisfy (1.4.6): the linear differential equation $\mathcal{L}[u_j] = 0$ and the boundary conditions (linear) $B(u_j) = 0$ for $j = 1, 2, \ldots, k$, then any linear combination $v := \sum_{i=1}^{\ell} \beta_i u_i$, $\ell \leq k$, satisfies the same equation and boundary condition: (1.4.6).

1.4.1 Exercises

Problem 1.1 Consider the problem

$$u_{xx} + u = 0, \quad x \in (0, \ell); \qquad u(0) = u(\ell) = 0.$$

Clearly, the function $u(x) \equiv 0$ is a solution. Is this solution unique? Does the answer depend on ℓ?

Problem 1.2 Consider the problem

$$u_{xx} + u_x = f(x), \quad x \in (0, \ell); \qquad u(0) = u'(0) = \frac{1}{2}[u'(\ell) + u(\ell)].$$

a) Is the solution unique? (f is a given function).
b) Under what condition on f a solution exists?

Problem 1.3 Suppose $u_i,\ i = 1, 2, \ldots, N$ are N solutions of the linear differential equation $\mathcal{L}[u] = F$, where $F \neq 0$. Under what condition on the constant coefficients $c_i,\ i = 1, 2, \ldots, N$ is the linear combination $\sum_{i=1}^{N} c_i u_i$ also a solution of this equation?

Problem 1.4 Consider the nonlinear ODE $u_x = u(1 - u)$.

a) Show that $u_1(x) \equiv 1$ and $u_2(x) = 1 - 1/(1 + e^x)$ both are solutions, but $u_1 + u_2$ is not a solution.
b) For which value of c_1 is $c_1 u_1$ a solution? What about $c_2 u_2$?

Problem 1.5 Show that each of the following equations has a solution of the form $u(x, y) = f(ax + by)$ for a proper choice of constants $a,\ b$. Find the constants for each example.

a) $u_x + 3u_y = 0$.
b) $3u_x - \pi u_y = 0$.
c) $2u_x + eu_y = 0$.

Problem 1.6

a) Consider the equation $u_{xx} + 2u_{xy} + u_{yy} = 0$. Write the equation in the coordinates $s = x,\ t = x - y$.
b) Find the general solution of the equation.
c) Consider the equation $u_{xx} - 2u_{xy} + 5u_{yy} = 0$. Write it in the coordinates $s = x + y$ and $t = 2x$.

Problem 1.7

a) Show that for $n = 1, 2, 3, \ldots,\quad u_n(x, y) = \sin(n\pi x)\sinh(n\pi y)$ satisfies

$$u_{xx} + u_{yy} = 0, \quad u(0, y) = u(1, y) = u(x, 0) = 0.$$

b) Find a linear combination of u_ns that satisfies $u(x, 1) = \sin 2\pi x - \sin 3\pi x$.
c) Solve the Dirichlet problem

$$u_{xx} + u_{yy} = 0, \quad u(0, y) = u(1, y) = 0, u(x, 0) = 2\sin \pi x,$$
$$u(x, 1) = \sin 2\pi x - \sin 3\pi x.$$

1.5 Some Equations of Mathematical Physics

Throughout the book, we focus on the study of some of the basic PDEs of mathematical physics. These equations involve differential operators as the *gradient* and *Laplacian*, acting on $C^{(2)}(\Omega),\quad \Omega \subseteq \mathbb{R}^n$:

$$\nabla u := \left(\frac{\partial u}{\partial x_1}, \frac{\partial u}{\partial x_2}, \ldots, \frac{\partial u}{\partial x_n}\right), \qquad \nabla^2 u := \Delta_n = \frac{\partial^2 u}{\partial x_1^2} + \frac{\partial^2 u}{\partial x_2^2} + \cdots + \frac{\partial^2 u}{\partial x_n^2}. \tag{1.5.1}$$

These equations describe fundamental physical phenomena as follows:

$$\begin{aligned}
\nabla^2 u &= f(\mathbf{x}), && \text{The stationary diffusion}\\
u_t - k\nabla^2 u &= f(\mathbf{x}, t), && \text{The diffusion process}\\
u_{tt} - c^2\nabla^2 u &= f(\mathbf{x}, t), && \text{The wave propagation}\\
u_t + \beta \cdot \nabla u - \alpha\nabla^2 u &= f(\mathbf{x}, t), && \text{Convection–diffusion.}
\end{aligned} \tag{1.5.2}$$

Here f is a given function. In the special case, when $f \equiv 0$, i.e. when Eq. (1.5.2) are *homogeneous*, the first equation is called *the Laplace equation* and, being time-independent, has some special features: besides the fact that it describes the steady-state heat transfer and the standing wave equations (loosely speaking, the time-independent versions of the other two equations), the Laplace equation arises in describing several physical phenomena such as *electrostatic potential* in regions with no electric charge, *gravitational potential* in the regions with no mass distribution, in elasticity, etc.

1.5.1 The Poisson Equation

The *Poisson equation* is a BVP of the form

$$\begin{cases} -\Delta u(x) = f(x), & x \in \Omega \subset \mathbb{R}^n, \quad n \geq 2 \\ u(x) = 0, & x \in \partial\Omega \quad \text{(Dirichlet data).} \end{cases} \tag{1.5.3}$$

Below are some common phenomena modeled by the Poisson's equation.

- *Electrostatics*: To describe the components of the Maxwell's equations, associating the electric- and potential fields $E(x)$ and $\varphi(x)$ to the charge density ρ, roughly speaking, we have

$$\left.\begin{cases} \nabla \cdot E = \rho, \quad \text{in } \Omega & \text{(Coulomb's law)} \\ \nabla \times E = 0, & \text{(Faraday's law)} \implies E = \nabla\varphi \end{cases}\right\} \implies \Delta\varphi = \rho,$$

  and with a Dirichlet boundary condition $\varphi = g$ on $\partial\Omega$.
- *Fluid mechanics*: The velocity field u of a rotation-free fluid flow satisfies $\nabla \times u = 0$ and hence, u is a, so-called, gradient field: $u = \nabla\varphi$, with φ being a scalar potential field. The rotation-free incompressible fluid flow satisfies $\nabla \cdot u = 0$, which yields the Laplace's equation $\Delta\varphi = 0$ for its potential. At a solid boundary, this problem will be associated with homogeneous Neumann boundary condition, due to the fact that in such a boundary, the normal velocity is zero.
- Statistical physics: The random motion of particles inside a container Ω until they hit the boundary is described by the probability $u(x)$ of a particle starting at the point $x \in \Omega$ winding up to stop at some point on $\partial\Omega$, where $u(x) = 1$ means that it is certain and $u(x) = 0$ means that it never happens. It turns out that u solves the Laplace's equation $\Delta u = 0$, with discontinuity at the boundary: $u = 0$ on $\partial\Omega_0$ and $u = 1$ on $\partial\Omega_1$ where $\partial\Omega = \partial\Omega_0 \cup \partial\Omega_1$. Poisson's equation is of vital

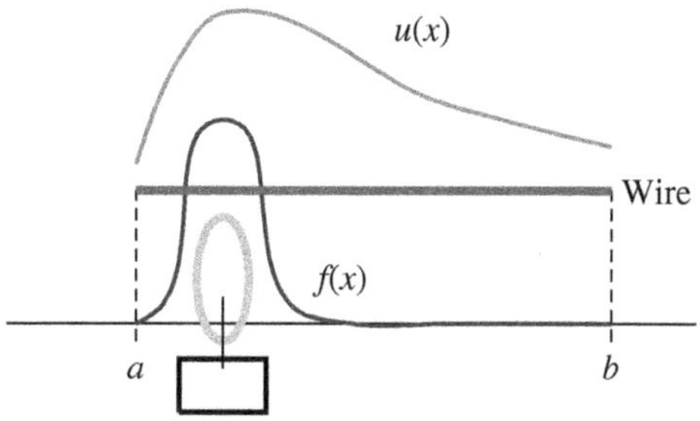

Figure 1.3 A heat-conducting one-dimensional wire.

importance in describing, the coupling of nonlinearity aspects, in the system of gas kinetic/dynamic equations.

1.5.2 The Heat Equation

1.5.2.1 A Model Problem for the Stationary Heat Equation in 1*d*

Below we model the heat conduction in a thin heat-conducting wire stretched between the two endpoints of an interval $[a, b]$ that is subject to a heat source of intensity $f(x)$, as in Figure 1.3. We are interested in the stationary distribution of temperature $u(x)$ in the wire.

To this end, let $q(x)$ denote the heat flux in the direction of the positive x-axis in the wire $a < x < b$. Conservation of energy in the stationary case requires that the net heat through the endpoints of an arbitrary subinterval (x_1, x_2) of (a, b) is equal to the heat produced in (x_1, x_2) per unit time:

$$q(x_2) - q(x_1) = \int_{x_1}^{x_2} f(x)\, dx.$$

By the *Fundamental Theorem of Calculus*,

$$q(x_2) - q(x_1) = \int_{x_1}^{x_2} q'(x)\, dx,$$

Hence, we conclude that

$$\int_{x_1}^{x_2} q'(x)\, dx = \int_{x_1}^{x_2} f(x)\, dx.$$

Since x_1 and x_2 are arbitrary, assuming that the integrands are continuous, yields

$$q'(x) = f(x), \quad \text{for} \quad a < x < b, \tag{1.5.4}$$

which expresses conservation of energy in differential equation form. We need an additional equation that relates the heat flux q to the temperature gradient u' called a constitutive equation. The simplest *constitutive equation* for heat flow is *Fourier's law*:

$$q(x) = -c(x)u'(x), \tag{1.5.5}$$

which states that heat flows from warm regions to cold regions at a rate proportional to the temperature gradient $u'(x)$. The constant of proportionality is the *coefficient of heat conductivity* $c(x)$, which we assume to be a positive function in $[a, b]$. Combining (1.5.4) and (1.5.5) gives the *stationary heat equation in one dimension*:

$$-(c(x)u'(x))' = f(x), \quad \text{for} \quad a < x < b. \tag{1.5.6}$$

To define a solution u uniquely, the differential equation is complemented by *boundary conditions* imposed at the boundary points $x = a$ and $x = b$. A common example is the homogeneous *Dirichlet conditions* $u(a) = u(b) = 0$, corresponding to keeping the temperature at zero at the endpoints of the wire. The result is a *two-point BVP*:

$$\begin{cases} -(c(x)u'(x))' = f(x), & \text{in} \quad (a, b), \\ u(a) = u(b) = 0. \end{cases} \tag{1.5.7}$$

The boundary condition $u(a) = 0$ may be replaced by $-c(a)u'(a) = q(a) = 0$, corresponding to prescribing zero heat flux, or insulating the wire, at $x = a$. Later, we also consider nonhomogeneous boundary conditions of the form $u(a) = u_a$ or $q(a) = g$ where u_a and g may be different from zero. For other types of boundary conditions, see Trinities (Section 1.2).

The time-dependent heat equation in (1.5.2) describes the diffusion of thermal energy in a homogeneous material, where $u = u(\mathbf{x}, t)$ is the temperature at a position $\mathbf{x}$ at time t and $k = k(\mathbf{x})$ is called *thermal diffusivity* or *heat conductivity* (corresponding to $c(x)$ in (1.5.5)–(1.5.7)) of the material.

Remark 1.3 The heat equation can be used to model the heat flow in solids and fluids, in the latter case, however, it does *not* take into account the convection phenomenon; and provides a reasonable model only if phenomena such as macroscopic currents in the fluid are not present (or negligible). Further, the heat equation is not a fundamental law of physics, and it does not give reliable answers at very low or very high temperatures.

Since temperature is related to heat, which is a form of energy, the basic idea in deriving the heat equation is to use the *law of conservation of energy*. Below we derive the general form of the heat equation in arbitrary dimension.

1.5.2.2 Fourier's Law of Heat Conduction, Derivation of the Heat Equation

Let $\Omega \subset \mathbb{R}^d$, $d = 1, 2, 3$, be a fixed spatial domain with boundary $\partial\Omega$. The rate of change of thermal energy with respect to time in Ω is equal to the net flow of energy across the boundary of Ω plus the rate at which heat is generated within Ω.

Let $u(\mathbf{x}, t)$ denote the temperature at the position $\mathbf{x} = (x, y, z) \in \Omega$ and at time t. We assume that the solid is at rest and that it is rigid, so that the only energy present is thermal energy and the density $\rho(\mathbf{x})$ is independent of the time t and temperature u. Let $\mathcal{E}$ denote the energy per unit mass. Then the amount of thermal energy in Ω is given by

$$\int_\Omega \rho\ \mathcal{E}\ d\mathbf{x},$$

and the time rate (time derivative) of change of thermal energy in Ω is:

$$\frac{d}{dt}\int_\Omega \rho\ \mathcal{E}\ d\mathbf{x} = \int_\Omega \rho\ \frac{\partial \mathcal{E}}{\partial t}\ d\mathbf{x}.$$

Let $\mathbf{q} = (q_x, q_y, q_z)$ denote the heat flux vector and $\mathbf{n} = (n_x, n_y, n_z)$ denote the outward unit normal to the boundary $\partial\Omega$, at the point $\mathbf{x} \in \partial\Omega$. Then $\mathbf{q} \cdot \mathbf{n}$ represents the flow of heat per unit cross-sectional area per unit time crossing a surface element. Thus,

$$-\int_{\partial\Omega} \mathbf{q} \cdot \mathbf{n}\ dS$$

is the amount of heat per unit time flowing into Ω across the boundary $\partial\Omega$. Here, dS represents the element of surface area. The minus sign reflects the fact that if more heat flows out of the domain Ω than in, the energy in Ω decreases. Finally, in general, the heat production is determined by external sources that are independent of the temperature. In some cases, (such as an air conditioner controlled by a thermostat), it depends on temperature itself, but not on its derivatives. Hence, in the presence of a source (or sink), we denote the corresponding rate at which heat is produced per unit volume by $f = f(\mathbf{x}, t, u)$ so that the source term becomes

$$\int_\Omega f(\mathbf{x}, t, u)\ d\mathbf{x}.$$

Now, the law of conservation of energy takes the form

$$\int_\Omega \rho\ \mathcal{E}_t\ d\mathbf{x} + \int_{\partial\Omega} \mathbf{q} \cdot \mathbf{n}\ dS = \int_\Omega f(\mathbf{x}, t, u)\ d\mathbf{x}. \tag{1.5.8}$$

Applying the Gauss divergence theorem to the integral over $\partial\Omega$, we get

$$\int_\Omega (\rho\ \mathcal{E}_t + \nabla \cdot \mathbf{q} - f)\ d\mathbf{x} = 0, \tag{1.5.9}$$

where ∇ denotes the divergence operator. In the sequel, we shall use the following simple result:

Lemma 1.1 *Let h be a continuous function satisfying $\int_\Omega h(\boldsymbol{x})\ d\mathbf{x} = 0$ for every domain $\Omega \subset \mathbb{R}^d$. Then $h \equiv 0$.*

Proof: Let us assume to the contrary that there exists a point $\mathbf{x}_0 \in \mathbb{R}^d$ where $h(\mathbf{x}_0) \neq 0$. Assume without loss of generality that $h(\mathbf{x}_0) > 0$. Since h is continuous, there exists a domain (maybe very small) Ω_0, containing $\mathbf{x}_0$, and an $\varepsilon > 0$, such that $h(\mathbf{x}) > \varepsilon$, for all $\mathbf{x} \in \Omega_0 \subset \Omega$. Therefore, we have $\int_{\Omega_0} h(\mathbf{x})\, d\mathbf{x} > \varepsilon \mathrm{Vol}(\Omega_0) > 0$, which contradicts the assumption.

From (1.5.9), using Lemma 1.1, we conclude that

$$\rho\, \mathcal{E}_t = -\nabla \cdot \mathbf{q} + f. \tag{1.5.10}$$

This is the basic form of our heat conduction law. The functions $\mathcal{E}$ and q are unknown and additional information of an empirical nature is needed to determine the equation for the temperature u. First, for many materials, over a fairly wide but not too large temperature range, the function $\mathcal{E} = \mathcal{E}(u)$ depends nearly linearly on u, so that

$$\mathcal{E}_t = \lambda u_t. \tag{1.5.11}$$

Here, λ, called the *specific heat*, is assumed to be constant. Next, we relate the temperature u to the heat flux $\mathbf{q}$. Here, we use *Fourier's law* but, first, to be specific, we describe the simple facts supporting Fourier's law:

i. Heat flows from regions of high temperature to regions of low temperature.
ii. The rate of heat flux is small or large accordingly as temperature changes between neighboring regions are small or large.

To describe these quantitative properties of heat flux, we postulate a linear relationship between the rate of heat flux and the rate of temperature change. Recall that if $\mathbf{x}$ is a point in the heat-conducting medium and $\mathbf{n}$ is a unit vector specifying a direction at $\mathbf{x}$, then the rate of heat flow at $\mathbf{x}$ in the direction $\mathbf{n}$ is $\mathbf{q} \cdot \mathbf{n}$ and the rate of change of the temperature is $\partial u/\partial \mathbf{n} = \nabla u \cdot \mathbf{n}$, the directional derivative of the temperature. Since $\mathbf{q} \cdot \mathbf{n} > 0$ requires $\nabla u \cdot \mathbf{n} < 0$, and vice versa (from calculus the direction of maximal growth of a function is given by its gradient), our linear relation takes the form $\mathbf{q} \cdot \mathbf{n} = -\kappa \nabla u \cdot \mathbf{n}$, with $\kappa = \kappa(\mathbf{x}) > 0$. Since $\mathbf{n}$ specifies any direction at $\mathbf{x}$, this is equivalent to the assumption

$$\mathbf{q} = -\kappa \nabla u, \tag{1.5.12}$$

which is *Fourier's law*. The positive function κ is called the *heat conduction (or Fourier) coefficient*. Let now $\sigma = \kappa/\lambda\rho$ and $F = f/\lambda\rho$ and insert (1.5.11) and (1.5.12) into (1.5.10) to get the final form of the heat equation:

$$u_t = \nabla \cdot (\sigma \nabla u) + F. \tag{1.5.13}$$

The quantity σ is referred to as the *thermal diffusivity (or diffusion) coefficient*. If we assume that σ is constant, then the final form of the heat equation would be

$$u_t = \sigma \nabla^2 u + F, \quad \text{or} \quad u_t = \sigma\, \Delta u + F. \tag{1.5.14}$$

1.5.3 The Wave Equation

The third equation in (1.5.2) is the wave equation: $u_{tt} - c^2\nabla^2 u = F$. Here, u represents a wave traveling through an n-dimensional medium; c is the speed of propagation of the wave in the medium and $u(\mathbf{x}, t)$ is the amplitude of the wave at position $\mathbf{x}$ and time t. The wave equation provides a mathematical model for a number of problems involving different physical processes as, e.g. in the following examples (i)–(vi):

(i) Vibration of a stretched string, such as a violin string (one-dimensional).
(ii) Vibration of a column of air, such as a clarinet (one-dimensional).
(iii) Vibration of a stretched membrane, such as a drumhead (two-dimensional).
(iv) Waves in an incompressible fluid, such as water (two-dimensional).
(v) Sound waves in air or other elastic media (three-dimensional).
(vi) Electromagnetic waves, such as light waves and radio waves (three-dimensional).

Note that in (i), (iii) and (iv), u represents the transverse displacement of the string, membrane, or fluid surface; in (ii) and (v), u represents the longitudinal displacement of the air; and in (vi), u is any of the components of the electromagnetic field. For detailed discussions and a derivation of the equations modeling (i)–(vi), see, e.g. Folland [62], Strauss [129], and Taylor [134]. We should point out, however, that in most cases, the derivation involves making some simplifying assumptions. Hence, the wave equation gives only an approximate description of the actual physical process, and the validity of the approximation will depend on whether certain physical conditions are satisfied. For instance, in example (i), the vibration should be small enough so that the string is not stretched beyond its limits of elasticity. In example (vi), it follows from Maxwell's equations, the fundamental equations of electromagnetism, that the wave equation is satisfied *exactly* in regions containing no electrical charges or current, which of course cannot be guaranteed under normal physical circumstances and can only be approximately justified in the real world. So an attempt to derive the wave equation corresponding to each of these examples from physical principles is beyond the scope of these notes. Nevertheless, to give an idea, below we shall derive the wave equation for a vibrating string.

1.5.3.1 The Vibrating String, Derivation of the Wave Equation in 1*d*

Consider a perfectly elastic and flexible string stretched along the segment $[0, L]$ of the x-axis, moving perpendicular to its equilibrium position. Let $\rho_0(x)$ denote the density of the string in the equilibrium position and $\rho(x, t)$ the density at time t. In an arbitrary small interval $[x, x + dx]$, the mass will satisfy, see Figure 1.4.

$$m = \int_x^{x+dx} \rho_0(x)\, dx = \int_x^{x+dx} \rho(x, t)\sqrt{1 + u_x^2}\, dx. \tag{1.5.15}$$

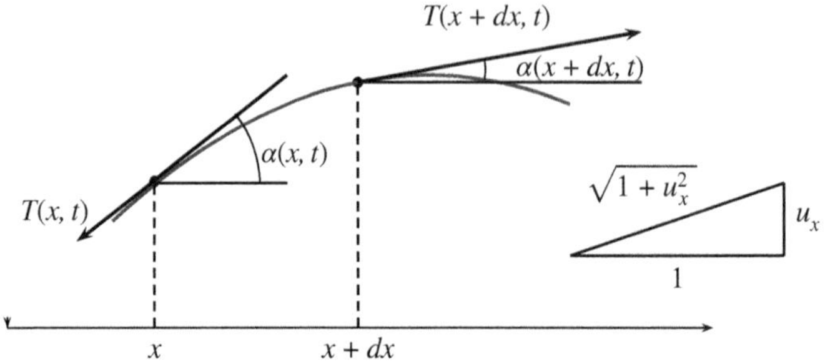

Figure 1.4 A vibrating string.

Thus, using Lemma 1.1, (1.5.15) gives the conservation of mass:

$$\rho_0(x) = \rho(x,t)\sqrt{1+u_x^2}. \tag{1.5.16}$$

Now we use the tensions $T(x,t)$ and $T(x+dx,t)$, at the endpoints of an element of the string and determine the forces acting on the small interval $[x, x+dx]$. Since we assumed that the string moves only vertically, the forces in the horizontal direction should be in balance: i.e.

$$T(x+dx,t)\cos\alpha(x+dx,t) - T(x,t)\cos\alpha(x,t) = 0. \tag{1.5.17}$$

Dividing (1.5.17) by dx and letting $dx \to 0$, we thus obtain

$$\frac{\partial}{\partial x}(T(x,t)\cos\alpha(x,t)) = 0, \tag{1.5.18}$$

Hence,

$$T(x,t)\cos\alpha(x,t) = \tau(t), \tag{1.5.19}$$

where $\tau(t) > 0$ because it is the magnitude of the horizontal component of the tension.

On the other hand, the vertical motion is determined by the fact that the time rate of change of linear momentum is given by the sum of the forces acting in the vertical direction. Hence, using (1.5.16), the momentum of the small element $[x, x+dx]$ is given by

$$\int_x^{x+dx} \rho_0(x)u_t\,dx = \int_x^{x+dx} \rho(x,t)\sqrt{1+u_x^2}\,u_t\,dx, \tag{1.5.20}$$

with the time rate of change:

$$\frac{d}{dt}\int_x^{x+dx} \rho_0 u_t\,dx = \int_x^{x+dx} \rho_0 u_{tt}\,dx. \tag{1.5.21}$$

There are two kinds of forces acting on the segment $[x, x+dx]$ of the string: (i) the forces due to tension that keep the string taut and whose horizontal components are in balance, and (ii) the forces acting along the whole length of the string, such as weight. Thus, using (1.5.19), the net tension force acting on the ends of the string element $[x, x + dx]$ is

$$\begin{aligned} F_\tau &:= T(x + dx, t) \sin \alpha(x + dx, t) - T(x, t) \sin \alpha(x, t) \\ &= \tau \left(\frac{\sin \alpha(x + dx, t)}{\cos \alpha(x + dx, t)} - \frac{\sin \alpha(x, t)}{\cos \alpha(x, t)} \right) \\ &= \tau(\tan \alpha(x + dx, t) - \tan \alpha(x, t)) \\ &= \tau(u_x(x + dx, t) - u_x(x, t)). \end{aligned} \tag{1.5.22}$$

Further, the weight of the string acting downward is

$$F_g = -\int \rho\, g\, dS = -\int_x^{x+dx} \rho\, g \sqrt{1 + u_x^2}\, dx = -\int_x^{x+dx} \rho_0 g\, dx. \tag{1.5.23}$$

Next, for an external load, with density $f(x, t)$, acting on the string (e.g. when a violin string is bowed), we have

$$F_f = \int \rho\, f\, dS = \int_x^{x+dx} \rho_0 f(x, t)\, dx. \tag{1.5.24}$$

Finally, one should model the friction forces acting on the string segment. We shall assume a linear law of friction of the form:

$$F_{\text{fr}} := -\int \sigma \rho u_t\, dS = -\int_x^{x+dx} \sigma \rho \sqrt{1 + u_x^2}\, u_t\, dx = -\int_x^{x+dx} \sigma \rho_0 u_t\, dx. \tag{1.5.25}$$

Now applying Newton's second law yields

$$\begin{aligned} \int_x^{x+dx} \rho_0 u_{tt}\, dx = F_\tau + F_g + F_f + F_{\text{fr}} &= \tau[u_x(x + dx, t) - u_x(x, t)] \\ &\quad - \int_x^{x+dx} \sigma \rho_0 u_t\, dx + \int_x^{x+dx} \rho_0 (f - g)\, dx. \end{aligned} \tag{1.5.26}$$

Dividing (1.5.26) by dx and letting $dx \to 0$, we obtain the equation

$$\rho_0 u_{tt} = \tau u_{xx} - \sigma \rho_0 u_t + \rho_0 (f - g). \tag{1.5.27}$$

Letting $c^2 = \tau / \rho_0$ and $F = f - g$, we end up with the following concise form:

$$u_{tt} + \sigma u_t = c^2 u_{xx} + F. \tag{1.5.28}$$

Equation (1.5.28) describes the vibration of the considered string once it is set into motion. The *smallness* assumption here results in a single linear equation for u. Due to the presence of the friction term σu_t, Eq. (1.5.28) is often referred to as the *damped one-dimensional wave equation*. If friction is negligible, then we can let $\sigma = 0$ and get the *inhomogeneous wave equation*

$$u_{tt} = c^2 u_{xx} + F. \tag{1.5.29}$$

In the absence of external forces and when the weight of the string is negligible, we may take $F \equiv 0$ to get the *one-dimensional wave equation*:

$$u_{tt} = c^2 u_{xx}. \tag{1.5.30}$$

Note that since u has the unit of length ℓ, u_{tt} has the unit of acceleration and u_{xx} the unit of ℓ^{-1}, hence, c has the unit of velocity.

1.5.4 Exercises

Problem 1.8 Show that $u(x, y) = \log(x^2 + y^2)$ satisfies Laplace's equation $u_{xx} + u_{yy} = 0$ for $(x, y) \neq (0, 0)$.

Problem 1.9 Show that $u(x, y, z) = (x^2 + y^2 + z^2)^{-1/2}$ satisfies Laplace's equation $u_{xx} + u_{yy} + u_{zz} = 0$, for $(x, y, z) \neq (0, 0, 0)$.

Problem 1.10 Show that $u(r, \theta) = Br^n \sin(n\theta)$ satisfies the Laplace equation in polar coordinates:

$$u_{rr} + \frac{1}{r}u_r + \frac{1}{r^2}u_{\theta\theta} = 0.$$

Problem 1.11 Verify that

$$u = \frac{-2y}{x^2 + y^2 + 2x + 1}, \qquad v = \frac{x^2 + y^2 - 1}{x^2 + y^2 + 2x + 1}$$

both satisfy the Laplace equation, and sketch the curves u = constant and v = constant. Show that

$$u + iv = \frac{i(z - 1)}{z + 1}, \quad \text{where } z = x + iy.$$

Problem 1.12 Show that $u(x, t) = t^{-1/2} \exp(-x^2/4kt)$ satisfies the heat equation $u_t = ku_{xx}$, for $t > 0$.

Problem 1.13 Show that $u(x, y, t) = t^{-1} \exp[-(x^2 + y^2)/4kt]$ satisfies the heat equation $u_t = k(u_{xx} + u_{yy})$, for $t > 0$.

Problem 1.14 The spherically symmetric form of the heat conduction equation is given by

$$u_{rr} + \frac{2}{r}u_r = \frac{1}{\kappa}u_t.$$

Show that $v = ru$ satisfies the standard one-dimensional heat equation.

Problem 1.15 Show that the equation

$$\theta_t = \kappa\theta_{xx} - h(\theta - \theta_0)$$

can be reduced to the standard heat conduction equation by writing $u = e^{ht}(\theta - \theta_0)$. How do you interpret the term $h(\theta - \theta_0)$?

Problem 1.16 Use the substitution $\xi = x - vt,\ \ \eta = t$ to transform the one-dimensional convection–diffusion equation

$$u_t = ku_{xx} - vu_x,$$

into a heat equation for $\tilde{u}(\xi, \eta) = u(\xi + v\eta, \eta)$.

Problem 1.17 If $f \in C[0, 1]$, let $u(x, t)$ satisfy

$$\begin{cases} u_t = u_{xx}, & 0 < x < 1, \quad t > 0, \\ u(0, t) = u(1, t) = 0, & t \geq 0, \\ u(x, 0) = f(x), & 0 \leq x \leq 1. \end{cases}$$

Derive the identity $2u(u_t - u_{xx}) = (u^2)_t - (2uu_x)_x + 2u_x^2$.

Problem 1.18 Find the possible values of a and b in the expression $u = \cos at \sin bx$, such that it satisfies the wave equation $u_{tt} = c^2u_{xx}$.

Problem 1.19 Taking $u = f(x + \alpha t)$, where f is any function, find the values of α that will ensure u satisfies the wave equation $u_{tt} = c^2u_{xx}$.

Problem 1.20 The spherically symmetric version of the wave equation $u_{tt} = c^2u_{xx}$ takes the form

$$u_{tt} = c^2(u_{rr} + 2u_r/r).$$

Show, by putting $v = ru$, that it has a solution of the form

$$v = f(ct - r) + g(ct + r).$$

Problem 1.21 Let $\xi = x - ct$ and $\eta = x + ct$. Use the chain rule to show that

$$u_{tt} - c^2 u_{xx} = -4c^2 u_{\xi\eta}.$$

Problem 1.22 Show that the solution of the initial value problem

$$u_{tt} = c^2 u_{xx}, \qquad u(x,0) = f(x), \qquad u_t(x,0) = g(x),$$

satisfies *d'Alembert's* formula:

$$u(x,t) = \frac{1}{2}[f(x-ct) + f(x+ct)] + \frac{1}{2c}\int_{x-ct}^{x+ct} g(y)\, dy.$$

2

Mathematical Tools

This chapter is devoted to the most fundamental mathematical environments and tools that are necessary in our studies in the book. To establish a framework, we give a brief introduction of the concept of vector spaces, also referred to as linear spaces. We introduce the main function spaces with corresponding norms and state the basic inequalities of vital importance. We continue with an abstract framework and derive the fundamental solution of the Poisson's equation based on Green's functions. The Poisson's equation is also a milestone in the study of heat and wave equations. Finally, we establish the well-posedness of a canonical form for partial differential equations (PDEs) based on Riesz representation (Lax–Milgram) theorem.

2.1 Vector Spaces

Definition 2.1 A set V of functions or vectors is called a *linear space*, or a *vector space*, if for all $u, v, w \in V$ and all $\alpha, \beta \in \mathbb{R}$ (real numbers), we have

$$\begin{array}{ll} \text{(i)} & u + \alpha v \in V, \\ \text{(ii)} & (u+v)+w = u+(v+w), \\ \text{(iii)} & u+v = v+u, \\ \text{(iv)} & \exists\ 0 \in V \quad \text{such that} \quad u+0 = 0+u = u, \\ \text{(v)} & \forall\ u \in V,\ \exists\ (-u) \in V, \quad \text{such that} \quad u+(-u) = 0, \\ \text{(vi)} & (\alpha+\beta)u = \alpha u + \beta u, \\ \text{(vii)} & \alpha(u+v) = \alpha u + \alpha v, \\ \text{(viii)} & \alpha(\beta u) = (\alpha\beta)u. \end{array} \tag{2.1.1}$$

Definition 2.2 A subset $U \subset V$ of a vector space V is a *subspace* of V if

$$\alpha u + \beta v \in U, \quad \forall u, v \in U, \quad \text{and} \quad \forall \alpha, \beta \in \mathbb{R}.$$

An Introduction to the Finite Element Method for Differential Equations, First Edition.
Mohammad Asadzadeh.

Observe that (iii) and (i), with $\alpha = 1$ and $v = (-u)$ implies (iv): that 0 (zero vector) is an element of every vector space.

Definition 2.3 *(Scalar product)* A scalar product (inner product) is a real valued operator on $V \times V$, viz. $(u, v): V \times V \to \mathbb{R}$ (or $\langle u, v\rangle: V \times V \to \mathbb{R}$) such that for all $u, v, w \in V$ and all $\alpha \in \mathbb{R}$,

$$\begin{array}{lll} \text{(i)} & (u, v) = (v, u) & \text{(symmetry)} \\ \text{(ii)} & (u + \alpha v, w) = (u, w) + \alpha(v, w) & \text{(bi-linearity)} \\ \text{(iii)} & (v, v) \geq 0 \quad \forall v \in V & \text{(positivity)} \\ \text{(iv)} & (v, v) = 0 \Longleftrightarrow v = 0. & \end{array} \tag{2.1.2}$$

Definition 2.4 A vector space W is called an inner product, or scalar product, space if W is associated with a *scalar product* $(\cdot, \cdot)$, defined on $W \times W$.

Example 2.1 In the one-dimensional case, the space of continuous functions on an interval $[a, b]$: $C([a, b])$; the space of k-times continuously differentiable functions on $[a, b]$: $C^{(k)}((a, b))$; the space of all polynomials of degree $\leq q$ on $[a, b]$: $\mathcal{P}^{(q)}(a, b)$, all are examples of inner product spaces associated with a scalar product defined by

$$(u, v) := \int_a^b u(x)v(x)\, dx. \tag{2.1.3}$$

Note that the space of all polynomials of degree $= q$ on $[a, b]$ is not a vector space (let $q = 1$ and $u = x$, $v = -x + 1$, then $u + v$ is not of degree 1).

Definition 2.5 *(Orthogonality)* Two real-valued functions $u(x)$ and $v(x)$ are called orthogonal if $(u, v) = 0$. The orthogonality is denoted as $u \perp v$.

Definition 2.6 Let $\mathbf{x} = (x_1, x_2, \ldots, x_n) \in \mathbb{R}^n$ be a vector in $\mathbb{R}^n$. The *norm* of $\mathbf{x}$ is defined as its Euclidean length:

$$||\mathbf{x}|| = |\mathbf{x}| = (x_1^2 + x_2^2 + \cdots + x_n^2)^{1/2}.$$

Definition 2.7 *(The L_2-space and the L_2-norm)* The space of all square integrable functions over a given domain Ω in $\mathbb{R}^n$ is known as the $L_2(\Omega)$-space. If $u \in L_2(\Omega)$, then the L_2-norm of u associated with the above scalar product is defined by

$$\begin{aligned} ||u||_{L_2(\Omega)} &:= \sqrt{(u, u)} = \left(\int_\Omega |u(x)|^2 \, dx \right)^{1/2}, \\ L_2(\Omega) &:= \{u : \Omega \to \mathbb{R};\quad ||u||_{L_2(\Omega)} < \infty\}. \end{aligned} \tag{2.1.4}$$

In the L_2 case, we usually suppress the subscript and write $\|u\|$ for $\|u\|_{L_2(\Omega)}$. General L_p-spaces and norms, $1 \leq p \leq \infty$, are defined below.

Example 2.2 For the one-dimensional case of the interval $[a, b]$ above, we simply have

$$\|u\| = \left(\int_a^b |u(x)|^2 \, dx \right)^{1/2}. \tag{2.1.5}$$

In a direct connection with the inner product and L_2-norm, it is more natural to state and prove the two most useful inequalities (their general forms as well as other central inequalities are introduced in Section 2.3).

Cauchy–Schwarz' inequality. We have that

$$|(u, v)| \leq \|u\| \|v\|.$$

Proof: Using the inequality $2ab \leq a^2 + b^2$, we have

$$2|(u, w)| \leq \|u\|^2 + \|w\|^2. \tag{2.1.6}$$

We let $w = (v/\|v\|^2)\,(u, v)$, then

$$2|(u, w)| = 2\left|\left(u, (u, v)\frac{v}{\|v\|^2}\right)\right| \leq \|u\|^2 + |(u, v)|^2 \frac{\|v\|^2}{\|v\|^4} \tag{2.1.7}$$

Thus,

$$2\frac{|(u, v)|^2}{\|v\|^2} \leq \|u\|^2 + |(u, v)|^2 \frac{\|v\|^2}{\|v\|^4}, \tag{2.1.8}$$

which, multiplying by $\|v\|^2$, gives

$$2|(u, v)|^2 \leq \|u\|^2 \, \|v\|^2 + |(u, v)|^2. \tag{2.1.9}$$

Hence,

$$|(u, v)|^2 \leq \|u\|^2 \, \|v\|^2, \tag{2.1.10}$$

which completes the proof. □

Remark 2.1 The Cauchy–Schwarz, inequality can be stated and proved without specific reference to the L_2-norm. It is simply valid with the notation $\|\cdot\| := \sqrt{(\cdot, \cdot)}$. It is evident for the usual scalar product defined on an Euclidean space, viz.

$$(u, v) = \|u\| \|v\| \cos(u, v).$$

Example 2.3 As a special case, we have when $V = \mathbb{R}^n$ and $\mathbf{x},\ \mathbf{y} \in \mathbb{R}^n$,

$$(\mathbf{x}, \mathbf{y}) := \mathbf{x}^T \cdot \mathbf{y} = \sum_{j=1}^{n} x_j y_j \leq \left(\sum_{j=1}^{n} x_j^2\right)^{1/2} \left(\sum_{j=1}^{n} y_j^2\right)^{1/2} = \|\mathbf{x}\|\,\|\mathbf{y}\|.$$

Using the Cauchy–Schwarz' inequality, we can prove the
Triangle inequality:

$$\|u + v\| \leq \|u\| + \|v\|.$$

Proof: By the definition of L_2-norm and Cauchy–Schwarz' Inequality, we have

$$\begin{aligned}\|u + v\|^2 &= (u + v, u + v) = \|u\|^2 + 2(u, v) + \|v\|^2 \\ &\leq \|u\|^2 + 2\|u\|\,\|v\| + \|v\|^2 = (\|u\| + \|v\|)^2,\end{aligned}$$

and the proof is complete. □

Remark 2.2 In the sequel, we keep the notion of the scalar product $(\cdot,\ \cdot)$ for integration over the domains $\Omega\ \ (\subset \mathbb{R}^n)$, $n \geq 1$ and use $\langle u, v\rangle$ for the integration over $\partial\Omega$, the boundary of Ω. Hence, for the scalar product of two integrable functions f and g defined on a set Ω, we have

$$(f, g) := \int_{\Omega} f(x)g(x)\, dx, \quad \langle f, g\rangle := \int_{\partial\Omega} f(x)g(x)\, d\sigma.$$

2.1.1 Linear Independence, Basis, and Dimension

In this part, we include a short review of the concepts of linear independence, linear dependence, basis, and dimension. A more elaborate approach to these concepts is the subject of a linear algebra course. The easiest way described for the geometric vectors in 2 or 3d: Loosely speaking three geometric vectors in $\mathbb{R}^3$ are linearly independent if they are not lying in the same plane. Such vectors may be used to construct a basis in $\mathbb{R}^3$. This is generalized to $\mathbb{R}^n$ as well as to arbitrary vector space V. Let $\{\mathbf{x}_1, \mathbf{x}_2, \dots, \mathbf{x}_k\}$ be a set of k vectors in V. A linear combination of these vectors is a vector of the form

$$\alpha_1\mathbf{x}_1 + \alpha_2\mathbf{x}_2 + \cdots + \alpha_k\mathbf{x}_k, \quad \alpha_j \in \mathbb{R}, \quad j = 1, 2, \dots, k.$$

The set of all such linear combination forms a subspace of V which we denote by $S = \text{span}\ \{\mathbf{x}_1, \mathbf{x}_2, \dots, \mathbf{x}_k\}$.

Definition 2.8 A set of vectors $\mathbf{x}_1, \mathbf{x}_2, \dots, \mathbf{x}_n$ is said to span V, if every element in V can be written as a linear combination of $\mathbf{x}_1, \mathbf{x}_2, \dots, \mathbf{x}_n$. That is, $V = \text{span}\ \{\mathbf{x}_1, \mathbf{x}_2, \dots, \mathbf{x}_n\}$.

Example 2.4 $\mathbb{R}^n$ is the span of

$$\mathbf{e}_1 = (1, 0, \dots, 0), \ \mathbf{e}_2 = (0, 1, \dots, 0), \ \dots, \ \mathbf{e}_n = (0, 0, \dots, 1).$$

An arbitrary vector $\mathbf{x} \in \mathbb{R}^n$ can be written as

$$\mathbf{x} = (x_1, x_2, \dots, x_n) = x_1\mathbf{e}_1 + x_2\mathbf{e}_2 + \cdots + x_n\mathbf{e}_n.$$

Definition 2.9 A set $S = \{\mathbf{x}_1, \mathbf{x}_2, \dots, \mathbf{x}_k\}$ of vectors in a vector space V (or in $\mathbb{R}^n$) is said to be linearly independent if, with the scalars $\alpha_1, \alpha_2, \dots, \alpha_k$,

$$\alpha_1\mathbf{x}_1 + \alpha_2\mathbf{x}_2 + \cdots + \alpha_k\mathbf{x}_k = 0 \quad \Longrightarrow \quad \alpha_1 = \alpha_2 = \cdots = \alpha_k = 0. \tag{2.1.11}$$

S is said to be linearly dependent if there exists constants $\alpha_1, \alpha_2, \dots, \alpha_k$, not all zero, such that $\alpha_1\mathbf{x}_1 + \alpha_2\mathbf{x}_2 + \cdots + \alpha_k\mathbf{x}_k = 0$.

Example 2.5 The vectors $\mathbf{e}_1, \ \mathbf{e}_2, \dots, \mathbf{e}_n \in \mathbb{R}^n$ are linearly independent, since if

$$\alpha_1\mathbf{e}_1 + \alpha_2\mathbf{e}_2 + \cdots + \alpha_n\mathbf{e}_n = 0,$$

then $\alpha_1 = \alpha_2 = \cdots = \alpha_n = 0$.

Remark 2.3 It should be emphasized that for any set of vectors $\{\mathbf{x}_1, \dots, \mathbf{x}_k\}$, Eq. (2.1.11) always holds if we choose the scalars $\alpha_1 = \cdots = \alpha_k = 0$. The point is whether or not it is possible to satisfy (2.1.11) with at least one of the scalars not equal to zero.

The procedure of determining that the set S is linearly dependent or linearly independent is as follows:

(I) Write down Eq. (2.1.11) (left identity) as an homogeneous system for the coefficients $\alpha_1, \alpha_2, \dots, \alpha_k$.
(II) If the homogeneous system in (I) has only the trivial solution $\alpha_1 = \alpha_2 = \cdots = \alpha_k = 0$, then the set of vectors $S = \{\mathbf{x}_1, \mathbf{x}_2, \dots, \mathbf{x}_k\}$ is linearly independent; if it has nontrivial solution, then the set of vectors is linearly dependent.

Example 2.6 Consider the vectors $\mathbf{x}_1 = (1, 0, 1, 2)$, $\mathbf{x}_2 = (0, 1, 1, 2)$, $\mathbf{x}_3 = (1, 1, 1, 3) \in \mathbb{R}^4$. To determine whether $S = \{\mathbf{x}_1, \mathbf{x}_2, \mathbf{x}_3\}$ is linearly independent or linearly dependent, we write the homogeneous equation

$$\alpha_1\mathbf{x}_1 + \alpha_2\mathbf{x}_2 + \alpha_3\mathbf{x}_3 = 0,$$

in the system form and solve it for $\alpha_1, \ \alpha_2$, and α_3:

$$\alpha_1 + \alpha_3 = 0, \ \ \alpha_2 + \alpha_3 = 0, \ \ \alpha_1 + \alpha_2 + \alpha_3 = 0, \ \ 2\alpha_1 + 2\alpha_2 + 3\alpha_3 = 0$$

which has only the trivial solution

$$\alpha_1 = \alpha_2 = \alpha_3 = 0.$$

Hence, $S = \{\mathbf{x}_1, \mathbf{x}_2, \mathbf{x}_3\}$ is linearly independent.

Example 2.7 Consider the vectors $\mathbf{x}_1 = (1, 2, -1)$, $\mathbf{x}_2 = (1, -2, 1)$, $\mathbf{x}_3 = (-3, 2, -1)$, $\mathbf{x}_4 = (2, 0, 0) \in \mathbb{R}^3$.

The equation

$$\alpha_1 \mathbf{x}_1 + \alpha_2 \mathbf{x}_2 + \alpha_3 \mathbf{x}_3 + \alpha_4 \mathbf{x}_4 = 0,$$

yields the homogeneous equation system

$$\alpha_1 + \alpha_2 - 3\alpha_3 + 2\alpha_4 = 0, \quad 2\alpha_1 - 2\alpha_2 + 2\alpha_3 = 0, \quad -\alpha_1 + \alpha_2 - \alpha_3 = 0$$

consisting of three equations with four unknowns, which has always a nontrivial solution. Hence, the vectors $\mathbf{x}_1$, $\mathbf{x}_2$, $\mathbf{x}_3$, and $\mathbf{x}_4$ are linearly dependent. In fact, we have infinitely many nontrivial solutions. Here are two of them:

$$\begin{aligned} &\alpha_1 = 1, \quad \alpha_2 = 2, \quad \alpha_3 = 1, \quad \alpha_4 = 0 \\ &\alpha_1 = 1, \quad \alpha_2 = 1, \quad \alpha_3 = 0, \quad \alpha_4 = -1. \end{aligned}$$

Example 2.8 Consider the polynomials

$$p_1(x) = x^2 + x + 2, \quad p_2(x) = 2x^2 + x, \quad p_3(x) = 3x^2 + 2x + 2.$$

To find out whether $S = \{p_1(x), p_2(x), p_3(x)\}$ is linearly dependent or linearly independent, we set up the equation

$$\alpha_1 p_1(x) + \alpha_2 p_2(x) + \alpha_3 p_3(x) = 0,$$

and solve for α_1, α_2, α_3. The resulting homogeneous system is

$$\alpha_1 + 2\alpha_2 + 3\alpha_3 = 0, \quad \alpha_1 + \alpha_2 + 2\alpha_3 = 0, \quad 2\alpha_1 + 2\alpha_3 = 0,$$

which has infinitely many solutions (verify). A particular solution is $\alpha_1 = \alpha_1 = 1$, $\alpha_3 = -1$, which yields

$$p_1(x) + p_2(x) - p_3(x) = 0.$$

Thus, S is linearly dependent.

Below are some basic definitions, lemmas, and theorems (without proofs) that are of crucial importance in the study of the finite dimensional vector spaces.

Definition 2.10 A set of vectors $S = \{\mathbf{x}_1, \mathbf{x}_2, \dots, \mathbf{x}_n\}$ in V is said to be a basis for V if S spans V and S is linearly independent.

Definition 2.11 A linear space V is said to have dimension n, we write $\dim V = n$, if n is the maximum number of linear independent vectors in V. If there is no such n, i.e. if one finds subsets of V with arbitrary number of linear independent vectors, then V is said to be infinite dimensional.

Example 2.9

- $\mathbb{R}^n$ has dimension n, since the set $\{\mathbf{e}_i\}_{i=1}^n$ is linearly independent and spans $\mathbb{R}^n$.

- The set of polynomials $\mathcal{P}^{(q)}$ has dimension $q+1$, due to the fact that the monomials $p_1(x)=1, p_2(x)=x, \dots, p_n(x)=x^q$ are linearly independent and span $\mathcal{P}^{(q)}$.
- The space $C[a,b]$ is infinite dimensional, since we can find arbitrary number of linear independent functions in $C[a,b]$. For example the monomials $p_1(x)=1, p_2(x)=x, \dots, p_n(x)=x^q$ are continuous and linearly independent functions for every q.

Lemma 2.1 *Assume that $\boldsymbol{x}_1, \boldsymbol{x}_2, \dots, \boldsymbol{x}_k$ are linearly independent vectors in V. If $\boldsymbol{x} \in V$ and $\boldsymbol{x} \notin$ span $\{\boldsymbol{x}_1, \boldsymbol{x}_2, \dots, \boldsymbol{x}_k\}$, then $\boldsymbol{x}_1, \boldsymbol{x}_2, \dots, \boldsymbol{x}_k, \boldsymbol{x}$ are linearly independent.*

Theorem 2.1 *Assume that* $\dim V = n > 0$. *Then, there is at least one set consisting of n linearly independent vectors in V. Every such set is a basis of V.*

Theorem 2.2 *Assume that V is the span of the vectors $\boldsymbol{x}_1, \boldsymbol{x}_2, \dots, \boldsymbol{x}_n$. Then every combination of more than n different vectors in V is linearly dependent.*

Theorem 2.3 *Assume that V is finite dimensional. Then all bases for V have the same number of elements which is the same number as the dimension of V.*

Theorem 2.4 *Every element in V can be, uniquely, represented as a linear combination of a given basis vectors. This representation is different (has different coefficients α_i) for different bases.*

2.2 Function Spaces

2.2.1 Spaces of Differentiable Functions

In Chapter 1, to formulate some basic PDEs, we introduced certain function spaces, e.g. the space of continuous functions. Accurate finite element approximations for PDEs rely on the maximal available regularity of their exact solutions: smoothness properties that are specified through belonging to certain class of function spaces. Below we introduce some function spaces that are fundamental in the study of PDEs.

Definition 2.12 Let $\Omega \subset \mathbb{R}^n$ be a bounded open set, then $C^{(k)}(\overline{\Omega})$ is the set of all functions $u \in C^{(k)}(\Omega)$ such that $D^\alpha u$ can be extended from Ω to its closure $\overline{\Omega}$, for all multiindex $\alpha = (\alpha_1, \dots, \alpha_n)$, $|\alpha| \le k$. The space $C^{(k)}(\overline{\Omega})$ is equipped with supremum norm defined by

$$\|u\|_{C^{(k)}(\overline{\Omega})} := \sum_{|\alpha| \le k} \sup_{x \in \Omega} |D^\alpha u(x)|.$$

Hence, e.g. for $k = 1$, we have

$$\|u\|_{C^{(1)}(\overline{\Omega})} := \sum_{|\alpha|\leq 1} \sup_{x\in\Omega} |D^\alpha u(x)| = \sup_{x\in\Omega} |u(x)| + \sum_{i=1}^{n} \sup_{x\in\Omega} \left|\frac{\partial u}{\partial x_i}(x)\right| .$$

For simplicity, for a bounded domain $\Omega \subset \mathbb{R}^n$, we define

$$C(\overline{\Omega}) := C^0(\overline{\Omega}) = \{u: \|u\|_{C(\overline{\Omega})} = \|u\|_{C^0(\overline{\Omega})} < \infty\}$$

where

$$\|u\|_{C(\overline{\Omega})} := \|u\|_{C^0(\overline{\Omega})} = \sup_{x\in\Omega} |u(x)| = \max_{x\in\Omega} |u(x)|.$$

For convenience, the frequently used spaces $C^{(k)}, \quad k = 0, 1, 2, \infty$ are denoted by $C^0, \ C^1, \ C^2$, and C^∞, respectively.

Example 2.10 Consider the open interval $\Omega = (a, b) \subset \mathbb{R}$. The function $u(x) = 1/(x - a) \in C^{(k)}(a, b), \quad \forall k$. However, since $\lim_{x\to a} u(x) = \infty$, u (and its derivatives) is not continuous in $\overline{\Omega} = [a, b]$. Hence for any $k \geq 0$, $u \notin C^{(k)}[a, b]$.

Definition 2.13 Let $u \in C(\Omega)$, $\Omega \subset \mathbb{R}^n$ we define the *support* of u as the closed set

$$\text{supp } u := \overline{\{x \in \Omega, u(x) \neq 0\}}.$$

The set of all functions $u \in C^{(k)}(\Omega)$ whose support is a bounded subset of Ω is denoted by $C_0^{(k)}(\Omega)$. We define $C_0^\infty(\Omega)$ as

$$C_0^\infty(\Omega) := \cap_{k\geq 0} C_0^{(k)}(\Omega). \tag{2.2.1}$$

Example 2.11 The function u is defined in $\mathbb{R}^n$ by

$$u(x) = \begin{cases} e^{-\frac{1}{a-|x|}}, & |x| < a, \\ 0, & \text{else.} \end{cases} \tag{2.2.2}$$

Note that $\text{supp } u = \{x \in \mathbb{R}^n: |x| \leq a\}$ and $u \in C_0^\infty(\mathbb{R}^n)$.

2.2.2 Spaces of Integrable Functions

Below we introduce a class of *Lebesgue integrable* functions on an open set $\Omega \subset \mathbb{R}^n$ (or $\Omega = \mathbb{R}^n$). For a real number $p \geq 1$, we define the space $L_p(\Omega)$ by

$$L_p(\Omega) := \left\{u: \int_\Omega |u(x)|^p \, dx < \infty\right\}, \qquad 1 \leq p < \infty,$$

$$L_\infty(\Omega) := \{ u: \text{ess.} \sup_{x\in\Omega} |u(x)| < \infty\}.$$

In this setting, two functions $u,\ v \in L_p(\Omega)$ are identical if $u = v$ except on a set of measure zero, then we call $u = v$ *almost everywhere* (and denote it by $u = v$ a.e.). The $L_p(\Omega)$ is associated with the norm

$$\|u\|_{L_p(\Omega)} := \left(\int_\Omega |u(x)|^p \, dx\right)^{1/p}, \qquad 1 \leq p < \infty$$
$$\|u\|_{L_\infty(\Omega)} := \text{ess.}\sup_{x\in\Omega} |u(x)|,$$

where the latter is called the *L-infinity* norm which is also known as *maximum-norm* in case the domain Ω is bounded.

Remark 2.4 In PDEs modeling density of particles $L_1(\Omega)$-norm corresponds to a measure for the mass. Of particular interest is the $L_2(\Omega)$-norm (which can be related to measuring the energy) defined as

$$\|u\|_{L_2(\Omega)} := \left(\int_\Omega |u(x)|^2 \, dx\right)^{1/2},$$

and the scalar product

$$(u, v) := \int_\Omega u(x)v(x) \, dx,$$

viz. $\|u\|_{L_2(\Omega)} = (u, u)^{1/2} \geq 0$ (with equality only if $u \equiv 0$).

2.2.3 Weak Derivative

Let Ω be an open subset of $\mathbb{R}^n$ and assume that $u \in C^{(k)}(\Omega)$, for some integer k. Then, we have that

$$\int_\Omega D^\alpha u(x)v(x) \, dx = (-1)^{|\alpha|} \int_\Omega u(x) \, D^\alpha v(x) \, dx, \quad |\alpha| \leq k, \quad \forall v \in C_0^\infty(\Omega).$$

This motivates the following definition:

Definition 2.14 Let $\omega \subset \Omega$ be a bounded and open subset of Ω with $\overline{\omega} \subset \Omega$. Suppose that $u \in L_1(\omega)$, $\forall \omega$ (i.e. u is locally integrable on Ω). Suppose, further, that there exists a locally integrable function χ_α on Ω such that

$$\int_\Omega \chi_\alpha(x)v(x) \, dx = (-1)^{|\alpha|} \int_\Omega u(x) \, D^\alpha v(x) \, dx, \quad \forall v \in C_0^\infty(\Omega),$$

then χ_α is called a *weak derivative* of u of order $|\alpha|$, and we write $D^\alpha u = \chi_\alpha$.

Obviously, for sufficiently smooth u (e.g. $u \in C^{(k)}(\Omega)$), χ_α, $|\alpha| \leq k$, coincides with its corresponding classical partial derivative:

$$\frac{\partial^{|\alpha|} u}{\partial x_1{}^{\alpha_1} \partial x_2{}^{\alpha_2} \dots \partial x_n^{\alpha_n}}.$$

Example 2.12 Consider $u(x) = (1 - |x|)_+$ defined on $\mathbb{R}$. Note that u is not differentiable at 0 and ± 1, but it is locally integrable on $\Omega := \mathbb{R}$ and may have a weak derivative. Let now $v \in C_0^\infty(\Omega) = C_0^\infty(-\infty, \infty)$, then

$$\begin{aligned}\int_{-\infty}^{\infty} u(x)v'(x)\,dx &= \int_{-\infty}^{\infty} (1-|x|)_+\, v'(x)\,dx = \int_{-1}^{1} (1-|x|)v'(x)\,dx \\ &= \int_{-1}^{0} (1+x)v'(x)\,dx + \int_{0}^{1} (1-x)v'(x)\,dx \\ &= (1+x)v(x)|_{-1}^{0} - \int_{-1}^{0} v(x)\,dx + (1-x)v(x)|_0^1 + \int_0^1 v(x)\,dx \\ &= \int_{-1}^{0} (-1)v(x)\,dx + \int_0^1 (1)v(x)\,dx \equiv \int_{-\infty}^{\infty} \chi(x)v(x)\,dx,\end{aligned}$$

where

$$\chi(x) = \begin{cases} 0, & x \notin [-1,1], \\ -1, & x \in (-1,0), \\ +1, & x \in (0,1). \end{cases}$$

Thus, the piecewise constant function $\chi(x)$ is the first weak derivative of the continuous linear function u, and $\chi = u' = Du$.

2.2.4 Sobolev Spaces

Definition 2.15 Let Ω be an open subset of $\mathbb{R}^n$. For a nonnegative integer k and $p \in [1, \infty]$, we define the *Sobolev space* of order k by

$$W_p^k(\Omega) := \{u \in L_p(\Omega) \colon D^\alpha u \in L_p(\Omega),\ \ |\alpha| \le k\}. \tag{2.2.3}$$

The corresponding *Sobolev norms* are

$$\|u\|_{W_p^k(\Omega)} := \left(\sum_{|\alpha| \le k} \|D^\alpha u\|_{L_p(\Omega)}^p \right)^{1/p}, \quad 1 \le p < \infty \tag{2.2.4}$$

and

$$\|u\|_{W_\infty^k(\Omega)} := \sum_{|\alpha| \le k} \|D^\alpha u\|_{L_\infty(\Omega)}. \tag{2.2.5}$$

We also define the *seminorms* as

$$|u|_{W_p^k(\Omega)} := \left(\sum_{|\alpha| = k} \|D^\alpha u\|_{L_p(\Omega)}^p \right)^{1/p}, \quad 1 \le p < \infty. \tag{2.2.6}$$

Thus, for $1 \le p < \infty$ we may write

$$\|u\|_{W_p^k(\Omega)} = \left(\sum_{j=0}^{k} |u|_{W_p^j(\Omega)}^p \right)^{1/p}. \tag{2.2.7}$$

Further in the L_∞-case, letting

$$|u|_{W^k_\infty(\Omega)} := \sum_{|\alpha|=k} \|D^\alpha u\|_{L_\infty(\Omega)}, \tag{2.2.8}$$

we write

$$\|u\|_{W^k_\infty(\Omega)} = \sum_{j=0}^{k} |u|_{W^j_\infty(\Omega)}. \tag{2.2.9}$$

Remark 2.5 For $k = 0$, $|\cdot|_{W^k_p(\Omega)}$ is the usual L_p-norm. The wording *seminorm* is used when $k \geq 1$.

2.2.5 Hilbert Spaces

The special cases $p = 1, 2, \infty$, and $k = 1, 2$ are the most frequently used function spaces, where in particular $p = 2$ and $k = 1, 2$ called the *Hilbert spaces* and denoted by $H^k(\Omega)$, $\Omega \subset \mathbb{R}^n$, viz.

$$H^1(\Omega) := \left\{ u \in L_2(\Omega) \left| \frac{\partial u}{\partial x_j} \in L_2(\Omega), \quad j = 1, \dots, n \right. \right\} \tag{2.2.10}$$

$$\|u\|_{H^1(\Omega)} := \left(\|u\|^2_{L_2(\Omega)} + \sum_{j=1}^{n} \left\| \frac{\partial u}{\partial x_j} \right\|^2_{L_2(\Omega)} \right)^{1/2},$$

$$|u|_{H^1(\Omega)} := \left(\sum_{j=1}^{n} \left\| \frac{\partial u}{\partial x_j} \right\|^2_{L_2(\Omega)} \right)^{1/2}.$$

and

$$H^2(\Omega) := \left\{ u \in L_2(\Omega) : \frac{\partial u}{\partial x_j} \in L_2(\Omega), \quad j = 1, \dots, n, \quad \text{and} \right.$$

$$\left. \frac{\partial^2 u}{\partial x_i \partial x_j} \in L_2(\Omega), \quad i,j = 1, \dots, n \right\}.$$

$$\|u\|_{H^2(\Omega)} := \left(\|u\|^2_{L_2(\Omega)} + \sum_{j=1}^{n} \left\| \frac{\partial u}{\partial x_j} \right\|^2_{L_2(\Omega)} + \sum_{i,j=1}^{n} \left\| \frac{\partial^2 u}{\partial x_i \partial x_j} \right\|^2_{L_2(\Omega)} \right)^{1/2},$$

$$|u|_{H^2(\Omega)} := \left(\sum_{i,j=1}^{n} \left\| \frac{\partial^2 u}{\partial x_i \partial x_j} \right\|^2_{L_2(\Omega)} \right)^{1/2}.$$

Further, solving the *homogeneous problems*, (e.g. a Dirichlet boundary value problem [BVP]), we shall use a particular Sobolev space: $H^1_0(\Omega)$, which is defined below.

Definition 2.16 The closer of $C^\infty_0(\Omega)$ in the norm $\|u\|_{H^1(\Omega)}$ is the set of all $u \in H^1(\Omega)$ such that u is the $H^1(\Omega)$ limit of a sequence $\{u_m\}^\infty_{m=1} \subset C^\infty_0(\Omega)$. This closer defines a special space denoted by $H_0^1(\Omega)$, which for sufficiently smooth $\partial\Omega$, can be identified as

$$H_0^1(\Omega) := \{u \in H^1(\Omega) | u = 0 \text{ on } \partial\Omega\}.$$

We emphasis that $H_0^1(\Omega)$ is a Hilbert space with the same norm as $H^1(\Omega)$.

Finally, we state (without proof) a very crucial theorem in PDE namely *Sobolev embedding theorem*. For a proof, we refer the reader to Adams [1], Sobolev spaces.

Theorem 2.5 (Sobolev embedding theorem) *We have that*

$$W_p^m(\Omega) \subset C^{(l)}(\overline{\Omega}), \quad for \ m - l > \frac{n}{p}, \quad 1 \le p < \infty,$$

and the identity operator ID: $W_p^m(\Omega) \to C^{(l)}(\overline{\Omega})$ *is a bounded linear operator.*

2.3 Some Basic Inequalities

Below we introduce the most important inequalities used in our estimates. First, we state (without proofs) the general forms of the triangle and Cauchy–Schwarz' inequalities that we derived above. To this end we define the concept of *conjugate exponents*:

Definition 2.17 If p and q are positive numbers such that $p + q = pq$, or equivalently

$$\frac{1}{p} + \frac{1}{q} = 1, \tag{2.3.1}$$

then we call p and q being a pair of conjugate exponents. We have $1 \le p, q \le \infty$, and in particular $p = q = 2$ is a conjugate exponent pair.

Theorem 2.6 (Minkowski's and Hölder's inequalities) *Let p and q be conjugate exponents,* $1 < p < \infty$. *Then for* $u,\ v \in L_p(\Omega)$, *we have the Minkowski's inequality:*

$$\|u + v\|_{L_p(\Omega)} \le \|u\|_{L_p(\Omega)} + \|v\|_{L_p(\Omega)}. \tag{2.3.2}$$

If $u \in L_p(\Omega)$ *and* $v \in L_q(\Omega)$, *then we have the Hölder's inequality:*

$$\int_\Omega u(x)v(x)\, dx \le \|u\|_{L_p(\Omega)} \|v\|_{L_q(\Omega)}. \tag{2.3.3}$$

For a proof, we refer to Rudin [121].

Next, we prove the Poincaré inequality.

Theorem 2.7 (Poincaré inequality) *Assume that u, the solution of a homogeneous Dirichlet BVP, and $|\nabla u|$ are square integrable functions in a bounded domain $\Omega \subset \mathbb{R}^d$. Then, there exists a constant C_Ω, independent of u such that*

$$\|u\| \le C_\Omega \|\nabla u\|. \tag{2.3.4}$$

Proof: We give separate argument for $d = 1$ and $d = 2$ (the argument for the one-dimensional case is easily generalized to two dimensions). For a general proof, see, e.g. Brenner–Scott [37].

(I) *The 1d case.* We restate the theorem in 1d. Let u and u' be square integrable functions in an interval $[0, L]$. There exists a constant C_L, independent of u, but depends on L, such that if $u(0) = 0$, then

$$\int_0^L u(x)^2\, dx \le C_L \int_0^L u'(x)^2\, dx, \quad \text{i.e. } \|u\| \le \sqrt{C_L}\,\|u'\|. \tag{2.3.5}$$

For $x \in [0, L]$, using Cauchy–Schwarz' inequality, we may write

$$\begin{aligned} u(x) &= \int_0^x u'(y)\, dy \le \int_0^x |u'(y)|\, dy = \int_0^x |u'(y)| \cdot 1\, dy \\ &\le \left(\int_0^L |u'(y)|^2\, dy\right)^{1/2} \cdot \left(\int_0^L 1^2\, dy\right)^{1/2} = \sqrt{L}\left(\int_0^L |u'(y)|^2\, dy\right)^{1/2}. \end{aligned}$$

Thus, squaring both sides and integrating over $[0, L]$, we get

$$\int_0^L u(x)^2\, dx \le \int_0^L L\left(\int_0^L |u'(y)|^2\, dy\right)\, dx = L^2 \int_0^L |u'(y)|^2\, dy. \tag{2.3.6}$$

Hence,

$$\|u\| \le L\,\|u'\|. \tag{2.3.7}$$

(II) *The 2d case.* Let φ satisfy $\Delta\varphi = 1$ in Ω, and $2|\nabla\varphi| \le C_\Omega$ in Ω, (it is easy to construct such a function φ, e.g. take $\varphi(x, y) = \frac{1}{4}(x^2 + y^2)$), then by Green's formula and using the boundary condition on u, we get

$$\|u\|^2 = \int_\Omega u^2 \Delta\varphi = -\int_\Omega 2u(\nabla u \cdot \nabla\varphi) \le C_\Omega \|u\|\ \|\nabla u\|. \tag{2.3.8}$$

Thus, we have the desired result.

$$\|u\| \le C_\Omega \|\nabla u\|.$$

□

Remark 2.6 In the one-dimensional case, the constant $C_L = L^2$ indicates that the Poincare inequality is valid for arbitrary bounded intervals, but not for unbounded intervals. If $u(x_0) \neq 0$ for $x_0 \in [0, L]$, then by a similar argument as above, we get the following version of the one-dimensional Poincare inequality:

$$||u||^2_{L_2(0,L)} \leq 2L(u(x_0)^2 + \|u'\|^2_{L_2(0,L)}). \tag{2.3.9}$$

We note that in (2.3.4), as in the one-dimensional case, we have $C_\Omega \approx \text{diam}(\Omega)$, and therefore (2.3.4) is valid as long as Ω is a bounded domain.

Further, we state the trace inequality that estimates the L_p-norm of a function at the boundary of a sufficiently smooth domain Ω, i.e. $\|u\|_{L_p(\partial\Omega)}$ by a product of its $L_p(\Omega)$- and $W^1_p(\Omega)$-norm (for a proof, see [37]):

Theorem 2.8 (Trace theorem) *Let $u \in W^1_p(\Omega)$, then there is a constant C such that, for $1 \leq p \leq \infty$, we have*

$$\|u\|_{L_p(\partial\Omega)} \leq C\|u\|^{1-1/p}_{L_p(\Omega)} \|u\|^{1/p}_{W^1_p(\Omega)}. \tag{2.3.10}$$

In particular, for $p = 2$, we have that

$$\|u\|^2_{L_2(\partial\Omega)} \leq C\|u\|_{L_2(\Omega)} \|u\|_{H^1(\Omega)}. \tag{2.3.11}$$

An important *inequality* to use is a consequence of another inequality. They together comprise the following:

Lemma 2.2 (Grönwall's lemma) *Suppose that α, β, and φ are real valued functions in an interval I and φ is nonnegative and continuous with*

$$\varphi(t) \leq \alpha(t) + \int_0^t \beta(s)\varphi(s)\, ds, \quad t > 0.$$

(a) If β is nonnegative, then

$$\varphi(t) \leq \alpha(t) + \int_0^t \alpha(s)\beta(s) \, \exp\left(\int_s^t \beta(r)\, dr\right) ds, \quad t > 0.$$

(b) If, in addition, α is nondecreasing, then

$$\varphi(t) \leq \alpha(t) \exp\left(\int_0^t \beta(s)\, ds\right).$$

Proof:

(a) We define

$$\psi(s) = \exp\left(-\int_0^s \beta(r)\, dr\right) \int_0^s \beta(r)\varphi(r)\, dr, \quad s > 0.$$

Then, by the fundamental theorem of calculus, we may differentiate ψ:

$$\psi'(s) = \left(\varphi(s) - \int_0^s \beta(r)\varphi(r)\, dr\right) \beta(s) \exp\left(-\int_0^s \beta(r)\, dr\right), \quad s > 0.$$

Thus, using the assumption of nonnegativity of α, and integrating the above derivative, we get

$$\psi(t) \leq \int_0^t \alpha(s)\beta(s)\,ds \exp\left(-\int_0^s \beta(r)\,dr\right), \quad s > 0. \tag{2.3.12}$$

Now, by the definition of ψ we have, using (2.3.12), that

$$\begin{aligned} \textstyle\int_0^t \beta(s)\varphi(s)\,ds &= \exp\left(\int_0^s \beta(r)\,dr\right)\psi(t) \\ &\leq \int_0^t \alpha(s)\beta(s)\exp\left(\int_0^t \beta(r)\,dr - \int_0^s \beta(r)\,dr\right)\,dr, \end{aligned} \tag{2.3.13}$$

which completes the proof of part (a).

(b) If α is nondecreasing, then $\alpha(s) \leq \alpha(t)$ for $s \leq t$, part (a), and the fundamental theorem of calculus imply that

$$\begin{aligned} \varphi(t) &\leq \alpha(t) + \left(-\alpha(t) \times \exp\left(\int_s^t \beta(r)\,dr\right)\Big|_{s=0}^{s=t}\right) \\ &= \alpha(t)\exp\left(\int_0^t \beta(r)\,dr\right), \quad t > 0. \end{aligned} \tag{2.3.14}$$

□

Finally, we shall make use of the following *Sobolev inequality* which is a direct consequence of the *Sobolev embedding theorem.*

Theorem 2.9 (Sobolev inequality) *There exists a constant* $C = C_{m,n,p}$ *such that*

$$\|u\|_{C^l(\overline{\Omega})} \leq C_{m,n,p}\|u\|_{W_p^m(\Omega)}, \quad \textit{for } m - l > n/p, \ 1 \leq p < \infty.$$

2.4 Fundamental Solution of PDEs[1]

One classical approach to solve PDEs is through Fourier techniques, e.g. multidimensional Fourier series, Laplace, and Fourier transforms and separation of variables technique: a procedure designed to fit more idealized models. For "real-world" problems, this however is, if applicable, less efficient, and efforts in this direction often lead to cumbersome calculus procedures. An alternative and more general approach is through using Green's functions. Below we shall introduce a general framework and compute the Green's functions for the Laplacian in some common geometries using the fundamental solution (another approach is the: *method of images*, see, e.g. [123]). To this end, we derive fundamental

[1]This section maybe skipped in a first reading.

solution for the Poisson equation through deriving rotationally symmetric solution $u(\mathbf{x}) = v(r)$, $r = |\mathbf{x}|$, $\mathbf{x} = (x_1, x_2, \dots, x_n)$ for the Laplace's equation:

$$\Delta u(\mathbf{x}) = v''(r) + \frac{n-1}{r} v'(r) = 0.$$

This yields the following ordinary differential equation for v:

$$\frac{v''}{v'} = -\frac{n-1}{r},$$

so that

$$\log\, v' = -(n-1) \log\, r + C.$$

Hence,

$$v(r) = \begin{cases} \dfrac{p}{r^{n-2}} + q, & \text{for } n \geq 3 \\ p \log\, r + q, & \text{for } n = 2. \end{cases}$$

The fundamental solution $\Phi(\mathbf{x})$ for the Laplace's equation is defined by setting $q = 0$. However, to use $\Phi(\mathbf{x})$ to determine the fundamental solution for the Poisson's equation, we choose p that normalizes $\Phi(\mathbf{x})$:

$$\Phi(\mathbf{x}) = \begin{cases} \dfrac{1}{n(n-2)|B_1(n)|} \dfrac{1}{|\mathbf{x}|^{n-2}}, & \text{for } n \geq 3 \\ -\dfrac{1}{2\pi} \log |\mathbf{x}|, & \text{for } n = 2, \end{cases}$$

where $B_1(n)$ is the volume of the unit ball in $\mathbb{R}^n$.

Remark 2.7 Note that Φ is integrable on bounded sets (even though it is not defined at $\mathbf{x} = 0$) and has integrable singularity at $\mathbf{x} = 0$. Nevertheless, the singularity of $-\Delta\Phi$ is not integrable, but it is the singular Dirac's δ function.

Now, we formulate the fundamental solution for the Poisson's equation in $\mathbb{R}^n$.

Theorem 2.10 *Suppose that $f \in C^2(\mathbb{R}^n)$ has compact support. Then the solution for the Poisson's equation*

$$-\Delta u(\boldsymbol{x}) = f(\boldsymbol{x}) \quad \boldsymbol{x} \in \mathbb{R}^n, \tag{2.4.1}$$

is given by the convolution product of f with the fundamental solution Φ of the corresponding Laplace's equation:

$$u(\boldsymbol{x}) = (\Phi * f)(\boldsymbol{x}) = \int_{\mathbb{R}^n} \Phi(\boldsymbol{x} - \mathbf{y}) f(\mathbf{y})\, d\mathbf{y}.$$

We skip the proof of this theorem and refer the reader to [123].

2.4.1 Green's Functions

We shall give a swift introduction to a general framework. A more thorough study is via distribution theory which is beyond the scope of this text. Green's functions provide us with an integral representation for solution of the BVPs:

Definition 2.18 Let u: $\mathbb{R}^n \to \mathbb{R}$ and consider the problem

$$\mathcal{L}u = f \text{ in } \Omega, \quad Bu = g \text{ on } \partial\Omega, \tag{2.4.2}$$

where $\mathcal{L}$ is a linear differential operator. Then, in *distribution sense*, the fundamental solution $\Phi(\mathbf{x}, \mathbf{y})$ for (2.4.2) satisfies

$$\mathcal{L}\Phi(\mathbf{x}, \mathbf{y}) = \delta(\mathbf{x} - \mathbf{y}), \quad \mathbf{x},\ \mathbf{y} \in \mathbb{R}^n. \tag{2.4.3}$$

Note that here Φ is defined on $\mathbb{R}^n \times \mathbb{R}^n$.

To see the advantage of this representation, we assume that for each $\mathbf{x} \in \mathbb{R}^n$ the function Φ is integrable in $\mathbf{y}$. Let now

$$v(\mathbf{x}) = \int_{\mathbb{R}^n} \Phi(\mathbf{x}, \mathbf{y}) f(\mathbf{y})\, d\mathbf{y} = (\Phi(\mathbf{x}, \mathbf{y}), f(\mathbf{y})). \tag{2.4.4}$$

Then,

$$\mathcal{L}v(\mathbf{x}) = \mathcal{L}(\Phi(\mathbf{x}, \mathbf{y}),\ f(\mathbf{y})) = (\delta(\mathbf{x} - \mathbf{y}),\ f(\mathbf{y})) = f(\mathbf{x}).$$

Thus, v satisfies the equation $\mathcal{L}v = f$ in $\mathbb{R}^n$.

The remaining issue is that, in general, $v(\mathbf{x})$ does not satisfy the boundary condition $Bv = g$. So, as in elementary calculus, we need to construct a solution by adding the solution $w(\mathbf{x})$ of the corresponding homogeneous equation ($\mathcal{L}w = 0$) to $v(\mathbf{x})$, so that $u(\mathbf{x}) = v(\mathbf{x}) + w(\mathbf{x})$ and $Bu = g$. This imposes the boundary condition on w, viz. $Bw = g - Bv$. Thus, knowing the fundamental solution, the inhomogeneous BVP (2.4.2) is reduced to the homogeneous BVP

$$\mathcal{L}w = 0 \text{ in } \Omega, \quad Bw = g - Bv \text{ on } \partial\Omega.$$

It is the boundary condition for w that can be smoothly tackled using the Green's function for the operator $\mathcal{L}$ with the boundary operator B. In this setting, the Green's function $G(\mathbf{x}, \mathbf{y})$ is defined as the solution of the problem

$$\begin{cases} \mathcal{L}G(\mathbf{x}, \mathbf{y}) = \delta(\mathbf{x} - \mathbf{y}), & \mathbf{x} \in \Omega, \\ BG(\mathbf{x}, \mathbf{y}) = 0, & \mathbf{x} \in \partial\Omega, \end{cases} \tag{2.4.5}$$

for each $\mathbf{y} \in \overline{\Omega}$. Note that unlike (2.4.3), here $\mathbf{x} \in \Omega$. We construct G using the fundamental solution that defines a $\mathbf{y}$-parametric function $\Phi^{\mathbf{y}}(\mathbf{x})$,

$$G(\mathbf{x}, \mathbf{y}) := \Phi(\mathbf{x}, \mathbf{y}) - \Phi^{\mathbf{y}}(\mathbf{x}),$$

then,

$$\begin{cases} \mathcal{L}\Phi^{\mathbf{y}}(\mathbf{x}) = 0, & \mathbf{x} \in \Omega, \\ B\Phi^{\mathbf{y}}(\mathbf{x}) = B\Phi(\mathbf{x}, \mathbf{y}), & \mathbf{x} \in \partial\Omega. \end{cases} \tag{2.4.6}$$

Thus, we may write the solution of (2.4.2) as $u = v + w$ where

$$\begin{cases} v(\mathbf{x}) = \displaystyle\int_\Omega G(\mathbf{x}, \mathbf{y}) f(\mathbf{y})\, d\mathbf{y} & \text{and} \\ \mathcal{L}w = 0 \text{ in } \Omega & \text{with } Bw = g \text{ on } \partial\Omega. \end{cases} \tag{2.4.7}$$

We use the above approach to solve the Dirichlet BVP:

Theorem 2.11 (Green's function for Laplace's equation) *Let $u \in C^2(\overline{\Omega})$ be the solution for*

$$-\Delta u = f \text{ in } \Omega, \quad u = g \text{ on } \partial\Omega, \tag{2.4.8}$$

then

$$u(\boldsymbol{x}) = \int_\Omega G(\boldsymbol{x}, \boldsymbol{y}) f(\boldsymbol{y})\, d\boldsymbol{y} - \int_{\partial\Omega} \frac{\partial G}{\partial \boldsymbol{\nu}_{\boldsymbol{y}}}(\boldsymbol{x}, \boldsymbol{y}) g(\boldsymbol{y})\, d\sigma_{\boldsymbol{y}}. \tag{2.4.9}$$

Here G is the Green's function given as

$$G(\boldsymbol{x}, \boldsymbol{y}) = \Phi(\boldsymbol{x} - \boldsymbol{y}) - \varphi^{\boldsymbol{y}}(\boldsymbol{x}),$$

where $\Phi(\boldsymbol{x} - \boldsymbol{y})$ is the fundamental solution and $\varphi^{\boldsymbol{y}}(\boldsymbol{x})$ is the solution for the problem

$$\begin{cases} \Delta\varphi^{\boldsymbol{y}}(\boldsymbol{x}) = 0, & \text{in } \Omega \\ \Delta\varphi^{\boldsymbol{y}}(\boldsymbol{x}) = \Phi(\boldsymbol{x} - \boldsymbol{y}), & \text{on } \partial\Omega. \end{cases} \tag{2.4.10}$$

For a proof, see [123].

2.5 The Weak/Variational Formulation

Below we prepare for a well-posedness approach for the solution of the PDEs of the second order. To this end, we shall derive the variational formulation (VF) for the Poisson equation, also known as weak formulation. Under certain assumptions, to solve a BVP is equivalent (have the same solution) to its corresponding VF which in turn is equivalent to an associated minimization problem (MP):

$$\text{BVP "} \Longleftrightarrow \text{" VF} \Longleftrightarrow \text{MP}.$$

The implication $\Leftarrow$ in the first equivalence " $\Longleftrightarrow$ " is subject to a regularity requirement on the solution up to the order of the underlying PDE. In Section 2.6, we state and prove the above equivalences for the one-dimensional case.

Green's formula. In the study of the elliptic (second-order) PDEs, certain version of the *Green's formula* plays a central role. This is the higher-dimensional form of the integration by parts: a simple weak derivation (see below), that yields, e.g. *weak formulation* for elliptic Dirichlet or Neumann BVPs:

$$\begin{cases} -\Delta u(x) = f(x), & x \in \Omega \subset \mathbb{R}^n, \quad n \geq 2 \\ u(x) = 0, & x \in \partial\Omega \quad \text{(Dirichlet data),} \quad \text{or} \\ \mathbf{n}(x) \cdot \nabla u(x) = 0, & x \in \partial\Omega \quad \text{(Neumann data).} \end{cases} \tag{2.5.1}$$

Below we state, without proof, a version of the Green's formula that we shall use to derive *variational formulations:*

Lemma 2.3 (Green's formula) *Let $\Omega \subset \mathbb{R}^n$, $u \in C^2(\Omega)$, and $v \in C^1(\Omega)$, then*

$$\int_\Omega (\Delta u)v \, dx \, dy = \int_{\partial\Omega} (\nabla u \cdot \boldsymbol{n})v \, ds - \int_\Omega \nabla u \cdot \nabla v \, dx \, dy, \tag{2.5.2}$$

The proof can be found in any book in calculus of several variables (a simple proof in two-dimensions is given in chapter of piecewise polynomials in $2d$).

Remark 2.8 For the Green's formula, for Problem (2.5.1), associated with the *Dirichlet boundary data*

$$u(x) = 0, \quad x \in \partial\Omega, \tag{2.5.3}$$

we choose the test function $v \in C_0^1(\Omega)$ where the boundary condition is strongly imposed by the test function. This reduces (2.5.2) to

$$\int_\Omega (\Delta u)v \, dx \, dy = -\int_\Omega \nabla u \cdot \nabla v \, dx \, dy. \tag{2.5.4}$$

Whereas, the Green's formula associated with the *Neumann boundary data*

$$\mathbf{n}(x) \cdot \nabla u(x) = g(x), \quad x \in \partial\Omega, \tag{2.5.5}$$

will be of the form

$$\int_\Omega (\Delta u)v \, dx \, dy = \int_{\partial\Omega} gv \, ds - \int_\Omega \nabla u \cdot \nabla v \, dx \, dy, \tag{2.5.6}$$

i.e. the boundary data $\mathbf{n}(x) \cdot \nabla u(x) = g(x)$ is weakly imposed in the trial space.

Now, we define an abstract form of the inner product:

Definition 2.19 Let V be a *vector space* of function on Ω and define a *bilinear form* on V by $a(\cdot,\cdot)\colon V \times V \to \mathbb{R}$, i.e. for $\alpha, \beta, x, y \in \mathbb{R}$ and $u, v, w \in V$, we have

$$\begin{cases} a(\alpha u + \beta v, w) = \alpha \cdot a(u, w) + \beta \cdot a(v, w) \\ a(u, xv + yw) = x \cdot a(u, v) + y \cdot a(u, w). \end{cases} \tag{2.5.7}$$

Using a bilinear form as $a(u, v) = (\nabla u, \nabla v)$ with

$$(w, \chi) = \int_{\Omega} w(x)\chi(x)\, dx, \quad \text{and} \quad \langle \eta, \zeta \rangle = \int_{\partial\Omega} \eta(x)\zeta(x)\, dx, \tag{2.5.8}$$

we may combine (2.5.1) with either (2.5.3) or (2.5.5) and obtain the weak formulations: find $u \in H_0^1(\Omega)$ so that

$$a(u, v) = (f, v) \quad \forall v \in H_0^1(\Omega). \tag{2.5.9}$$

Or, find $u \in H^1(\Omega)$, with the Neumann data $\nabla u \cdot \mathbf{n} = g$, so that

$$a(u, v) = (f, v) + \langle g, v \rangle \quad \forall v \in H^1(\Omega). \tag{2.5.10}$$

Relations (2.5.9) and (2.5.10) are the *variational formulations* for the Poisson equation (2.5.1) with the corresponding boundary conditions, respectively.

Definition 2.20 A function $L: V \to \mathbb{R}$ is called a linear form on V, if for α, $\beta \in \mathbb{R}$ and $u,\ v \in V$,

$$L(\alpha u + \beta v) = \alpha L(u) + \beta L(v). \tag{2.5.11}$$

Now, if we define a linear form as

$$L_D(v) = (f, v) \quad \text{or} \quad L_N(v) = (f, v) + \langle g, v \rangle, \tag{2.5.12}$$

then we have a general form of *variational formulations*, viz.

$$a(u, v) = L(v), \quad \forall v \in \mathcal{H}(\Omega), \tag{2.5.13}$$

where for the Dirichlet problem $L = L_D$ with $\mathcal{H}(\Omega) = H_0^1(\Omega)$ and in the case of Neumann data, $L = L_N$ with $\mathcal{H}(\Omega) = H^1(\Omega)$.

2.6 A Framework for Analytic Solution in 1*d*

Assume that a horizontal elastic bar which occupies the interval $I := [0, 1]$, is fixed at the end-points. Let $u(x)$ denote the displacement of the bar at a point $x \in I$, $a(x)$ be the *modulus of elasticity*, and $f(x)$ a given *load function*. We showed, in (1.5.7), that u satisfies the following BVP

$$(\text{BVP})_1 \qquad \begin{cases} -(a(x)u'(x))' = f(x), & 0 < x < 1, \\ u(0) = u(1) = 0. \end{cases} \tag{2.6.1}$$

Equation (2.6.1) is of Poisson's type modeling also the stationary heat flux. Below we demonstrate a 1*d*-version of the *Green's function* approach. For simplicity, we assume $a(x) = 1$ and consider

$$\begin{cases} -u''(x) = f(x), & 0 < x < 1, \\ u(0) = u(1) = 0. \end{cases} \tag{2.6.2}$$

The solution to this problem is obtained by integration:

$$-u'(x) = \int_0^x f(y)\,dy + C_1, \quad -u(x) = \int_0^x \int_0^z f(y)\,dy\,dz + C_1 x + C_2.$$

Reversing the order of integration, we have that

$$\int_0^x \int_0^z f(y)\,dy\,dz = \int_0^x \int_y^x f(y)\,dz\,dy = \int_0^x (x-y)f(y)\,dy.$$

Thus,

$$u(x) = -\int_0^x (x-y)f(y)\,dy - C_1 x - C_2. \tag{2.6.3}$$

Using the boundary data, we get

$$C_1 = 0, \quad C_2 = -\int_0^1 (1-y)f(y)\,dy.$$

Inserting in (2.6.3), we end up with

$$\begin{aligned} u(x) &= \int_0^x (y-x)f(y)\,dy + \int_0^1 x(1-y)f(y)\,dy \\ &= \int_0^x (y-x)f(y)\,dy + \int_0^x x(1-y)f(y)\,dy + \int_x^1 x(1-y)f(y)\,dy \\ &= \int_0^x y(1-x)f(y)\,dy + \int_x^1 x(1-y)f(y)\,dy. \end{aligned}$$

This can be rewritten in a compact form as

$$u(x) = \int_0^1 G(x,y)f(y)\,dy, \quad 0 \le x \le 1, \tag{2.6.4}$$

where $G(x,y)$ is an integral kernel given by

$$G(x,y) = \begin{cases} y(1-x), & y \le x \\ x(1-y), & x \le y. \end{cases} \tag{2.6.5}$$

Here, $G(x,y)$ is the Green's function for our BVP. Hence, we have defined an integral operator as

$$(\mathcal{G}f)(x) = \int_0^1 G(x,y)f(y)\,dy, \quad 0 \le x \le 1.$$

Here, we have that the differential operator $\mathcal{L} = -\dfrac{d^2}{dx^2}: V \to C[0,1]$ acts on functions $u \in V$, where V is the space of $C^2[0,1]$ functions satisfying the boundary conditions in (2.6.2). Whereas the integral operator $\mathcal{G}: C[0,1] \to V$, operates the other way, acting on continuous functions to reach twice differentiable functions that satisfy (2.6.2). Hence, the integral operator $\mathcal{G}$ is the inverse of the differential operator $\mathcal{L}$.

Here, is a list of some properties of the Green's Function $G: [0,1] \times [0,1] \to \mathbb{R}$:

(i) G is nonnegative: $G(x,y) \geq 0$;
(ii) G is symmetric: $G(x,y) = G(y,x)$;
(iii) G is continuous; and
(iv) G is differentiable, except on the diagonal $x = y$. At the diagonal $\partial G/\partial y$ has a jump discontinuity:

$$\left[\frac{\partial G}{\partial y}\right] = -1.$$

There are counterparts of the properties (i)–(iv) for differential operator $\mathcal{L}$.

We also notice that $\frac{\partial^2 G}{\partial y^2} = 0$ except on the diagonal, where $\partial G/\partial y$ may be interpreted to have a negative slope. Hence, we may use the Dirac δ function to write

$$-\frac{\partial^2 G}{\partial y^2}(x,y) = \delta(x-y).$$

We may relate the Green's function to the fundamental solution $\Phi(x,y) = -\frac{1}{2}|x-y|$ (roughly speaking, a solution of (2.6.1) for $f(x) = \delta(x)$). Thus, if

$$G(x,y) = -\frac{1}{2}|x-y| + \varphi^y(x),$$

then

$$-\frac{d^2 G}{dc^2}\varphi^y(x) = 0, \quad \forall\ y \in [0,1]$$

and since $G(x,y)$ satisfies the homogeneous boundary conditions, hence $\varphi^y(x)$ satisfies the boundary conditions:

$$\varphi^y(x) - \Phi(x-y) = 0, \quad \text{for } x = 0 \text{ and } x = 1. \tag{2.6.6}$$

Thus, we may modify the fundamental solution to obtain the Green's function, where for each y we construct a solution $u = \varphi^y$ for the homogeneous equation $u'' = 0$ that satisfies the boundary conditions of (2.6.6).

This, however, is an abstract approach: a procedure in the realm of advanced theoretical study. A less abstract theoretical study is the eigenfunction expansion based on the separation of variables and related Fourier techniques.

We assume that the reader is familiar with the Fourier techniques and hence focus on the approximation procedure based on finite element methods. To this end we introduce the variational formulation in 1d.

2.6.1 The Variational Formulation in 1*d*

We shall assume that $a(x)$ is continuous in $(0,1)$ and bounded for $0 \leq x \leq 1$. Let $v(x)$ and its derivative $v'(x), x \in I$, be square integrable functions, that is $v,\ v' \in L_2(0,1)$, and recall the L_2-based Sobolev space, referred to as the *Hilbert space*

$$H_0^1(0,1) = \left\{ v\colon \int_0^1 (v(x)^2 + v'(x)^2)\, dx < \infty, \quad v(0) = v(1) = 0 \right\}. \tag{2.6.7}$$

As a consequence of the Poincaré inequality, the space $H_0^1(0,1)$ is identically defined as

$$H_0^1(0,1) = \left\{ v\colon \int_0^1 v'(x)^2\, dx < \infty, \quad v(0) = v(1) = 0 \right\}, \tag{2.6.8}$$

as well. It is this latter version (where often H is replaced by $\mathcal{H}$) that we shall use. When the domain is obvious, we suppress it from the involved function spaces, e.g. $(0,1)$ (or Ω) from $H^1(0,1)$ (or $H^1(\Omega)$).

Solving PDEs by Fourier techniques is based on multiplying the Fourier series representation (obtained, e.g. by the separation of variables technique) by adequate (orthogonal) eigenfunctions and integrating over the domain. In this way, we extract an information from the structure of the PDE and the data function, i.e. a Fourier coefficient. We perform this procedure in the variational formulation as follows:

We multiply the equation in $(\text{BVP})_1$ by a so-called test function $v \in H_0^1(0,1)$ and integrate over $(0,1)$ to obtain

$$-\int_0^1 (a(x)u'(x))'v(x)\, dx = \int_0^1 f(x)v(x)\, dx. \tag{2.6.9}$$

Using integration by parts, we get

$$-[a(x)u'(x)v(x)]_0^1 + \int_0^1 a(x)u'(x)v'(x)\, dx = \int_0^1 f(x)v(x)\, dx. \tag{2.6.10}$$

Now, since $v(0) = v(1) = 0$, we have thus obtained the *variational formulation* for the Problem (2.6.1) as follows: find $u \in H_0^1(0,1)$ such that

$$(\text{VF})_1 \qquad \int_0^1 a(x)u'(x)v'(x)\, dx = \int_0^1 f(x)v(x)\, dx, \quad \forall v \in H_0^1. \tag{2.6.11}$$

In other words, we have shown that if u satisfies $(\text{BVP})_1$, then u also satisfies the $(\text{VF})_1$ above. We write this as $(\text{BVP})_1 \Longrightarrow (\text{VF})_1$. Now, the question is whether the reverse implication is true, i.e. under which conditions can we deduce the implication $(\text{VF})_1 \Longrightarrow (\text{BVP})_1$? It appears that this question has an affirmative answer, *provided that the solution u to* $(\text{VF})_1$ *is twice differentiable.* Then, in the modulo of this regularity requirement, the two problems are indeed equivalent. We prove this in the following theorem.

Theorem 2.12 *The following two properties are equivalent*

(i) u satisfies $(BVP)_1$
(ii) u is twice differentiable and satisfies $(VF)_1$.

Proof: We have already shown $(\text{BVP})_1 \Longrightarrow (\text{VF})_1$.

It remains to show that $(\text{VF})_1 \Longrightarrow (\text{BVP})_1$:

Integrating by parts on the left-hand side in (2.6.11), *assuming that u is twice differentiable*, and using $f \in C[0,1]$, $a \in C^1(0,1)$ and $v(0) = v(1) = 0$ we return to the relation (2.6.9):

$$-\int_0^1 (a(x)u'(x))'v(x)\,dx = \int_0^1 f(x)v(x)\,dx, \quad \forall v \in H_0^1 \tag{2.6.12}$$

which can be rewritten as

$$\int_0^1 \{-(a(x)u'(x))' - f(x)\}v(x)\,dx = 0, \quad \forall v \in H_0^1. \tag{2.6.13}$$

To show that u satisfies $(\mathrm{BVP})_1$ is equivalent to *claim* that (2.6.13) implies

$$-(a(x)u'(x))' - f(x) \equiv 0, \quad \forall x \in (0,1). \tag{2.6.14}$$

Suppose not. Then there exists at least one point $\xi \in (0,1)$, such that

$$-\big(a(\xi)u'(\xi)\big)' - f(\xi) \neq 0, \tag{2.6.15}$$

where we may assume, without loss of generality, that

$$-(a(\xi)u'(\xi))' - f(\xi) > 0 \quad (\text{or} < 0). \tag{2.6.16}$$

Thus, by continuity, $\exists\delta > 0$, such that in a δ-neighborhood of ξ,

$$g(x) := -(a(x)u'(x))' - f(x) > 0, \quad \forall\, x \ \in I_\delta := (\xi - \delta, \xi + \delta). \tag{2.6.17}$$

Now, take the test function v in (2.6.13) as the continuous linear *hat-function* $v^*(x) > 0$, with $v^*(\xi) = 1$ and the support I_δ, see Figure 2.1. Then $v^* \in H_0^1$ and

$$\int_0^1 \{-(a(x)u'(x))' - f(x)\}v^*(x)\,dx = \int_{I_\delta} \underbrace{g(x)}_{>0}\ \underbrace{v^*(x)}_{>0}\ dx > 0.$$

This contradicts (2.6.13). Thus, our claim is true. Note further that in $(\mathrm{VF})_1$, $u \in H_0^1$ implies that $u(0) = u(1) = 0$ and hence, we have also the boundary conditions and the proof is complete. □

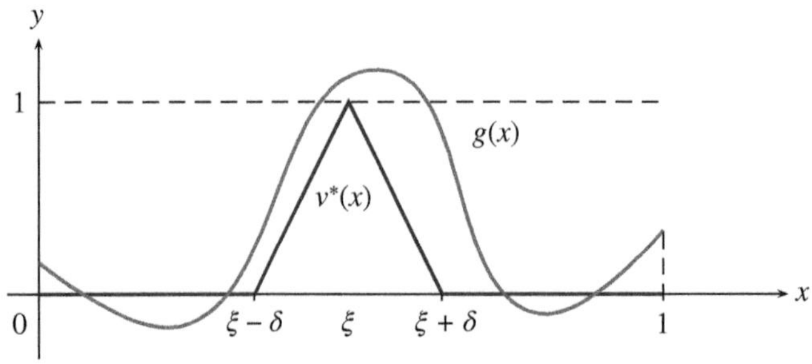

Figure 2.1 The hat function $v^*(x)$ over the interval $(\xi - \delta, \xi + \delta)$.

Corollary 2.1

(i) If $f(x)$ is continuous and $a(x)$ is continuously differentiable, i.e. $f \in C(0,1)$ and $a \in C^1(0,1)$, then $(BVP)_1$ and $(VF)_1$ have the same solution.
(ii) If $a(x)$ is discontinuous and $f \in L_2(0,1)$, then $(BVP)_1$ is not always well defined, but $(VF)_1$ has still a meaning. Therefore, $(VF)_1$ covers a larger set of data than $(BVP)_1$.
(iii) More important in $(VF)_1$, $u \in C^1(0,1)$, while $(BVP)_1$ is formulated for u having two derivatives, e.g. $u \in C^2(0,1)$.

2.6.2 The Minimization Problem in 1*d*

For the Problem (2.6.1), we may formulate yet another equivalent problem:

Find $u \in H_0^1$ such that $F(u) \le F(w)$, $\forall w \in H_0^1$, where $F(w)$ is the total potential energy of the displacement $w(x)$, given by

$$(\mathrm{MP})_1 \quad F(w) = \underbrace{\frac{1}{2}\int_0^1 a(w')^2\,dx}_{\text{Internal (elastic) energy}} - \underbrace{\int_0^1 fw\,dx}_{\text{Load potential}}. \tag{2.6.18}$$

This means that the solution u minimizes the energy functional $F(w)$.

Below we show that the MP is equivalent to the variational formulation $(\mathrm{VF})_1$ and hence also to the BVP $(\mathrm{BVP})_1$.

Theorem 2.13 *The following two properties are equivalent*

(a) u satisfies the variational formulation $(VF)_1$
(b) u is the solution for the minimization problem $(MP)_1$

i.e.

$$\int_0^1 au'v'\,dx = \int_0^1 fv\,dx, \quad \forall v \in H_0^1 \Longleftrightarrow F(u) \le F(w), \quad \forall w \in H_0^1. \tag{2.6.19}$$

Proof: ($\Longrightarrow$): First, we show that the variational formulation $(\mathrm{VF})_1$ implies the minimization problem $(\mathrm{MP})_1$. To this end, for $w \in H_0^1$, we let $v = w - u$, then, since H_0^1 is a vector space and $u \in H_0^1$, hence, $v \in H_0^1$ and

$$\begin{aligned} F(w) = F(u+v) &= \frac{1}{2}\int_0^1 a((u+v)')^2\,dx - \int_0^1 f(u+v)\,dx \\ &= \underbrace{\frac{1}{2}\int_0^1 2au'v'\,dx}_{(i)} + \underbrace{\frac{1}{2}\int_0^1 a(u')^2\,dx}_{(ii)} + \tfrac{1}{2}\textstyle\int_0^1 a(v')^2\,dx \\ &\quad - \underbrace{\int_0^1 fu\,dx}_{(iii)} - \underbrace{\int_0^1 fv\,dx}_{(iv)}. \end{aligned}$$

Now using $(\mathrm{VF})_1$ we have (i)–(iv)=0. Further by the definition of the functional F, (ii)–(iii)=$F(u)$. Thus,

$$F(w) = F(u) + \frac{1}{2}\int_0^1 a(x)(v'(x))^2\, dx, \tag{2.6.20}$$

and since $a(x) > 0$ we get $F(w) \geq F(u)$, thus, we have proved "$\Longrightarrow$" part.

($\Leftarrow$): Next, we show that the minimization problem $(\mathrm{MP})_1$ implies the variational formulation $(\mathrm{VF})_1$. To this end, assume that $F(u) \leq F(w)\ \ \forall w \in H_0^1$, and for an arbitrary function $v \in H_0^1$, set $g_v(\varepsilon) = F(u + \varepsilon v)$, then by $(\mathrm{MP})_1$, g_v (as a function of ε) has a *minimum* at $\varepsilon = 0$. In other words, $\frac{\partial}{\partial\varepsilon} g_v(\varepsilon)|_{\varepsilon=0} = 0$. We have that

$$\begin{aligned} g_v(\varepsilon) &= F(u + \varepsilon v) = \frac{1}{2}\int_0^1 a((u+\varepsilon v)')^2\, dx - \int_0^1 f(u+\varepsilon v)\, dx \\ &= \frac{1}{2}\int_0^1 \{a(u')^2 + a\varepsilon^2(v')^2 + 2a\varepsilon u'v'\}\, dx - \int_0^1 fu\, dx - \varepsilon\int_0^1 fv\, dx. \end{aligned}$$

The derivative $dg_v(\varepsilon)/d\varepsilon$, of $g_v(\varepsilon)$ is

$$\frac{dg_v(\varepsilon)}{d\varepsilon} = \frac{1}{2}\int_0^1 \{2a\varepsilon(v')^2 + 2au'v'\}\, dx - \int_0^1 fv\, dx, \tag{2.6.21}$$

where $\frac{dg_v(\varepsilon)}{d\varepsilon}|_{(\varepsilon=0)} = 0$, yields

$$\int_0^1 au'v'\, dx - \int_0^1 fv\, dx = 0, \tag{2.6.22}$$

which is our desired variational formulation $(\mathrm{VF})_1$. Hence, we conclude that $F(u) \leq F(w),\ \forall w \in H_0^1 \Longrightarrow (\mathrm{VF})_1$, and the proof is complete. □

We summarize the two theorems in short as

Corollary 2.2

$$(BVP)_1 \ \text{“}\Longleftrightarrow\text{”}\ (VF)_1 \Longleftrightarrow (MP)_1.$$

Recall that "$\Longleftrightarrow$" is a conditional equivalence, requiring u to be twice differentiable, for the reverse implication.

2.6.3 A Mixed Boundary Value Problem in 1*d*

Obviously, changing the boundary conditions would require changes in the variational formulation. This can be seen, e.g. in deriving the variational formulation corresponding to the following mixed BVP: find u such that

$$(\mathrm{BVP})_2 \qquad \begin{cases} -(a(x)u'(x))' = f(x), & 0 < x < 1 \\ u(0) = 0, & a(1)u'(1) = g_1. \end{cases} \tag{2.6.23}$$

As usual, we multiply the equation by a suitable test function $v(x)$, and integrate over the interval $(0, 1)$. Note that here the test function should satisfy only one boundary condition: $v(0) = 0$. This is due to the fact that now $u(1)$ is not given, and to get a value of u at $x = 1$, we need to supply a test function at $x = 1$. Therefore, the proper choice for a function space is now

$$\tilde{H}_0^1 = \left\{ v\colon \int_0^1 (v(x)^2 + v'(x)^2)\, dx < \infty, \text{such that } v(0) = 0 \right\}. \tag{2.6.24}$$

So $v(1) \neq 0$ and we have, multiplying the equation by such a v and integrating over $I = (0, 1)$, that

$$-\int_0^1 (a(x)u'(x))'v(x)\, dx = \int_0^1 f(x)v(x)\, dx, \quad \forall v \in \tilde{H}_0^1. \tag{2.6.25}$$

Integrating by parts, yields

$$-[a(x)u'(x)v(x)]_0^1 + \int_0^1 a(x)u'(x)v'(x)\, dx = \int_0^1 f(x)v(x)\, dx, \tag{2.6.26}$$

and using the boundary data $a(1)u'(1) = g_1$ and $v(0) = 0$ we get

$$\int_0^1 a(x)u'(x)v'(x)\, dx = \int_0^1 f(x)v(x)\, dx + g_1 v(1), \quad \forall v \in \tilde{H}_0^1. \tag{2.6.27}$$

Hence, (2.6.27) yields the variational formulation: find $u \in \tilde{H}_0^1$ such that

$$(\text{VF})_2 \qquad \int_0^1 a(x)u'(x)v'(x)\, dx = \int_0^1 f(x)v(x)\, dx + g_1 v(1), \quad \forall v \in \tilde{H}_0^1.$$

Now, as in the case of *Dirichlet* problem $(\text{BVP})_1$, we want to show that

Theorem 2.14 *$(BVP)_2 \Longleftrightarrow (VF)_2$, in the sense that the two problems have the same solution (also here, with the smoothness requirement in $\Longleftarrow$ part).*

Proof: That $(\text{BVP})_2 \Longrightarrow (\text{VF})_2$, is already shown along (2.6.25)–(2.6.27). To prove $(\text{VF})_2 \Longrightarrow (\text{BVP})_2$: that a solution of the variational problem $(\text{VF})_2$ is also a solution of the two-point BVP $(\text{BVP})_2$, we have to prove the following claims:

(i) the solution of $(\text{VF})_2$ satisfies the differential equation
(ii) the solution of $(\text{VF})_2$ satisfies the boundary conditions

A reversed integration by parts on the left-hand side of $(\text{VF})_2$ yields

$$\int_0^1 a(x)u'(x)v'(x)\, dx = [a(x)u'(x)v(x)]_0^1 - \int_0^1 (a(x)u'(x))'v(x)\, dx, \tag{2.6.28}$$

and since $v(0) = 0$, we get

$$\int_0^1 a(x)u'(x)v'(x)\, dx = a(1)u'(1)v(1) - \int_0^1 (a(x)u'(x))'v(x)\, dx. \tag{2.6.29}$$

Thus, the variational formulation $(\mathrm{VF})_2$ can be rewritten as

$$-\int_0^1 (a(x)u'(x))'v(x)\,dx + a(1)u'(1)v(1) = \int_0^1 f(x)v(x)\,dx + g_1 v(1). \tag{2.6.30}$$

Equation (2.6.30) is valid for every $v \in \tilde{H}_0^1(0,1)$, including the special class of test functions $v(x)$ that satisfies, in addition to $v(0) = 0$, also $v(1) = 0$, i.e. as in the Dirichlet case: $-(au')' = f,\ \ u(0) = u(1) = 0$. This is simply because $H_0^1(0,1) \subset \tilde{H}_0^1(0,1)$. Consequently, choosing $v(1) = 0$, (2.6.30) is reduced to

$$-\int_0^1 (a(x)u'(x))'v(x)\,dx = \int_0^1 f(x)v(x)\,dx, \quad \forall v \in H_0^1. \tag{2.6.31}$$

Now, as in the Dirichlet case, the variational formulation (2.6.31) gives the differential equation in (2.6.23) and hence the claim (i) is true. Inserting $(au')' = f$ into (2.6.30), we get $g_1 v(1) = a(1)u'(1)v(1)$. Choosing $v(1) \neq 0$, e.g. $v(1) = 1$, gives the boundary condition $g_1 = a(1)u'(1)$. The boundary condition $u(0) = 0$ follows directly from the definition of $\tilde{H}_0^1(0,1)$, and the proof is complete. □

Remark 2.9

(i) The Dirichlet boundary condition is called an essential boundary condition and is *"strongly imposed"* in the trial/test function space: Enforced explicitly to the trial and test functions in (VF).

(ii) Neumann and Robin boundary conditions are called natural boundary conditions. These boundary conditions are automatically satisfied, within the corresponding variational formulations, and are therefore *weakly imposed.*

Now, we are prepared to give an approach to the existence of unique solution for the BVPs. Even this approach will be presented in the one-dimensional case showing that the equivalent MP has a unique solution.

2.7 An Abstract Framework

Example 2.13 Consider the simple one-dimensional BVP:

$$\text{(BVP):} \qquad \begin{cases} -u''(x) = f(x), & 0 < x < 1 \\ u(0) = u(1) = 0, \end{cases} \tag{2.7.1}$$

Let $V = \mathcal{H}_0^1$ and define

$$a(u,v) := (u,v) := \int_0^1 u'(x)v'(x)\,dx, \tag{2.7.2}$$

then $(\cdot,\cdot)$ is *symmetric*, i.e. $(u,v)=(v,u)$, *bilinear* (obvious), and *positive definite* in the sense that

$$(u,u)\geq 0, \quad \text{and } \ (u,u)=0 \Longleftrightarrow u\equiv 0.$$

Note that, here,

$$(u,u)=\int_0^1 u'(x)^2\,dx=0 \Longleftrightarrow u'(x)=0,$$

thus $u(x)$ is constant and since $u(0)=u(1)=0$ we have $u(x)\equiv 0$.

Further, for $f\in L_2(0,1)$, let

$$\ell(v)=\int_0^1 fv\,dx, \quad \forall v\in \mathcal{H}_0^1. \tag{2.7.3}$$

Then our (VF) can be restated as follows: Find $u\in\mathcal{H}_0^1$ such that

$$a(u,v)=\ell(v), \quad \forall v\in\mathcal{H}_0^1. \tag{2.7.4}$$

Generalizing Example 2.13, *(e.g. in a Hilbert space defined below)*, to a bilinear form $a(\cdot,\cdot)$, and a linear form $L(\cdot)$, we get the following (see (2.5.13)) *abstract problem*: Find $u\in V$, such that

$$a(u,v)=L(v) \quad \forall v\in V. \tag{2.7.5}$$

Definition 2.21 Let $\|\cdot\|_V$ be a norm corresponding to a scalar product $(\cdot,\cdot)_V$ defined on $V\times V$. Then the bilinear form $a(\cdot,\cdot)$ is called *coercive* (V-elliptic), and $a(\cdot,\cdot)$ and $L(\cdot)$ are continuous, if there are constants c_i, $i=1,2,3$ such that

$$a(v,v)\geq c_1\|v\|_V^2, \quad \forall v\in V \quad \text{(coercivity)} \tag{2.7.6}$$

$$|a(u,v)|\leq c_2\|u\|_V\|v\|_V, \quad \forall u,v\in V \quad (a \text{ is bounded}) \tag{2.7.7}$$

$$|L(v)|\leq c_3\|v\|_V, \quad \forall v\in V \quad (L \text{ is bounded}). \tag{2.7.8}$$

Note that since L is linear, we have, using the relation (2.7.8) above that

$$|L(u)-L(v)|=|L(u-v)|\leq c_3\|u-v\|_V,$$

which shows that $L(u)\to L(v)$ in $\mathbb{R}$, as $u\to v$ in V. Thus, L is continuous. Similarly the relation $|a(u,v)|\leq c_1\|u\|_V\|v\|_V$ implies that the bilinear form $a(\cdot,\cdot)$ is continuous in each component.

Definition 2.22 The energy norm on V is given by $\|v\|_a=\sqrt{a(v,v)}, \quad v\in V$.

Recalling relations (2.7.6) and (2.7.7) above, the energy norm satisfies

$$c_1\|v\|_V^2\leq a(v,v)=\|v\|_a^2\leq c_2\|v\|_V^2. \tag{2.7.9}$$

Hence, the energy norm $\| \bullet \|_a$ is equivalent to the abstract $\| \bullet \|_V$ norm.

Example 2.14 For the scalar product

$$(u, v) = \int_0^1 u'(x)v'(x)\, dx, \quad \text{in } \mathcal{H}_0^1, \tag{2.7.10}$$

and the norm

$$\|u\| = \sqrt{(u, u)}, \tag{2.7.11}$$

relations (2.7.6) and (2.7.7) are valid with $c_1 = c_2 \equiv 1$:

For a general scalar product, we have that

(i) $(v, v) = \|v\|^2$ is an identity, and
(ii) The Cauchy–Schwarz' inequality $|(u, v)| \leq \|u\|\|v\|$ holds.

Below we further develop the abstract framework and provide general criteria to prove the existence of a unique solution for the PDEs.

Definition 2.23 A *Hilbert space* is a *complete* linear space with a scalar product.

To define a complete linear space, we first need to define a *Cauchy sequence* of *real* or *complex* numbers.

Definition 2.24 A sequence $\{z_k\}_{k=1}^{\infty} \subset \mathbb{C}$ is a Cauchy sequence if for every $\varepsilon > 0$, there is an integer $N > 0$, such that $m, n > N \Longrightarrow |z_m - z_n| < \varepsilon$.

Now we state, without proof, a classical theorem of analysis:

Theorem 2.15 *Every Cauchy sequence in* $\mathbb{C}$ *is convergent. More precisely, if* $\{z_k\}_{k=1}^{\infty} \subset \mathbb{C}$ *is a Cauchy sequence, then there is a* $z \in \mathbb{C}$*, such that for every* $\epsilon > 0$*, there is an integer* $M > 0$*, such that* $m \geq M \Longrightarrow |z_m - z| < \varepsilon$.

Definition 2.25 A linear space V (vector space) with the norm $\| \cdot \|$ is called *complete* if every Cauchy sequence in V is convergent. In other words, for every $\{v_k\}_{k=1}^{\infty} \subset V$ with the property that for every $\varepsilon_1 > 0$ there is an integer $N > 0$, such that $m, n > N \Longrightarrow \|v_m - v_n\| < \varepsilon_1$, (i.e. for every Cauchy sequence) there is a $v \in V$ such that for every $\varepsilon > 0$ there is an integer $M > 0$ such that $m \geq M \Longrightarrow \|v_m - v\| < \varepsilon$.

Theorem 2.16 *The space* $\mathcal{H}_0^1 := H_0^1(0, 1)$ *is a complete Hilbert space with the norm*

$$\|u\| = \sqrt{(u, u)} = \left(\int_0^1 u'(x)^2\, dx \right)^{1/2}. \tag{2.7.12}$$

Definition 2.26 We define a *functional* ℓ as a mapping from a (linear) function space V into $\mathbb{R}$, i.e.

$$\ell\colon\ V \to \mathbb{R}. \tag{2.7.13}$$

- A functional ℓ is called linear if

$$\begin{cases} \ell(u+v) = \ell(u) + \ell(v) & \text{for all } u, v \in V \\ \ell(\alpha u) = \alpha\, \ell(u) & \text{for all } u \in V \text{and } \alpha \in \mathbb{R}. \end{cases} \tag{2.7.14}$$

- A functional is called bounded if there is a constant C such that

$$|\ell(u)| \le C\ \|u\| \quad \text{for all } u \in V \quad (C \text{ is independent of } u)$$

Example 2.15 If $f \in L^2(0,1)$, i.e. $\int_0^1 f(x)^2\, dx$ is finite, then

$$\ell(v) = \int_0^1 f(x)v(x)\, dx \tag{2.7.15}$$

is a bounded linear functional.

2.7.1 Riesz and Lax–Milgram Theorems

Recalling

$$(u,v) = \int_0^1 u'(x)v'(x)\, dx \quad \text{and} \quad \ell(v) = \int_0^1 f(x)v(x)\, dx,$$

we may redefine our VF and MP in an abstract form as (V) and (M), respectively:

(V) Find $u \in \mathcal{H}_0^1$, such that $(u,v) = \ell(v)$ for all $v \in \mathcal{H}_0^1$.
(M) Find $u \in \mathcal{H}_0^1$, such that $F(u) = \min_{v\in\mathcal{H}_0^1} F(v)$ with $F(v) = \frac{1}{2}\|v\|^2 - \ell(v)$.

Theorem 2.17 *There exists a unique solution for the, equivalent, problems (V) and (M).*

Proof: That (V) and (M) are equivalent is shown above. To proceed, we note that there exists a real number σ such that $F(v) > \sigma$ for all $v \in \mathcal{H}_0^1$, (otherwise, it is not possible to minimize F): namely, we can write

$$F(v) = \frac{1}{2}\|v\|^2 - \ell(v) \ge \frac{1}{2}\|v\|^2 - \gamma\|v\|, \tag{2.7.16}$$

where γ is the constant bounding ℓ, i.e. $|\ell(v)| \le \gamma\|v\|$. But since

$$0 \le \frac{1}{2}(\|v\| - \gamma)^2 = \frac{1}{2}\|v\|^2 - \gamma\|v\| + \frac{1}{2}\gamma^2, \tag{2.7.17}$$

thus, evidently we have

$$F(v) \ge \frac{1}{2}\|v\|^2 - \gamma\|v\| \ge -\frac{1}{2}\gamma^2. \tag{2.7.18}$$

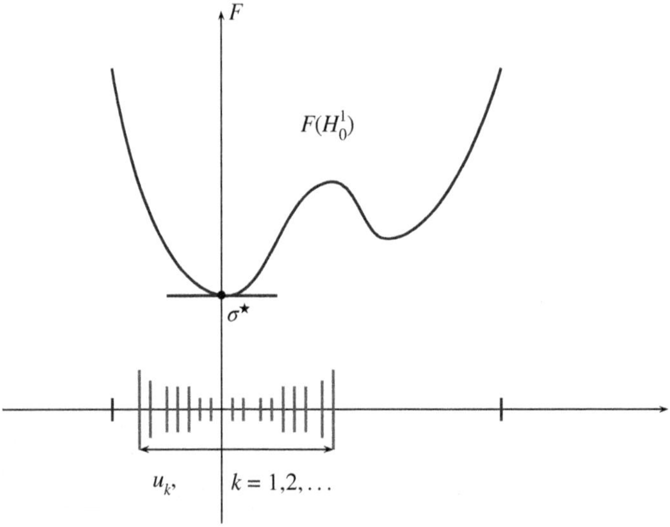

Figure 2.2 Illustrating the existence of a unique solution for (V) and (M).

Let now σ^* be the largest real number σ such that

$$F(v) \geq \sigma \quad \text{for all } v \in \mathcal{H}_0^1. \tag{2.7.19}$$

Take a sequence of functions $\{u_k\}_{k=0}^{\infty}$ (see Figure 2.2), such that

$$F(u_k) \to \sigma^*. \tag{2.7.20}$$

To show that there *exists* a *unique* solution for (V) and (M), we shall use the following two fundamental results:

(i) It is always possible to find a sequence $\{u_k\}_{k=0}^{\infty}$, such that $F(u_k) \to \sigma^*$ (because $\mathbb{R}$ is complete). This we already used in writing (2.7.20).
(ii) The parallelogram law (elementary linear algebra).

$$\|a+b\|^2 + \|a-b\|^2 = 2\|a\|^2 + 2\|b\|^2.$$

Using (ii) and the linearity of ℓ, we can write

$$\begin{aligned}
\|u_k - u_j\|^2 &= 2\|u_k\|^2 + 2\|u_j\|^2 - \|u_k+u_j\|^2 - 4\ell(u_k) - 4\ell(u_j) + 4\ell(u_k+u_j) \\
&= 2\|u_k\|^2 - 4\ell(u_k) + 2\|u_j\|^2 - 4\ell(u_j) - \|u_k+u_j\|^2 + 4\ell(u_k+u_j) \\
&= 4F(u_k) + 4F(u_j) - 8F\left(\frac{u_k+u_j}{2}\right),
\end{aligned}$$

where we have used the definition of $F(v) = \frac{1}{2}\|v\|^2 - \ell(v)$ with $v = u_k,\ \ u_j$, and $v = (u_k+u_j)/2$, respectively. In particular, by linearity of ℓ:

$$\begin{aligned}
-\|u_k+u_j\|^2 + 4\ell(u_k+u_j) &= -4\left\|\frac{u_k+u_j}{2}\right\|^2 + 8\ell\left(\frac{u_k+u_j}{2}\right) \\
&= -8F\left(\frac{u_k+u_j}{2}\right).
\end{aligned}$$

Now by (2.7.19) (applied to $(u_j + u_k)/2$) since $F(u_k) \to \sigma^*$ and $F(u_j) \to \sigma^*$, we have

$$\|u_k - u_j\|^2 \le 4F(u_k) + 4F(u_j) - 8\sigma^* \to 0, \quad \text{as} \quad k, j \to \infty.$$

Hence, we have shown that $\{u_k\}_{k=0}^{\infty}$ is a Cauchy sequence. Since $\{u_k\} \subset \mathcal{H}_0^1$ and $\mathcal{H}_0^1$ is Complete, thus $\{u_k\}_{k=1}^{\infty}$ is a convergent sequence. Therefore, there exists a $u \in \mathcal{H}_0^1$ such that $u = \lim_{k\to\infty} u_k$, then, by the continuity of F we get that

$$\lim_{k\to\infty} F(u_k) = F(u). \tag{2.7.21}$$

Now (2.7.20) and (2.7.21) yield $F(u) = \sigma^*$ and by (2.7.19) and the definition of σ^* we end up with

$$\text{(M)} \qquad F(u) \le F(v), \quad \forall v \in \mathcal{H}_0^1, \quad (\text{or} \quad F(u) < F(v), \quad v \ne u). \tag{2.7.22}$$

This in our minimization problem (M). And since (M) $\Longleftrightarrow$ (V) we conclude that *there is a unique* $u \in \mathcal{H}_0^1$, *such that* $\ell(v) = (u, v)$, $\forall v \in \mathcal{H}_0^1$. Note that the uniqueness follows from the last step of our minimization proof: $F(u + w) \ge F(u)$, $\forall w \in \mathcal{H}_0^1$. □

Summing up, we have proved that

Proposition 2.1 *Every bounded linear functional can be represented as a scalar product with a given function u. This u is the unique solution for both (V) and (M).*

Theorem 2.18 (Riesz representation theorem) *If V is a Hilbert space with the scalar product (u, v) and norm $\|u\| = \sqrt{(u,u)}$, and $\ell(v)$ is a bounded linear functional on V, then there is a unique $u \in V$ such that*

$$\ell(v) = (u, v), \quad \forall v \in V.$$

Theorem 2.19 (Lax–Milgram theorem) *(A generalized version of Riesz representation theorem) Assume that $\ell(v)$ is a bounded linear functional on V and $a(u, v)$ is bilinear bounded and elliptic in V, then there is a unique $u \in V$, such that*

$$a(u, v) = \ell(v), \quad \forall v \in V. \tag{2.7.23}$$

Remark 2.10 Here *bilinear* means that $a(u, v)$ satisfies the same properties as a scalar product, except for *symmetry*.

Remark 2.11 Note that if the source term $f \in L_2(\Omega)$ then (2.7.8) holds for $V = H^1(\Omega)$ and hence, in our definition of boundedness $c_3 = \|f\|_{L_2(\Omega)}$. The question is whether one can reach (2.7.8) by a somewhat weaker assumption on f than $f \in L_2(\Omega)$? It appears that a weaker assumption on the right- hand side f can be obtained using negative norm. More specifically, we define the $H^{-1}(\Omega)$ norm of f as

$$\|f\|_{H^{-1}(\Omega} = \sup_{v\in H_0^1(\Omega)} \frac{(f, v)}{\|v\|_V}, \quad V = H_0^1(\Omega). \tag{2.7.24}$$

Thus, using negative norm the equivalent norms $||v||_{H_0^1(\Omega)} \approx ||\nabla v||_{L_2(\Omega)}$,

$$||f||_{H^{-1}(\Omega)} \leq \frac{||f||_{L_2(\Omega)}||v||_{L_2(\Omega)}}{||\nabla v||_{L_2(\Omega)}} \leq C||f||_{L_2(\Omega)}. \tag{2.7.25}$$

Hence, (2.7.8) holds with $c_3 = ||f||_{H^{-1}(\Omega)}$. Therefore, it suffices that the right-hand side $f(x)$ in the Poisson's equation satisfies a weaker assumption, i.e. $f \in H^{-1}(\Omega)$ rather than $f \in L_2(\Omega)$.

Below we give some examples of verifying Lax–Milgram theorem and also applications of Lax–Milgram theorem proving existence of a unique solution for certain differential equations of interest.

Example 2.16 Verify the assumptions of the Lax–Milgram theorem and determine the constants of the assumptions for $I = (0, 1)$, $f \in L_2(I)$, $V = H^1(I)$, $||w||_V^2 = ||w||_{L_2(I)}^2 + ||w'||_{L_2(I)}^2$, and

$$a(v, w) = \int_I (uw + v'w')\, dx + v(0)w(0), \quad L(v) = \int_I fv\, dx.$$

Solution: It is trivial to show that $a(\cdot, \cdot)$ is bilinear and $L(\cdot)$ is linear. We have that

$$a(v, v) = \int_I v^2 + (v')^2\, dx + v(0)^2 \geq ||v||_V^2, \tag{2.7.26}$$

so that we can take $c_1 = 1$ in (2.7.6). Further, using Cauchy–Schwarz' inequalitys,

$$\begin{aligned}
|a(v, w)| &\leq \left|\int_I vw\, dx\right| + \left|\int_I v'w'\, dx\right| + |v(0)w(0)| \\
&\leq ||v||_{L_2(I)}||w||_{L_2(I)} + ||v'||_{L_2(I)}||w'||_{L_2(I)} + |v(0)||w(0)| \\
&\leq (||v||_{L_2(I)} + ||v'||_{L_2(I)})(||w||_{L_2(I)} + ||w'||_{L_2(I)}) + |v(0)||w(0)| \\
&\leq \sqrt{2}(||v||_{L_2(I)}^2 + ||v'||_{L_2(I)}^2)^{1/2} \cdot \sqrt{2}(||w||_{L_2(I)}^2 + ||w'||_{L_2(I)}^2)^{1/2} \\
&\quad + |v(0)||w(0)| \\
&= \sqrt{2}||v||_V \sqrt{2}||w||_V + |v(0)||w(0)|.
\end{aligned}$$

Now, we have that

$$v(0) = -\int_0^x v'(y)\, dy + v(x), \quad \forall x \in I, \tag{2.7.27}$$

and by the mean-value theorem for the integrals, there exists $\xi \in I$ so that $v(\xi) = \int_0^1 v(y)\, dy$. Choose $x = \xi$ in (2.7.27), then by Cauchy–Schwarz, inequality

$$\begin{aligned}
|v(0)| &= |-\int_0^{\xi} v'(y)\, dy + \int_0^1 v(y)\, dy| \\
&\leq \int_0^1 |v'|\, dy + \int_0^1 |v|\, dy \leq ||v'||_{L_2(I)} + ||v||_{L_2(I)} \leq \sqrt{2}||v||_V,
\end{aligned}$$

implies that

$$|v(0)||w(0)| \le 2||v||_V||w||_V,$$

and consequently

$$|a(u,w)| \le 2||v||_V||w||_V + 2||v||_V||w||_V = 4||v||_V||w||_V,$$

so that we can take $c_2 = 4$ in (2.7.7). Finally

$$|L(v)| = \left|\int_I fv\,dx\right| \le ||f||_{L_2(I)}||v||_{L_2(I)} \le ||f||_{L_2(I)}||v||_V,$$

taking $c_3 = ||f||_{L_2(I)}$ in (2.7.8), all the conditions in the Lax–Milgram theorem are fulfilled.

Example 2.17 Let $u \in H_0^1(\Omega)$ and consider the convection–diffusion problem

$$-\text{div}(\varepsilon\nabla u + \beta u) = f, \quad \text{in } \Omega \subset \mathbb{R}^2, \quad u = 0 \text{ on } \Gamma = \partial\Omega, \tag{2.7.28}$$

where Ω is a bounded convex polygonal domain, $\varepsilon > 0$ is constant, $\beta = (\beta_1(x), \beta_2(x))$ and $f = f(x)$. Determine conditions in the Lax–Milgram theorem that would guarantee existence of a unique solution for this problem. Prove a stability estimate for u i terms of $||f||_{L_2(\Omega)}$, ε and diam(Ω), and under the conditions that you derived.

Solution: Recall that

$$H_0^1(\Omega) := \{w\colon w \in L_2(\Omega),\ |\nabla w| \in L_2(\Omega),\quad w = 0 \ \text{ on } \ \partial\Omega\}.$$

(a) Multiply Eq. (2.7.28) by $v \in H_0^1(\Omega)$ and integrate over Ω to obtain

$$-\int_\Omega \text{div}(\varepsilon\nabla u + \beta u)v\,dx = \int_\Omega (\varepsilon\nabla u + \beta u)\cdot\nabla v\,dx = \int_\Omega fv\,dx.$$

Thus, the variational formulation for (2.7.28) is: Find $u \in H_0^1(\Omega)$ such that

$$a(u,v) = L(v), \quad \forall v \in H_0^1(\Omega), \tag{2.7.29}$$

where

$$a(u,v) = \int_\Omega (\varepsilon\nabla u + \beta u)\cdot\nabla v\,dx,$$

and

$$L(v) = \int_\Omega fv\,dx.$$

According to the Lax–Milgram theorem, for a unique solution for (2.7.29) we need to verify that there are positive constants $c_1,\ c_2,\ c_3$ such that the following relations hold true

(i) $\quad |a(v, w)| \le c_1 ||u||_{H^1(\Omega)} ||w||_{H^1(\Omega)}, \quad \forall v,\ w \in H_0^1(\Omega),$

(ii) $\quad a(v, v) \ge c_2 ||v||^2_{H^1(\Omega)}, \quad \forall v \in H_0^1(\Omega),$

(iii) $\quad |L(v)| \le c_3 ||v||_{H^1(\Omega)}, \quad \forall v \in H_0^1(\Omega).$

Now using Cauchy–Schwarz' inequality, we have

$$|L(v)| = \left| \int_\Omega fv\ dx \right| \le ||f||_{L_2(\Omega)} ||v||_{L_2(\Omega)} \le ||f||_{L_2(\Omega)} ||v||_{H^1(\Omega)},$$

hence (iii) follows with $c_3 = ||f||_{L_2(\Omega)}$, and consequently the first condition is that $f \in L_2(\Omega)$.

Further we have, using Cauchy–Schwarz' inequality, that

$$\begin{aligned} |a(v, w)| &\le \int_\Omega |\varepsilon \nabla v + \beta v| |\nabla w|\ dx \le \int_\Omega (\varepsilon |\nabla v| + |\beta| |v|) |\nabla w|\ dx \\ &\le \left(\int_\Omega (\varepsilon |\nabla v| + |\beta| |v|)^2\ dx \right)^{1/2} \left(\int_\Omega |\nabla w|^2\ dx \right)^{1/2} \\ &\le \sqrt{2} \max(\varepsilon, ||\beta||_\infty) \left(\int_\Omega (|\nabla v|^2 + v^2)\ dx \right)^{1/2} ||w||_{H^1(\Omega)} \\ &= \gamma ||v||_{H^1(\Omega)} ||w||_{H^1(\Omega)}, \end{aligned}$$

which, with $c_1 = \sqrt{2}\max(\varepsilon, ||\beta||_\infty)$, gives (i). Hence, the second condition is that $\beta \in L_\infty(\Omega)$.

Finally, if div $\beta \le 0$, then using Green's formula

$$\begin{aligned} a(v, v) &= \int_\Omega (\varepsilon |\nabla v|^2 + (\beta \cdot \nabla v) v)\ dx = \int_\Omega \left(\varepsilon |\nabla v|^2 + \left(\beta_1 \frac{\partial v}{\partial x_1} + \beta_2 \frac{\partial v}{\partial x_2} \right) v \right)\ dx \\ &= \int_\Omega \left(\varepsilon |\nabla v|^2 + \frac{1}{2} \left(\beta_1 \frac{\partial}{\partial x_1} (v)^2 + \beta_2 \frac{\partial}{\partial x_2} (v)^2 \right) \right)\ dx \\ &= \int_\Omega \left(\varepsilon |\nabla v|^2 - \frac{1}{2} (\text{div } \beta) v^2 \right)\ dx \ge \int_\Omega \varepsilon |\nabla v|^2\ dx. \end{aligned}$$

Now by the Poincare's inequality

$$\int_\Omega |\nabla v|^2\ dx \ge C \int_\Omega (|\nabla v|^2 + v^2)\ dx = C ||v||^2_{H^1(\Omega)},$$

for some constant $C = C(\text{diam}(\Omega))$, we have

$$a(v, v) \ge c_2 ||v||^2_{H^1(\Omega)}, \quad \text{with } c_2 = \varepsilon C(\text{diam}(\Omega)),$$

which means that (ii) is valid under the *condition* that div $\beta \le 0$.

From (ii), (2.7.29) (with $v = u$) and (iii), we conclude that

$$c_2 ||u||^2_{H^1(\Omega)} \le a(u, u) = L(u) \le c_3 ||u||_{H^1(\Omega)},$$

and hence, a *stability estimate* is given by

$$||u||_{H^1(\Omega)} \leq \frac{c_3}{c_2},$$

with $c_3 = ||f||_{L_2(\Omega)}$ and $c_2 = C\varepsilon$ defined above.

2.8 Exercises

Problem 2.1 Show that ℓ, defined in Example 2.15 above is linear.

Problem 2.2 Prove using Cauchy's and Poincare's inequalities that ℓ, defined as in Example 2.15, is bounded in $\mathcal{H}_0^1$.

Problem 2.3 Show that the functions x, e^x, xe^x, and e^{2x} are linearly independent on $\mathbb{R}$.

Problem 2.4 Is the vector $(4, 2, 1)$ a linear combination of the vectors $(1, 2, 1)$, $(-1, 1, 2)$, and $(-3, -3, 0)$.

Problem 2.5 Is the set of vectors

$$\{x^2 + 2x + 2,\ 2x^2 + 3x + 1, -x - 3\},$$

linearly dependent or linearly independent? If it is linearly dependent, write one of the vectors as a linear combination of the other two.

Problem 2.6 Find a basis for the subspace of $\mathbb{R}^4$ consisting of all vectors of the form

$$(a + b,\ b + c,\ a - b - 2c,\ b + c).$$

What is the dimension of this subspace?

Problem 2.7 Find a basis for the solution space of the homogeneous system

$$\begin{aligned} x_1 + 2x_2 - x_3 + x_4 + 2x_5 &= 0 \\ x_1 + x_2 + 2x_3 - 3x_4 + x_5 &= 0. \end{aligned}$$

What is the dimension of the solution space?

Problem 2.8 For what values of α is the set

$$\{x + 3,\ 2x + \alpha^2 + 2\}$$

linearly independent?

Problem 2.9 For what values of α is the vector $(\alpha^2,\ \alpha,\ 1)$ in the span of $\{(1,2,3),\ (1,1,1),\ (0,1,2)\}$?

Problem 2.10 Let $\mathbf{x}_1,\ \mathbf{x}_2$, and $\mathbf{x}_3$ be vectors in a vector space. Verify that

$$\text{span}\ \{\mathbf{x}_1,\ \mathbf{x}_2,\ \mathbf{x}_3\} = \text{span}\ \{\mathbf{x}_1+\mathbf{x}_2,\ \mathbf{x}_1-\mathbf{x}_3,\ \mathbf{x}_1+\mathbf{x}_3\}.$$

Problem 2.11 Verify that the assumptions of the Lax–Milgram theorem are satisfied for the following problems with appropriate assumptions on α and f.

$$\text{(I)} \quad \begin{cases} -u''+\alpha u = f, & \text{in } (0,1), \\ u(0)=u'(1)=0, & \alpha = 0 \text{ and } 1. \end{cases}$$

$$\text{(II)} \quad \begin{cases} -u''+\alpha u = f, & \text{in } (0,1), \\ u(0)=u(1) & u'(0)=u'(1)=0. \end{cases}$$

$$\text{(III)} \quad \begin{cases} -u'' = f, & \text{in } (0,1), \\ u(0)-u'(0)=u(1) & +u'(1)=0. \end{cases}$$

Problem 2.12 Let Ω be a bounded domain in $\mathbb{R}^d$ with boundary Γ, show that there is a constant C such that for all $v \in H^1(\Omega)$,

$$\|v\|_{L_2(\Gamma)} \leq C\|v\|_{H^1(\Omega)}, \tag{2.8.1}$$

where $\|v\|^2_{H^1(\Omega)} = \|v\|^2 + \|\nabla v\|^2$.
Hint: Use the following Green's formula

$$\int_\Omega v^2 \Delta\varphi = \int_\Gamma v^2 \partial_n \varphi - \int_\Omega 2v\nabla v \cdot \nabla\varphi, \tag{2.8.2}$$

with $\partial_n \varphi = 1$. Equation (2.8.1) is known as trace inequality, or trace theorem.

Problem 2.13 Let u be the solution of the following Neumann problem:

$$\begin{cases} -\Delta u = f, & \text{in } \Omega \subset \mathbb{R}^d, \\ -\partial_n u = ku, & \text{on } \Gamma = \partial\Omega,. \end{cases}$$

where $\partial_n u = n \cdot \nabla u$ with n being outward unit normal to Γ and $k \geq 0$.
(a) Show the stability estimate

$$\|u\|_\Omega \leq C_\Omega(\|u\|_\Gamma + \|\nabla u\|_\Omega).$$

(b) Use the estimate in (a) to show that $\|u\|_\Gamma \to 0$ as $k \to \infty$.

Problem 2.14 Using the trace inequality, show that the solution for the problem

$$\begin{cases} -\Delta u + u = 0, & \text{in } \Omega \\ \partial_n u = g, & \text{on } \Gamma, \end{cases}$$

satisfies the inequality

$$\|u\|^2 + \|\nabla u\|^2 \le C\|g\|^2_{L_2(\Gamma)}.$$

Problem 2.15 Consider the BVP

$$\begin{cases} \Delta u = 0, & \text{in } \Omega \subset \mathbb{R}^2, \\ \partial_n u + u = g, & \text{on } \Gamma = \partial\Omega, \quad n \text{ is outward unit normal to } \Gamma. \end{cases}$$

(a) Show the stability estimate

$$||\nabla u||^2_{L_2(\Omega)} + \frac{1}{2}||u||^2_{L_2(\Gamma)} \le \frac{1}{2}||g||^2_{L_2(\Gamma)}.$$

(b) Discuss, concisely, the conditions for applying the Lax–Milgram theorem to this problem.

Problem 2.16 Let $\Omega \subset \mathbb{R}^2$ be a smooth domain with the boundary $\Gamma := \partial\Omega$. Prove that there is no constant C such that

$$||u||_{L_2(\Gamma)} \le C||u||_{L_2(\Omega)}, \quad \forall v \in L_2(\Omega).$$

Problem 2.17 Consider the following Neumann BVP

$$\begin{cases} -\Delta u + u = f, & \text{in } \Omega \\ \partial_n u = 0, & \text{on } \Gamma, \end{cases} \tag{2.8.3}$$

where $\Omega := \{x \in \mathbb{R}^2 \colon |x| < 1\}$ and $f(x) = 1/|x|$.
Show that, in spite of the fact that $f \notin L_2(\Omega)$, the Lax–Milgram applies to this problem.

Problem 2.18 Let Ω be a square with boundary Γ. Consider the Hilbert space $H_0^2(\Omega)$ associated with the boundary condition $u = \frac{\partial u}{\partial n} = 0$ on Γ. Show that there is a constant C such that

$$\|u\|_{H^2(\Omega)} \le C\|\Delta u\|_{L_2(\Omega)}, \quad \forall u \in H_0^2(\Omega).$$

Hint:

1. Use the Green's formula to show that

$$\int_\Omega \left(\frac{\partial^2 u}{\partial x_1 \partial x_2}\right)^2 dx = \int_\Omega \frac{\partial^2 u}{\partial x_1^2} \cdot \frac{\partial^2 u}{\partial x_2^2}\, dx, \quad \forall u \in H_0^2(\Omega).$$

2. Note that $u = 0$ on Γ implies that $\frac{\partial u}{\partial s} = 0$ on Γ, where $\frac{\partial}{\partial s}$ is differentiation in a tangential direction to Γ.

3

Polynomial Approximation/Interpolation in 1*d*

In the Chapters 1 and 2, we introduced some differential equations and outlined analytic approaches for their solutions in some adequate mathematical environments. In this chapter, we mimic this procedure for discretized versions to these continuous problems in the one-dimensional case. More specifically, our objective is to present the finite element method (FEM) as an approximation technique for solution of differential equations using piecewise polynomials. Here, we give motivation and unifying ideas in using finite elements as a numerical strategy arising from the need of rendering a continuous problem to its discrete version. Loosely speaking, a continuous problem will have infinitely many unknowns (e.g. if one asks for the solution $u(x)$ at every point $x \in \mathbb{R}$), and it cannot be solved exactly on a computer. Therefore, the idea is to approximate the continuous problem using an adequate discrete formulation and get a discrete problem that possesses a finite number of unknowns. The more unknowns we keep, the better the accuracy of the approximation will be, but at a greater computational cost. So, a desirable approximation is a reasonably accurate one with fewer approximation points. To this approach, we introduce the intermediate procedure of the concept of polynomial interpolation and give a swift overture of the numerical integration.

3.1 Finite Dimensional Space of Functions on an Interval

Throughout the book, we shall frequently approximate the solutions for the differential equations by the functions of the following finite dimensional spaces.

I. $\mathcal{P}^{(q)}(a,b) := \{$The space of polynomials in x of degree $\leqq q$, $a \le x \le b\}$.
A possible basis for $\mathcal{P}^{(q)}(a,b)$ would be $\{x^j\}_{j=0}^{q} = \{1, x, x^2, x^3, \dots, x^q\}$. These are, in general, nonorthogonal polynomials (monomials) and may be orthogonalized by the Gram–Schmidt orthogonalization procedure. The dimension of $\mathcal{P}^{(q)}$ is $q+1$.

An Introduction to the Finite Element Method for Differential Equations, First Edition.
Mohammad Asadzadeh.

II. An example of orthogonal basis functions, on $(0, 1)$ or $(-1, 1)$, are the *Legendre polynomials*:

$$\tilde{P}_k(x) = \frac{(-1)^k}{k!}\frac{d^k}{dx^k}[x^k(1-x)^k] \quad \text{or} \quad P_n(x) = \frac{1}{2^n n!}\frac{d^n}{dx^n}[(x^2-1)^n],$$

respectively. The first four Legendre orthogonal polynomials on $(-1, 1)$ are

$$P_0(x) = 1, \quad P_1(x) = x, \quad P_2(x) = \frac{3}{2}x^2 - \frac{1}{2}, \quad P_3(x) = \frac{5}{2}x^3 - \frac{3}{2}x.$$

III. Periodic orthogonal bases on $[0, T]$ are usually represented by trigonometric polynomials (also known from Fourier series representation):

$$T^N := \left\{ f(x) \middle| f(x) = \sum_{n=0}^{N}\left[a_n \cos\left(\frac{2\pi}{T}nx\right) + b_n \sin\left(\frac{2\pi}{T}nx\right)\right]\right\}.$$

IV. A general form of polynomial basis functions on an interval is introduced in Section 3.5: namely the Lagrange basis $\{\lambda_i\}_{i=0}^q \subset \mathcal{P}^{(q)}(a, b)$, consisting of *continuous polynomials of degree* $\leq q$, associated to a set of $(q+1)$ distinct points $\xi_0 < \xi_1 < \cdots < \xi_q$ in $[a, b]$ and determined by the requirement that

$$\lambda_i(\xi_j) = \begin{cases} 1 & i = j, \\ 0 & i \neq j, \end{cases} \quad \text{or} \quad \lambda_i(x) = \prod_{j=0,(j\neq i)}^{q} \frac{x - \xi_j}{\xi_i - \xi_j}.$$

It is easy to verify that $\{\lambda_i\}_{i=0}^q$, $i = 0, 1, \ldots, q$; are linearly independent. A polynomial $P(x) \in \mathcal{P}^{(q)}(a, b)$, that has the value $p_i = P(\xi_i)$ at the nodes $x = \xi_i$ for $i = 0, 1, \ldots, q$, is uniquely, determined in terms of the corresponding Lagrange basis by

$$P(x) = p_0\lambda_0(x) + p_1\lambda_1(x) + \cdots + p_q\lambda_q(x). \tag{3.1.1}$$

Note that for each nodal-point $x = \xi_i$, we have associated a basis function $\lambda_i(x)$, $i = 0, 1, \ldots, q$. Thus, we have $q + 1$ basis functions, each being a polynomial of degree q, whence a $(q+1) \times (q+1)$ system.

Remark 3.1 Our goal is to approximate general functions by piecewise polynomials of Lagrange type. Then, for a given function f, the Lagrange coefficients p_i, $0 \leq i \leq q$, in (3.1.1) will be replaced by $f(\xi_i)$, and $f(x)$ will be approximated by its *Lagrange interpolant* defined as

$$f(x) \approx \sum_{i=0}^{q} f(\xi_i)\lambda_i(x) := \pi_q f(x), \tag{3.1.2}$$

with $f(\xi_i) = \pi_q f(\xi_i)$, $i = 0, 1, \ldots, q$. Detailed study of the Lagrange interpolant is the subject of Section 3.5.

We illustrate this, for the linear case, in Example 3.1.

Example 3.1 The linear Lagrange basis functions, $q = 1$, are given by (see Figures 3.1 and 3.2).

$$\lambda_0(x) = \frac{\xi_1 - x}{\xi_1 - \xi_0} \quad \text{and} \quad \lambda_1(x) = \frac{x - \xi_0}{\xi_1 - \xi_0}. \tag{3.1.3}$$

Choosing $\xi_0 = a$ and $\xi_1 = b$ in (3.1.3), we get the linear approximation (known as linear interpolant)

$$\pi_1 f(x) = f(a)\lambda_0(x) + f(b)\lambda_1(x),$$

where

$$\lambda_0(x) = \frac{b - x}{b - a} \quad \text{and} \quad \lambda_1(x) = \frac{x - a}{b - a},$$

i.e.

$$\pi_1 f(x) = f(a)\frac{b - x}{b - a} + f(b)\frac{x - a}{b - a}.$$

V. In FEMs, we approximate the solution of differential equations by (continuous or discontinuous) piecewise polynomials. Here is an example of the space of *continuous piecewise polynomials* on a partition of an interval into a collection of subintervals. Let $\mathcal{T}_h{:}0 = x_0 < x_1 < \cdots < x_M < x_{M+1} = 1$, with $h(x) = |I_j|$, $I_j := x_j - x_{j-1}$ and $j = 1, \dots, M + 1$, be a partition of $[0, 1]$ into $M + 1$ subintervals. Then, $V_h^{(q)}$ denotes the space of all continuous piecewise polynomial functions of degree $\leq q$ on $\mathcal{T}_h$ and

$$V_{0,h}^{(q)} = \{v{:}\ v \in V_h^{(q)},\quad v(0) = v(1) = 0\}.$$

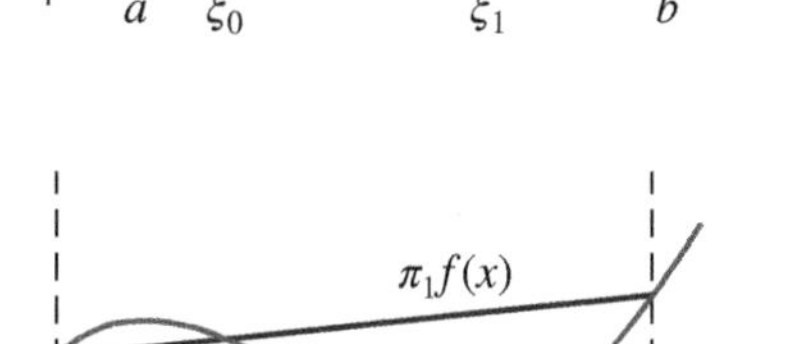

Figure 3.1 Linear Lagrange basis functions for $q = 1$.

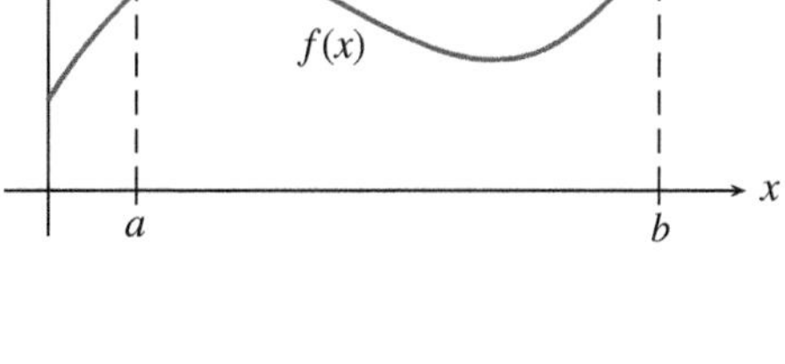

Figure 3.2 The linear interpolant $\pi_1 f(x)$ on a single interval.

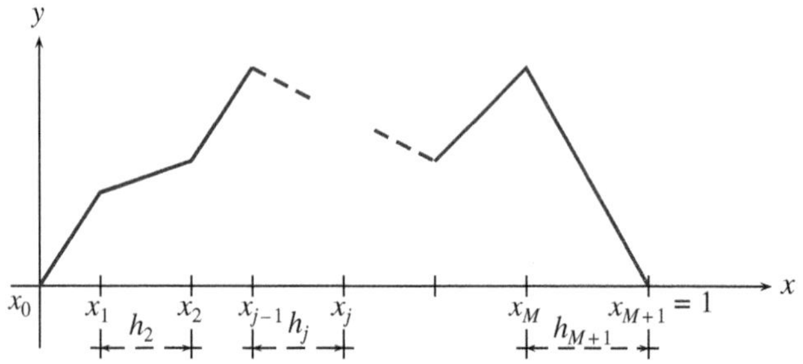

Figure 3.3 An example of a function in $V_{0,h}^{(1)}$.

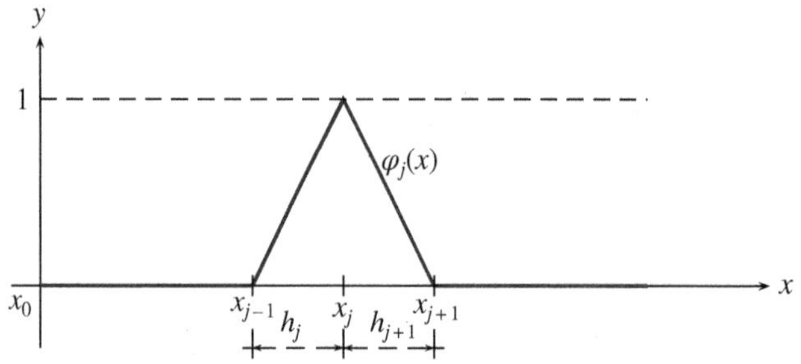

Figure 3.4 A general piecewise linear basis function $\varphi_j(x)$.

The standard basis for piecewise linears: $V_h := V_h^{(1)}$ is given by the so-called *hat-functions* $\varphi_j(x)$ with the additional property that $\varphi_j(x)$ is a continuous piecewise linear function, namely Figure 3.4, such that $\varphi_j(x_i) = \delta_{ij}$,

$$\delta_{ij} = \begin{cases} 1 & i = j, \\ 0 & i \neq j, \end{cases} \quad \text{i.e. } \varphi_j(x) = \begin{cases} \frac{x-x_{j-1}}{h_j} & x_{j-1} \le x \le x_j \\ \frac{x_{j+1}-x}{h_{j+1}} & x_j \le x \le x_{j+1} \\ 0 & x \notin [x_{j-1}, x_{j+1}], \end{cases}$$

where as in Figure 3.6, with obvious modifications for $j = 0$ and $j = M + 1$.

In the Section 3.2, we compare the behavior of two approximation procedures for an ordinary differential equation (ODE) and a partial differential equation (PDE) using monomials and piecewise linears, respectively.

3.2 An Ordinary Differential Equation (ODE)

An *initial value problem* (IVP): a model problem in population dynamics, with $u(t)$ denoting the size of the population at time t, can be written as

$$\dot{u}(t) = \lambda u(t), \quad 0 < t < T, \qquad u(0) = u_0, \tag{3.2.1}$$

where $\dot{u}(t) = \frac{du}{dt}$ and λ is a positive constant. For λ and $u_0 > 0$, this problem has the increasing analytic solution $u(t) = u_0 e^{\lambda t}$, which blows up as $t \to \infty$.

Generally, population dynamics equation, depending on many "parameters," is stated in vector form, and we have $\dot{\mathbf{u}}(t) = F(\mathbf{u}(t), t)$, where $\mathbf{u}(t) \in \mathbb{R}^n$ is a time-dependent vector in $\mathbb{R}^n$, with $\dot{\mathbf{u}} = \partial \mathbf{u}(t)/\partial t \in \mathbb{R}^n$ being its component-wise derivative with respect to $t \in \mathbb{R}^+$. Then $\mathbf{u}(t) = [u_1(t), u_2(t), \dots, u_n(t)]^T$, $\dot{\mathbf{u}}(t) = [\dot{u}_1(t), \dot{u}_2(t), \dots, \dot{u}_n(t)]^T$ and

$$F\colon \mathbb{R}^n \times \mathbb{R}^+ \to \mathbb{R}^n.$$

3.2.1 Forward Euler Method to Solve IVP

Below we give numerical solution of (3.2.1) using a *finite difference* method: We *discretize* the IVP (3.2.1) with the forward Euler method based on a partition of the interval $[0, T]$ into N subintervals, as in Figure 3.5, and an approximation of the derivative by a difference quotient at each subinterval $[t_k, t_{k+1}]$ letting $\dot{u}(t) \approx \frac{u(t_{k+1}) - u(t_k)}{t_{k+1} - t_k}$. Then an approximation of (3.2.1) is given by

$$\frac{u(t_{k+1}) - u(t_k)}{t_{k+1} - t_k} = \lambda \cdot u(t_k), \quad k = 0, \dots, N-1, \quad \text{and} \quad u(0) = u_0. \tag{3.2.2}$$

Hence, with $\Delta t_k = t_{k+1} - t_k$, we get

$$u(t_{k+1}) = (1 + \lambda \Delta t_k) u(t_k). \tag{3.2.3}$$

Starting with $k = 0$ and the data $u(0) = u_0$, the solution $u(t_k)$ would, iteratively, be computed at the subsequent points: $t_1, \ t_2, \dots, t_N = T$.

For a uniform partition, where all subintervals have the same length Δt, (3.2.3) would be of the form

$$u(t_{k+1}) = (1 + \lambda \Delta t) u(t_k), \quad k = 0, 1, \dots, N-1. \tag{3.2.4}$$

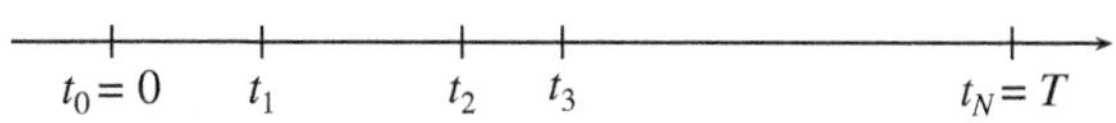

Figure 3.5 A partition of $[0, T]$.

Iterating we end up with

$$u(t_{k+1}) = (1 + \lambda\Delta t)^2 u(t_{k-1}) = \cdots = (1 + \lambda\Delta t)^{k+1} u_0. \tag{3.2.5}$$

Other finite difference methods for (3.2.1) are introduced in the study of the initial boundary value problems (BVPs) in Chapter 6. There are corresponding finite difference methods for PDEs. Our goal, however, is to study the *Galerkin FEMs*.

3.2.2 Variational Formulation for IVP

We multiply the IVP (3.2.1) with test functions v in a certain vector space V and integrate over $[0, T]$, to obtain

$$\int_0^T \dot{u}(t)v(t)\,dt = \lambda \int_0^T u(t)v(t)\,dt, \quad \forall v \in V, \tag{3.2.6}$$

or equivalently

$$\int_0^T (\dot{u}(t) - \lambda\, u(t))v(t)\,dt = 0, \quad \forall v(t) \in V, \tag{3.2.7}$$

which, interpreted as an inner product, means that

$$(\dot{u}(t) - \lambda\, u(t)) \perp v(t), \quad \forall v(t) \in V. \tag{3.2.8}$$

We refer to (3.2.6) as the *variational problem* for (3.2.1).

For Problem (3.2.6), it is natural to seek a solution in $C([0, T])$, or in

$$V := H^1(0, T) := \left\{ f\colon \int_0^T (f(t)^2 + \dot{f}(t)^2)\,dt < \infty \right\},$$

with H^1 consisting of all functions in $L_2(0, T)$ having also their first derivatives in $L_2(0, T)$.

Definition 3.1 If w is an approximation of u in the variational problem (3.2.6), then $\mathcal{R}(w(t)) := \dot{w}(t) - \lambda w(t)$ is called the residual error of $w(t)$.

In general for an approximate solution w for (3.2.1), we have that $\dot{w}(t) - \lambda w(t) \neq 0$; Otherwise, w and u would satisfy the same equation, and by uniqueness, we would get the exact solution ($w = u$). Our requirement is instead that w satisfies Eq. (3.2.1) in average, or in other words, we require that w satisfies (3.2.8):

$$\mathcal{R}(w(t)) \perp v(t), \quad \forall v(t) \in V. \tag{3.2.9}$$

We look for an approximate solution $U(t)$, called a *trial function* for (3.2.1), in a finite dimensional subspace $\hat{V}$ of V, say $\mathcal{P}^{(q)}$, the space of polynomials of degree $\leq q$:

$$\hat{V} := \mathcal{P}^{(q)} = \{U\colon U(t) = \xi_0 + \xi_1 t + \xi_2 t^2 + \cdots + \xi_q t^q\}. \tag{3.2.10}$$

Hence, to determine $U(t)$, we need to determine the coefficients $\xi_0, \xi_1, \dots, \xi_q$. We refer to $\hat{V}$ as the *trial space*. Note that $u(0) = u_0$ is given and therefore we may take $U(0) = \xi_0 = u_0$. It remains to find the real numbers $\xi_1, \dots, \xi_q$. These are coefficients of the q linearly independent monomials $t,\ t^2,\ \dots,\ t^q$. To this approach, we define the *test function space*:

$$V_0 := \mathcal{P}_0^{(q)} = \{v \in \mathcal{P}^{(q)} : v(0) = 0\}. \tag{3.2.11}$$

Thus, v can be written as $v(t) = c_1 t + c_2 t^2 + \cdots + c_q t^q$. Note that

$$\mathcal{P}_0^{(q)} = \text{span}[t,\ t^2,\ \dots,\ t^q]. \tag{3.2.12}$$

For an approximate solution U, we require its residual $R(U)$ to satisfy the orthogonality condition (3.2.9):

$$\mathcal{R}(U(t)) \perp v(t), \quad \forall v(t) \in \mathcal{P}_0^{(q)}.$$

Below we shall show how this condition yields the coefficient $\xi_1, \dots, \xi_q$ and determines $U(t)$. We also illustrate which are the better choice for the test function spaces V_h.

3.2.3 Galerkin Method for IVP

Given $u(0) = u_0$, find the approximate solution $U(t) \in \mathcal{P}^{(q)}(0,T)$, of (3.2.1) satisfying

$$\int_0^T \mathcal{R}(U(t))v(t)\,dt = \int_0^T \left(\dot{U}(t) - \lambda\, U(t)\right) v(t)\,dt = 0, \quad \forall v(t) \in \mathcal{P}_0^{(q)}. \tag{3.2.13}$$

Formally, this can be obtained writing a *wrong!* equation by replacing u by $U \in \mathcal{P}^{(q)}$ in (3.2.1),

$$\begin{cases} \dot{U}(t) = \lambda\, U(t), & 0 < t < T \\ U(0) = u_0, \end{cases} \tag{3.2.14}$$

then, multiplying (3.2.14) by a function $v(t) \in \mathcal{P}_0^{(q)}$ from the test function space and integrating over $[0, T]$.

Now since $U \in \mathcal{P}^{(q)}$, we can write $U(t) = u_0 + \sum_{j=1}^q \xi_j t^j$, then $\dot{U}(t) = \sum_{j=1}^q j\xi_j t^{j-1}$. Further, we have that $\mathcal{P}_0^{(q)}$ is spanned by $v_i(t) = t^i, i = 1, 2, \dots, q$. Therefore, it suffices to use these t^i values as test functions. Inserting these representations for $U,\ \dot{U}$, and $v = t^i,\ i = 1, 2, \dots, q$ into (3.2.13), we get (for simplicity we have put $T \equiv 1$)

$$\int_0^1 \left(\sum_{j=1}^q j\xi_j t^{j-1} - \lambda u_0 - \lambda \sum_{j=1}^q \xi_j t^j \right) \cdot t^i\,dt = 0, \quad i = 1, 2, \dots, q. \tag{3.2.15}$$

Moving the data to the right-hand side, this relation can be rewritten as

$$\int_0^1 \left(\sum_{j=1}^{q} (j\, t^{i+j-1} - \lambda\, t^{i+j}) \right) \xi_j \, dt = \lambda u_0 \int_0^1 t^i \, dt, \quad i = 1, 2, \ldots, q. \tag{3.2.16}$$

Performing the integration (ξ_j values are constants independent of t), we get

$$\sum_{j=1}^{q} \xi_j \left[j \frac{t^{i+j}}{i+j} - \lambda \frac{t^{i+j+1}}{i+j+1} \right]_{t=0}^{t=1} = \left[\lambda\, u_0 \frac{t^{i+1}}{i+1} \right]_{t=0}^{t=1}, \tag{3.2.17}$$

or equivalently

$$\sum_{j=1}^{q} \left(\frac{j}{i+j} - \frac{\lambda}{i+j+1} \right) \xi_j = \frac{\lambda}{i+1}\, u_0 \quad i = 1, 2, \ldots, q, \tag{3.2.18}$$

which is a linear system of equations (LSEs) with q equations and q unknowns $(\xi_1, \xi_2, \ldots, \xi_q)$; in the coordinates form. In the matrix form, (3.2.18) reads

$$\mathcal{A}\Xi = \mathbf{b}, \quad \text{with} \quad \mathcal{A} = (a_{ij})_{i,j=1}^{q}, \quad \Xi = (\xi_j)_{j=1}^{q}, \quad \text{and} \quad \mathbf{b} = (b_i)_{i=1}^{q}. \tag{3.2.19}$$

But the matrix $\mathcal{A}$ although invertible, is *ill-conditioned*, i.e. difficult to invert numerically with any accuracy. Mainly, because $\{t^i\}_{i=1}^q$ does not form an orthogonal basis and for large i and j the last two rows (columns) of $\mathcal{A}$, computed from $a_{ij} = \dfrac{j}{i+j} - \dfrac{\lambda}{i+j+1}$, are very close to each others resulting in very small value for the determinant of $\mathcal{A}$.

If we insist to use polynomial basis up to certain order, then instead of monomials, the use of Legendre orthogonal polynomials would yield a diagonal (sparse) coefficient matrix and make the problem both well conditioned and easier to solve. This, however, is rather involved. A better approach would be through the use of piecewise polynomial approximations (see, e.g. Chapter 6) on a partition of $[0, T]$ into subintervals, where we use *low order* polynomial approximations on each subinterval.

3.3 A Galerkin Method for (BVP)

We consider the Galerkin method for the following stationary ($\dot{u} = du/dt = 0$) heat equation in one dimension

$$\begin{cases} -\frac{d}{dx}\left(c(x)\, \frac{d}{dx} u(x) \right) = f(x), & 0 < x < 1; \\ u(0) = u(1) = 0. \end{cases} \tag{3.3.1}$$

For simplicity, we let $c(x) = 1$, then we have

$$-u''(x) = f(x), \quad 0 < x < 1; \qquad u(0) = u(1) = 0. \tag{3.3.2}$$

Now, let $\mathcal{T}_h : 0 = x_0 < x_1 < \cdots < x_M < x_{M+1} = 1$ be a partition of the interval $(0, 1)$ into the subintervals $I_j = (x_{j-1}, x_j)$, with length $|I_j| = h_j = x_j - x_{j-1}$, $j = 1, 2, \dots, M+1$. We define the finite dimensional space V_h^0 by

$$V_h^0 := \{v \in C(0,1): v \text{ is a piecewise linear function on } \mathcal{T}_h,\ v(0) = v(1) = 0\},$$

with the basis being the "hat-functions" $\{\varphi_j\}_{j=1}^M$. Due to the fact that u is known at the boundary points 0 and 1, it is not necessary to supply test functions corresponding to the values at $x_0 = 0$ and $x_{M+1} = 1$. However, in the case of given nonhomogeneous boundary data $u(0) = u_0 \neq 0$ and/or $u(1) = u_1 \neq 0$, to represent the trial function, one uses the basis functions to all internal nodes as well as those corresponding to the nonhomogeneous data (i.e. at $x = 0$ and/or $x = 1$).

Remark 3.2 If the Dirichlet boundary condition is given at only one of the boundary points, say $x_0 = 0$ and the other one satisfies, e.g. a Neumann condition as

$$-u''(x) = f(x), \quad 0 < x < 1; \qquad u(0) = b_0, \quad u'(1) = b_1, \tag{3.3.3}$$

then the test function representing $x = 0$: φ_0 will be unnecessary (no matter whether $b_0 = 0$ or $b_0 \neq 0$), whereas one needs to provide the *half-base* function φ_{M+1} at $x_{M+1} = 1$ (dashed in Figure 3.6). Again, φ_0 is present in representing the trial function U, with b_0 as its coefficient. Then, the term involving $b_0\varphi_0(x)$ being a data term is moved to the right-hand side of the equation.

To proceed with (3.3.2), we use the function space

$$V_0 = H_0^1(0,1) = \left\{ w\colon \int_0^1 (w(x)^2 + w'(x)^2)\, dx < \infty,\quad w(0) = w(1) = 0 \right\}.$$

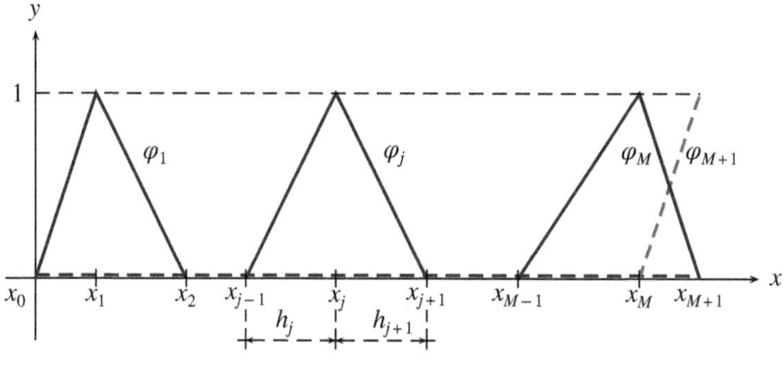

Figure 3.6 Piecewise linear basis functions.

A *variational formulation* for Problem (3.3.2), is based on multiply it by a test function in V_0 and integrating over $[0, 1]$: find $u \in V_0$ such that

$$\int_0^1 (-u''(x) - f(x))v(x)\,dx = 0, \quad \forall v \in V_0. \tag{3.3.4}$$

Integrating by parts, we get

$$-\int_0^1 u''(x)v(x)\,dx = \int_0^1 u'(x)v'(x)\,dx - [u'(x)v(x)]_0^1 \tag{3.3.5}$$

and since $v(x) \in V_0$; $v(0) = v(1) = 0$, we end up with

$$-\int_0^1 u''(x)v(x)\,dx = \int_0^1 u'(x)v'(x)\,dx. \tag{3.3.6}$$

Thus, the *variational formulation* is: find $u \in V_0$ such that

$$\int_0^1 u'(x)v'(x)\,dx = \int_0^1 f(x)v(x)\,dx, \quad \forall v \in V_0. \tag{3.3.7}$$

This is a guideline for the *finite element formulation*:

The Galerkin FEM for problem (3.3.2): find $U \in V_h^0$, such that

$$\int_0^1 U'(x)v'(x)\,dx = \int_0^1 f(x)v(x)\,dx, \quad \forall v \in V_h^0. \tag{3.3.8}$$

Remark 3.3 In this way, formally, the Galerkin approximation U of u and the L_2-projection Pu (defined below) of u are differing only on the requirement that U should satisfy the boundary conditions, whereas Pu does not need to.

We shall determine $\xi_j = U(x_j)$ *the approximate* values of $u(x)$ at the node points x_j, $1 \leq j \leq M$. To this end using basis functions $\varphi_j(x)$, we may write

$$U(x) = \sum_{j=1}^{M} \xi_j\, \varphi_j(x), \quad \text{which implies that } U'(x) = \sum_{j=1}^{M} \xi_j \varphi_j'(x). \tag{3.3.9}$$

Thus, we can write (3.3.8) as

$$\sum_{j=1}^{M} \xi_j \int_0^1 \varphi_j'(x)\, v'(x)\,dx = \int_0^1 f(x)v(x)\,dx, \quad \forall v \in V_h^0. \tag{3.3.10}$$

Since every $v \in V_h^0$ is a linear combination of the basis functions φ_i, it suffices to try with $v(x) = \varphi_i(x)$, for $i = 1, 2, \dots, M$: that is, to find ξ_j (constants), $1 \leq j \leq M$ such that

$$\sum_{j=1}^{M} \left(\int_0^1 \varphi_i'(x)\varphi_j'(x)\,dx \right) \xi_j = \int_0^1 f(x)\varphi_i(x)\,dx, \quad i = 1, 2, \dots, M. \tag{3.3.11}$$

This $M \times M$ system of equations can be written in the matrix form as

$$\mathbf{A}\xi = \mathbf{b}. \tag{3.3.12}$$

Here, $\mathbf{A}$ is called the *stiffness matrix* and $\mathbf{b}$ the *load vector*:

$$\mathbf{A} = \{a_{ij}\}_{i,j=1}^{M}, \quad a_{ij} = \int_0^1 \varphi_i'(x)\varphi_j'(x)\, dx, \tag{3.3.13}$$

$$\mathbf{b} = \begin{pmatrix} b_1 \\ b_2 \\ \vdots \\ b_M \end{pmatrix}, \quad \text{with } b_i = \int_0^1 f(x)\varphi_i(x)\, dx, \quad \text{and} \quad \xi = \begin{pmatrix} \xi_1 \\ \xi_2 \\ \vdots \\ \xi_M \end{pmatrix}. \tag{3.3.14}$$

To compute the entries a_{ij} of the stiffness matrix $\mathbf{A}$, in the case of approximating by continuous piecewise linears, first, we need to derive $\varphi_i'(x)$, viz.

$$\varphi_i(x) = \begin{cases} \frac{x-x_{i-1}}{h_i} & x_{i-1} \le x \le x_i \\ \frac{x_{i+1}-x}{h_{i+1}} & x_i \le x \le x_{i+1} \\ 0 & \text{else} \end{cases} \implies \varphi_i'(x) = \begin{cases} \frac{1}{h_i} & x_{i-1} < x < x_i \\ -\frac{1}{h_{i+1}} & x_i < x < x_{i+1} \\ 0 & \text{else.} \end{cases}$$

Stiffness matrix A:

If $|i-j| > 1$, then φ_i and φ_j have disjoint supports, see Figure 3.7, and

$$a_{ij} = \int_0^1 \varphi_i'(x)\varphi_j'(x)\, dx = 0.$$

As for $i = j$: we have that

$$a_{ii} = \int_{x_{i-1}}^{x_i} \left(\frac{1}{h_i}\right)^2 dx + \int_{x_i}^{x_{i+1}} \left(-\frac{1}{h_{i+1}}\right)^2 dx = \frac{\overbrace{x_i - x_{i-1}}^{h_i}}{h_i^2} + \frac{\overbrace{x_{i+1} - x_i}^{h_{i+1}}}{h_{i+1}^2}$$
$$= \frac{1}{h_i} + \frac{1}{h_{i+1}}.$$

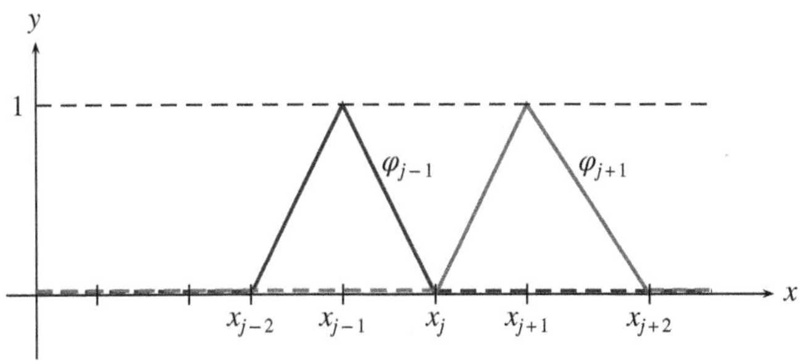

Figure 3.7 φ_{j-1} and φ_{j+1}.

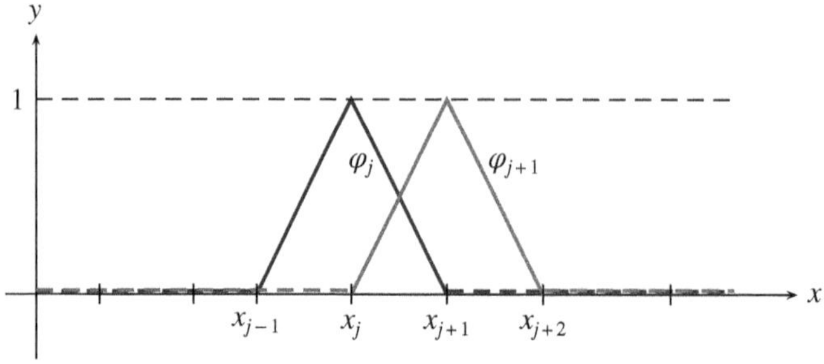

Figure 3.8 φ_j and φ_{j+1}.

It remains to compute a_{ij} for the case of (applicable!) $j = i \pm 1$: a straightforward calculation (see Figure 3.8) yields

$$a_{i,i+1} = \int_{x_i}^{x_{i+1}} \left(-\frac{1}{h_{i+1}}\right) \cdot \frac{1}{h_{i+1}}\, dx = -\frac{x_{i+1} - x_i}{h_{i+1}^2} = -\frac{1}{h_{i+1}}. \tag{3.3.15}$$

Obviously, $a_{i+1,i} = a_{i,i+1} = -\frac{1}{h_{i+1}}$. To *summarize*, we have

$$\begin{cases} a_{ij} = 0, & \text{if } |i-j| > 1, \\ a_{ii} = \frac{1}{h_i} + \frac{1}{h_{i+1}}, & i = 1, 2, \dots, M, \\ a_{i-1,i} = a_{i,i-1} = -\frac{1}{h_i}, & i = 2, 3, \dots, M. \end{cases} \tag{3.3.16}$$

By symmetry $a_{ij} = a_{ji}$, and we finally have the stiffness matrix for the stationary heat conduction as follows:

$$\mathbf{A} = \begin{bmatrix} \frac{1}{h_1} + \frac{1}{h_2} & -\frac{1}{h_2} & 0 & \cdots & 0 \\ -\frac{1}{h_2} & \frac{1}{h_2} + \frac{1}{h_3} & -\frac{1}{h_3} & 0 & 0 \\ 0 & \cdots & \cdots & \cdots & 0 \\ \cdots & 0 & \cdots & \cdots & -\frac{1}{h_M} \\ 0 & \cdots & 0 & -\frac{1}{h_M} & \frac{1}{h_M} + \frac{1}{h_{M+1}} \end{bmatrix}. \tag{3.3.17}$$

With a *uniform mesh*, i.e. $h_i = h,\ i = 1, \dots, M$ we get that

$$\mathbf{A}_{\text{unif}} = \frac{1}{h} \cdot \begin{bmatrix} 2 & -1 & 0 & \cdots & \cdots & 0 \\ -1 & 2 & -1 & 0 & \cdots & \cdots \\ 0 & -1 & 2 & -1 & 0 & \cdots \\ \cdots & \cdots & \cdots & \cdots & \cdots & 0 \\ \cdots & \cdots & 0 & -1 & 2 & -1 \\ 0 & \cdots & \cdots & 0 & -1 & 2 \end{bmatrix}. \tag{3.3.18}$$

As for the components of the load vector **b**, we have

$$b_i = \int_0^1 f(x)\varphi_i(x)\,dx = \int_{x_i-1}^{x_i} f(x)\frac{x - x_{i-1}}{h_i}\,dx + \int_{x_i}^{x_{i+1}} f(x)\frac{x_{i+1} - x}{h_{i+1}}\,dx.$$

3.3.1 An Equivalent Finite Difference Approach

To illustrate a finite difference approach, we reconsider the stationary heat equation (3.3.2):

$$-u''(x) = f(x), \quad 0 < x < 1; \tag{3.3.19}$$

and motivate for its boundary conditions. Equation (3.3.19) is linear in the unknown function u, with inhomogeneous source term f. There is some arbitrariness left in the problem, because any combination $C + Dx$ could be added to any solution. The sum would constitute another solution, since the second derivative of $C + Dx$ contributes nothing. Therefore, the uncertainty left by these two arbitrary constants C and D will be removed by adding a *boundary condition* at each end point of the interval, e.g.

$$u(0) = 0, \qquad u(1) = 0. \tag{3.3.20}$$

The result is a *two-point boundary-value problem*, describing not a transient but a steady-state phenomenon – the temperature distribution in a rod, for example with ends fixed at 0° and with a heat source $f(x)$.

As our goal is to solve a discrete problem, we cannot accept more than a finite amount of information about f, say its values at equally spaced points $x_1 = h$, $x_2 = 2h, \dots, x_n = nh$. And what we compute will be approximate values $u_1, u_2, \dots, u_n$ for the true solution u at these same points. At the end points $x_0 = 0$ and $x_{n+1} = 1 = (n + 1)h$, we are already given the correct boundary values $u_0 = 0$, $u_{n+1} = 0$.

The first question is: How do we replace the derivative d^2u/dx^2? Since every derivative is a limit of difference quotients, it can be approximated by stopping at a finite step size, and not permitting h (or Δx) to approach zero. For du/dx there are several alternatives:

$$\frac{du}{dx} \approx \frac{u(x+h) - u(x)}{h} \quad \text{or} \quad \frac{u(x) - u(x-h)}{h} \quad \text{or} \quad \frac{u(x+h) - u(x-h)}{2h}.$$

The last, because it is symmetric about x, is the most accurate. For the second derivative, we can write

$$\frac{d^2u}{dx^2} \approx \frac{u'(x) - u'(x-h)}{h}. \tag{3.3.21}$$

Replacing the approximations $u'(x) \approx \frac{u(x+h)-u(x)}{h}$ and $u'(x-h) \approx \frac{u(x)-u(x-h)}{h}$ in (3.3.21) we get

$$\begin{aligned}\frac{d^2u}{dx^2} &\approx \frac{(u(x+h) - u(x))/h - (u(x) - u(x-h))/h}{h} \\ &= \frac{u(x+h) - 2u(x) + u(x-h)}{h^2},\end{aligned} \tag{3.3.22}$$

which has the merit of being symmetric about x. To repeat the right-side approaches, the true value of d^2u/dx^2 as $h \to 0$, but we have to stop at a positive h.

At a typical mesh point $x_j = jh$, the differential equation $-d^2u/dx^2 = f(x)$ is now replaced by this discrete analogue (3.3.22); after multiplying by h^2,

$$-u_{j+1} + 2u_j - u_{j-1} = h^2 f(jh). \tag{3.3.23}$$

There are n equations of exactly this form, for every value $j = 1, 2, \dots, n$. The first and last equations include the expressions u_0 and u_{n+1}, which are not unknowns. Their values are the boundary conditions, and they are shifted to the right-hand side of the equation and contribute to the inhomogeneous terms (or at least, they might, if they were not known to be equal zero). It is easy to understand (3.3.23) as a steady-state equation, in which the flows $(u_j - u_{j+1})$ coming from the right and $(u_j - u_{j-1})$ coming from the left are balanced by the source $h^2 f(jh)$ at the center.

The structure of the n equations (3.3.23) can be better visualized in matrix form $\mathbf{A}u = \mathbf{b}$ as

$$\begin{bmatrix} 2 & -1 & 0 & \cdots & \cdots & 0 \\ -1 & 2 & -1 & 0 & \cdots & \cdots \\ 0 & -1 & 2 & -1 & 0 & \cdots \\ \cdots & \cdots & \cdots & \cdots & \cdots & 0 \\ \cdots & \cdots & 0 & -1 & 2 & -1 \\ 0 & \cdots & \cdots & 0 & -1 & 2 \end{bmatrix} \begin{bmatrix} u_1 \\ u_2 \\ u_3 \\ \vdots \\ \vdots \\ u_n \end{bmatrix} = h^2 \begin{bmatrix} f(h) \\ f(2h) \\ f(3h) \\ \vdots \\ \vdots \\ f(nh) \end{bmatrix}, \tag{3.3.24}$$

which, once again, gives the structure of our uniform stiffness matrix $\mathbf{A}_{\text{unif}}$ given in (3.3.18).

So we conclude that, for problem (3.3.2), the finite element and finite difference approximations, in a uniform mesh, are two identical approaches.

Remark 3.4 Returning back and comparing the discretizations made for the IVP: (3.2.1) and the BVP: (3.3.2) we realize that, unlike the matrix $\mathcal{A}$ for polynomial approximation of IVP in (3.2.18), $\mathbf{A}$ in (3.3.18) or (3.3.24) has a more desirable structure, e.g. $\mathbf{A}$ is a sparse, tridiagonal, and symmetric matrix. This is due to the fact that the basis functions $\{\varphi_j\}_{j=1}^M$ are *nearly orthogonal*. The most important property of $\mathbf{A}$ is that it is *positive definite*.

Definition 3.2 An $M \times M$ matrix $\mathbf{A}$ is called positive definite if

$$\forall \eta \in R^M, \quad \eta \neq 0, \ \eta^T \mathbf{A} \eta > 0, \quad \text{i.e.} \ \sum_{i,j=1}^{M} \eta_i a_{ij} \eta_j > 0. \tag{3.3.25}$$

We shall use the positive definiteness of $\mathbf{A}$ to show that (3.3.12) is uniquely solvable. To this end, we prove the following well-known result:

Proposition 3.1 *If a square matrix* $\mathbf{A}$ *is positive definite then* $\mathbf{A}$ *is* invertible *and hence* $\mathbf{A}\xi = \mathbf{b}$ *has a unique solution.*

Proof: Suppose $\mathbf{Ax} = \mathbf{0}$ then $\mathbf{x}^T\mathbf{Ax} = \mathbf{0}$, and since $\mathbf{A}$ is positive definite, then $\mathbf{x} \equiv \mathbf{0}$. Thus, $\mathbf{A}$ has full range, and we conclude that $\mathbf{A}$ is invertible. Since $\mathbf{A}$ is invertible $\mathbf{A}\xi = \mathbf{b}$ has a unique solution: $\xi = \mathbf{A}^{-1}\mathbf{b}$. □

Note, however, that it is a bad idea to invert a matrix to solve the linear system of equations (too involved!). Finally, we illustrate an example of the positive-definiteness argument for $\mathbf{A}_{\text{unif}}$.

Example 3.2 Assume $M = 2$ and let $\eta(x, y) = \begin{pmatrix} x \\ y \end{pmatrix}$, then

$$\begin{aligned}\eta^T\mathbf{A}_{\text{unif}}\eta &= (x, y)\begin{pmatrix} 2 & -1 \\ -1 & 2 \end{pmatrix}\begin{pmatrix} x \\ y \end{pmatrix} = (x, y)\begin{pmatrix} 2x - y \\ -x + 2y \end{pmatrix} \\ &= 2x^2 - xy - xy + 2y^2 = x^2 + y^2 + x^2 - 2xy + y^2 \\ &= x^2 + y^2 + (x - y)^2 \geq 0.\end{aligned} \tag{3.3.26}$$

Thus, $\mathbf{A}_{\text{unif}}$ is positive definite.

Summary: Roughly speaking, a systematic procedure for approximate solution for a differential equation would involve the following steps:

1. We need to approximate functions by polynomials agreeing with the functional values at certain points (nodes). This is the matter of *Interpolation techniques*, which we shall introduce in Section 3.5. We need to construct approximate solutions that, in average (in weak sense), satisfies the differential equation. This is the finite element approximation as being a discrete version of variational formulation. Throughout the book, we deal with the issue of Galerkin approximation, its efficiency, reliability, and other qualitative properties.
2. In the final system of equations: $\mathbf{A}\xi = \mathbf{b}$, entries of the coefficient matrix $\mathbf{A}$ as well as the components of the load vector $\mathbf{b}$ are integrals. For a more involved data function $f(x)$, and when approximating by higher-order polynomials and/or solving equations with variable coefficients, these integrals are not easy to compute. Therefore, we need to approximate different integrals over subintervals of a partition. This may be done using *quadrature rules*. In a simple case, one may use usual or composite *midpoint-*, *trapezoidal-*, or *Simpson's rules*. In more involved cases, one may employ *composite Gauss quadrature rules*. We briefly introduce the idea of quadrature rule at the end of this chapter.

3. Finally, we end up with LSEs of type (3.3.12). To solve LSE efficiently, we may use exact *Gauss – elimination* or the iteration procedures such as *Gauss–Seidel*, *Gauss–Jacobi*, or *Over-relaxation methods*. We discuss these concepts in Chapter 4.

3.4 Exercises

Problem 3.1 Prove that $V_0^{(q)} := \{v \in \mathcal{P}^{(q)}(0,1) : v(0) = 0\}$, is a subspace of $\mathcal{P}^{(q)}(0,1)$.

Problem 3.2 Consider the ODE

$$\dot{u}(t) = u(t), \quad 0 < t < 1; \qquad u(0) = 1.$$

Compute its Galerkin approximation in $\mathcal{P}^{(q)}(0,1)$, for $q = 1, 2, 3$, and 4.

Problem 3.3 Consider the ODE

$$\dot{u}(t) = u(t), \quad 0 < t < 1; \qquad u(0) = 1.$$

Compute the $L_2(0,1)$ projection of the exact solution u into $\mathcal{P}^3(0,1)$.

Problem 3.4 Compute the stiffness matrix and load vector in a finite element approximation of the BVP

$$-u''(x) = f(x), \quad 0 < x < 1, \qquad u(0) = u(1) = 0,$$

with $f(x) = x$ and $h = 1/4$.

Problem 3.5 We want to find a solution approximation $U(x)$ to

$$-u''(x) = 1, \quad 0 < x < 1, \qquad u(0) = u(1) = 0,$$

using the ansatz $U(x) = A \sin \pi x + B \sin 2\pi x$.

a. Calculate the exact solution $u(x)$.
b. Write down the residual $R(x) = -U''(x) - 1$
c. Use the orthogonality condition

$$\int_0^1 R(x) \sin \pi n x \, dx = 0, \quad n = 1, 2,$$

to determine the constants A and B.

d. Plot the error $e(x) = u(x) - U(x)$.

Problem 3.6 Consider the BVP

$$-u''(x) + u(x) = x, \quad 0 < x < 1, \qquad u(0) = u(1) = 0.$$

a. Verify that the exact solution of the problem is given by

$$u(x) = x - \frac{\sinh x}{\sinh 1}.$$

b. Let $U(x)$ be a solution approximation defined by

$$U(x) = A \sin \pi x + B \sin 2\pi x + C \sin 3\pi x,$$

where A, B, and C are unknown constants. Compute the residual function

$$R(x) = -U''(x) + U(x) - x.$$

c. Use the orthogonality condition

$$\int_0^1 R(x) \sin \pi n x \, dx = 0, \quad n = 1, 2, 3,$$

to determine the constants A, B, and C.

Problem 3.7 Let $U(x) = \xi_0 \phi_0(x) + \xi_1 \phi_1(x)$ be a solution approximation to

$$-u''(x) = x - 1, \quad 0 < x < \pi, \qquad u'(0) = u(\pi) = 0,$$

where ξ_i, $i = 0, 1$, are unknown coefficients and

$$\phi_0(x) = \cos \frac{x}{2}, \qquad \phi_1(x) = \cos \frac{3x}{2}.$$

a. Find the analytical solution $u(x)$.
b. Define the approximate solution residual $R(x)$.
c. Compute the constants ξ_i using the orthogonality condition

$$\int_0^\pi R(x)\, \phi_i(x) \, dx = 0, \quad i = 0, 1,$$

i.e. by projecting $R(x)$ onto the vector space spanned by $\phi_0(x)$ and $\phi_1(x)$.

Problem 3.8 Use the projection technique of Exercise 3.7 to solve

$$-u''(x) = 0, \quad 0 < x < \pi, \qquad u(0) = 0, \ u(\pi) = 2,$$

assuming that $U(x) = A \sin x + B \sin 2x + C \sin 3x + \frac{2}{\pi^2} x^2$, using test functions $\{\sin x, \ \sin 2x, \ \sin 3x\}$.

3.5 Polynomial Interpolation in 1*d*

In this section, we give a detailed study of the interpolation concept in one dimension, briefly introduced in Sections 3.1. We focus on the case of a real-valued function f, defined on an interval $I = [a, b]$. Interpolation in higher dimensions is a straightforward generalization of the content in this chapter.

Definition 3.3 An interpolant $\pi_q f$ of a function f, defined on an interval $I = [a, b]$, is a polynomial of degree $\leq q$ having the nodal values at $q+1$ distinct points $x_j \in [a, b]$, $j = 0, 1, \dots, q$, coinciding with those of f, i.e. $\pi_q f \in \mathcal{P}^{(q)}(a, b)$ and $\pi_q f(x_j) = f(x_j)$, $j = 0, \dots, q$.

Example 3.3 A typical application of the Lagrange basis functions is in finding a polynomial interpolant $\pi_q f \in \mathcal{P}^{(q)}(a, b)$ of a continuous function $f(x)$ on an interval $[a, b]$. The procedure is as follows:

Choose distinct interpolation nodes, e.g. ξ_i: $a = \xi_0 < \xi_1 < \cdots < \xi_q = b$ and let $\pi_q f(\xi_i) = f(\xi_i)$. Then, $\pi_q f \in \mathcal{P}^{(q)}(a, b)$, defined as the sum in (3.1.2), interpolates $f(x)$ at the nodes ξ_i, $i = 0, \dots, q$, and using Lagrange's formula (3.1.1), with $p_i = f(\xi_i)$, $i = 0, 1, \dots, q$, yields

$$\pi_q f(x) = f(\xi_0)\lambda_0(x) + f(\xi_1)\lambda_1(x) + \cdots + f(\xi_q)\lambda_q(x), \quad x \in [a, b].$$

Below we illustrate this definition through a simple example:

Example 3.4 *(Linear interpolation on an interval)* We start with the unit interval $I = [0, 1]$ and a continuous function f: $[0, 1] \to \mathbb{R}$. Let $q = 1$ and seek the linear interpolant of f on I, e.g. the linear function $\pi_1 f \in \mathcal{P}^{(1)}$, such that $\pi_1 f(0) = f(0)$ and $\pi_1 f(1) = f(1)$, viz., Figure 3.9. Thus, we seek the constants C_0 and C_1 in the following representation of $\pi_1 f \in \mathcal{P}^{(1)}$,

$$\pi_1 f(x) = C_0 + C_1 x, \quad x \in I, \tag{3.5.1}$$

where

$$\pi_1 f(0) = f(0) \quad \Longrightarrow \quad C_0 = f(0), \quad \text{and}$$

$$\pi_1 f(1) = f(1) \quad \Longrightarrow \quad C_0 + C_1 = f(1) \Longrightarrow C_1 = f(1) - f(0). \tag{3.5.2}$$

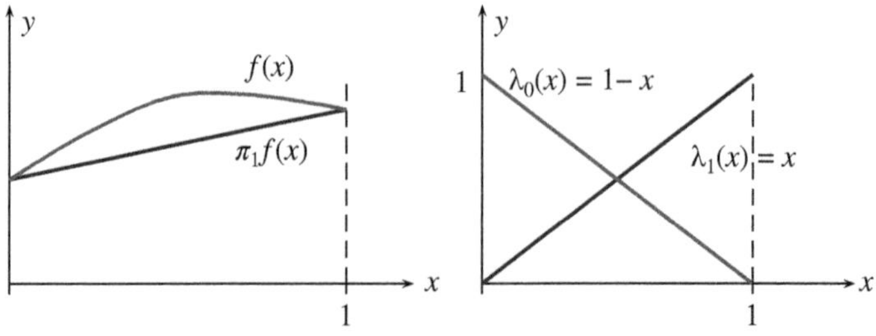

Figure 3.9 (a) Linear interpolation and (b) basis functions for $q = 1$.

Inserting C_0 and C_1 into (3.5.1), we get

$$\pi_1 f(x) = f(0) + (f(1) - f(0))x = f(0)(1-x) + f(1)x := f(0)\lambda_0(x) + f(1)\lambda_1(x).$$

In other words, $\pi_1 f(x)$ is represented in two different bases:

$$\pi_1 f(x) = C_0 \cdot 1 + C_1 \cdot x, \quad \text{with } \{1,\ x\} \text{ as the set of basis functions and}$$

$$\pi_1 f(x) = f(0)(1-x) + f(1)x, \quad \text{with } \{1-x,\ x\} \text{ as the set of basis functions.}$$

The functions $\lambda_0(x) = 1 - x$ and $\lambda_1(x) = x$ are linearly independent, since if

$$0 = \alpha_0(1-x) + \alpha_1 x = \alpha_0 + (\alpha_1 - \alpha_0)x, \quad x \in I, \tag{3.5.3}$$

then

$$\left.\begin{array}{lll} x = 0 & \Longrightarrow & \alpha_0 = 0 \\ x = 1 & \Longrightarrow & \alpha_1 = 0 \end{array}\right\} \Longrightarrow \alpha_0 = \alpha_1 = 0. \tag{3.5.4}$$

Remark 3.5 Note that if we define a scalar product on $\mathcal{P}^{(k)}(a,b)$ by

$$(p,q) = \int_a^b p(x)q(x)\,dx, \quad \forall p,q \in \mathcal{P}^{(k)}(a,b), \tag{3.5.5}$$

then we can easily verify that neither $\{1,x\}$ nor $\{1-x,x\}$ is an orthogonal basis for $\mathcal{P}^{(1)}(0,1)$, since $(1,x) := \int_0^1 1 \cdot x\,dx = [\frac{x^2}{2}] = \frac{1}{2} \neq 0$ and $(1-x,x) := \int_0^1 (1-x)x\,dx = \frac{1}{6} \neq 0$. To approximate with functions represented in nonorthogonal basis, the choice of the basis functions is crucial for the procedure and quality of the approximation.

With this in mind, it is natural to pose the following question.

Question 3.1 How well does $\pi_q f$ approximate f? In other words, how large/small will the error be in approximating $f(x)$ by $\pi_q f(x)$?

To answer this question, we need to estimate the difference between $f(x)$ and $\pi_q f(x)$. For instance, for $q = 1$, geometrically, the deviation of $f(x)$ from $\pi_1 f(x)$ (from being linear) depends on the *curvature* of $f(x)$, i.e. on how *curved* $f(x)$ is. In other words, on how *large* $f''(x)$ is, say, on an interval (a,b). To quantify the relationship between the size of the error $f - \pi_1 f$ and the size of f'', we recall the definitions of some measuring environments for functions: namely, the simple one-dimensional version of the previously defined L_p norms and spaces: i.e. $L_p(a,b)$-norms of f over (a,b):

$$\|f\|_{L_p(a,b)} := \left(\int_a^b |f(x)|^p\,dx \right)^{1/p}, \quad 1 \leq p < \infty,$$

$$\|f\|_{L_\infty(a,b)} := \max_{x \in [a,b]} |f(x)|,$$

and, for $1 \leq p \leq \infty$ the $L_p(a,b)$-space

$$L_p(a,b) := \{f\colon \|f\|_{L_p(a,b)} < \infty\}.$$

Below we shall answer the above question, first in the L_∞-norm, and then in the L_p-norm, (mainly for $p = 1, \ 2$.)

Theorem 3.1 *(L_∞-error estimates for linear interpolation in an interval) Assume that $f'' \in L_\infty(a,b)$. Then, for $q = 1$, i.e. only two interpolation nodes (e.g. the end-points of the interval), there are interpolation constants, C_i, $i = 1, 2, 3$, independent of the function f and the size of the interval $[a,b]$, such that*

(1) $\|\pi_1 f - f\|_{L_\infty(a,b)} \leq C_1(b-a)^2\|f''\|_{L_\infty(a,b)}$
(2) $\|\pi_1 f - f\|_{L_\infty(a,b)} \leq C_2(b-a)\|f'\|_{L_\infty(a,b)}$
(3) $\|(\pi_1 f)' - f'\|_{L_\infty(a,b)} \leq C_3(b-a)\|f''\|_{L_\infty(a,b)}$.

Proof: Note that every linear function $p(x)$ on $[a,b]$ can be written as a linear combination of the basis functions $\lambda_a(x)$ and $\lambda_b(x)$ where

$$\lambda_a(x) = \frac{b-x}{b-a} \quad \text{and} \quad \lambda_b(x) = \frac{x-a}{b-a}; \tag{3.5.6}$$

$$p(x) = p(a)\lambda_a(x) + p(b)\lambda_b(x). \tag{3.5.7}$$

Recall that the linear combination of $\lambda_a(x)$ and $\lambda_b(x)$ gives the basis functions $\{1, x\}$ for $\mathcal{P}^{(1)}$ (as in Figure 3.10):

$$\lambda_a(x) + \lambda_b(x) = 1, \qquad a\lambda_a(x) + b\lambda_b(x) = x. \tag{3.5.8}$$

Here, $\pi_1 f(x)$ being a linear function connecting the two points $(a, f(a))$ and $(b, f(b))$, (Figure 3.10) is represented by

$$\pi_1 f(x) = f(a)\lambda_a(x) + f(b)\lambda_b(x). \tag{3.5.9}$$

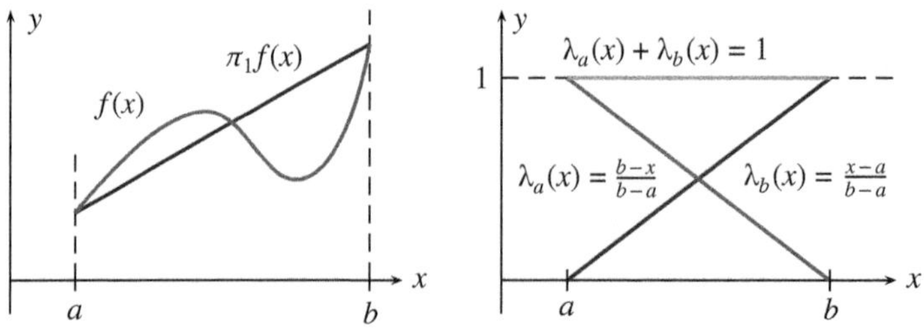

Figure 3.10 Linear Lagrange basis functions for $q = 1$.

By the Taylor expansion for $f(a)$ and $f(b)$ about $x \in (a, b)$, we can write

$$\begin{cases} f(a) = f(x) + (a - x)f'(x) + \frac{1}{2}(a - x)^2 f''(\eta_a), & \eta_a \in [a, x] \\ f(b) = f(x) + (b - x)f'(x) + \frac{1}{2}(b - x)^2 f''(\eta_b), & \eta_b \in [x, b]. \end{cases} \tag{3.5.10}$$

Inserting $f(a)$ and $f(b)$ from (3.5.10) into (3.5.9), it follows that

$$\pi_1 f(x) = \left[f(x) + (a - x)f'(x) + \frac{1}{2}(a - x)^2 f''(\eta_a)\right] \lambda_a(x)$$
$$+ \left[f(x) + (b - x)f'(x) + \frac{1}{2}(b - x)^2 f''(\eta_b)\right] \lambda_b(x).$$

Rearranging the terms, using (3.5.8) and the identity (which also follows from (3.5.8)) $(a - x)\lambda_a(x) + (b - x)\lambda_b(x) = 0$ we get

$$\begin{aligned} \pi_1 f(x) &= f(x)[\lambda_a(x) + \lambda_b(x)] + f'(x)[(a - x)\lambda_a(x) + (b - x)\lambda_b(x)] \\ &\quad + \frac{1}{2}(a - x)^2 f''(\eta_a)\lambda_a(x) + \frac{1}{2}(b - x)^2 f''(\eta_b)\lambda_b(x) \\ &= f(x) + \frac{1}{2}(a - x)^2 f''(\eta_a)\lambda_a(x) + \frac{1}{2}(b - x)^2 f''(\eta_b)\lambda_b(x). \end{aligned}$$

Consequently,

$$|\pi_1 f(x) - f(x)| = \left|\frac{1}{2}(a - x)^2 f''(\eta_a)\lambda_a(x) + \frac{1}{2}(b - x)^2 f''(\eta_b)\lambda_b(x)\right|. \tag{3.5.11}$$

To proceed, we note that for $a \leq x \leq b$ both $(a - x)^2 \leq (a - b)^2$ and $(b - x)^2 \leq (a - b)^2$, furthermore $\lambda_a(x) \leq 1$ and $\lambda_b(x) \leq 1$, $\forall\, x \in (a, b)$. Moreover, by the definition of the maximum norm both $|f''(\eta_a)| \leq \|f''\|_{L_\infty(a,b)}$ and $|f''(\eta_b)| \leq \|f''\|_{L_\infty(a,b)}$. Thus, we may estimate (3.5.11) as

$$|\pi_1 f(x) - f(x)| \leq \frac{1}{2}(a - b)^2 \cdot 1 \cdot \|f''\|_{L_\infty(a,b)} + \frac{1}{2}(a - b)^2 \cdot 1 \cdot \|f''\|_{L_\infty(a,b)}, \tag{3.5.12}$$

and hence

$$|\pi_1 f(x) - f(x)| \leq (a - b)^2 \|f''\|_{L_\infty(a,b)} \quad \text{corresponding to } C_1 = 1. \tag{3.5.13}$$

The other two estimates (2) and (3) are proved similarly. □

Remark 3.6 We can show that the optimal value of $C_1 = \frac{1}{8}$ (cf. Problem 3.19), i.e. the constant $C_1 = 1$ of the proof above is not the optimal one.

An analogue to Theorem 3.1 can be proved in the L_p-norm for $p = 1,\ 2$ using integral form of the Taylors' expansion:

Theorem 3.2 *Let $\pi_1 v(x)$ be the linear interpolant of the function $v(x)$ on (a,b). Then, assuming that v is twice continuously differentiable ($v \in C^2(a,b)$), there are interpolation constants c_i, $i = 1,2,3$, such that for $p = 1,\ 2, \infty$,*

$$\|\pi_1 v - v\|_{L_p(a,b)} \le c_1(b-a)^2\|v''\|_{L_p(a,b)}, \tag{3.5.14}$$

$$\|(\pi_1 v)' - v'\|_{L_p(a,b)} \le c_2(b-a)\|v''\|_{L_p(a,b)}, \tag{3.5.15}$$

$$\|\pi_1 v - v\|_{L_p(a,b)} \le c_3(b-a)\|v'\|_{L_p(a,b)}. \tag{3.5.16}$$

Proof: For $p = \infty$ this is just Theorem 3.1. For $p = 1,\ 2$, as mentioned above, use the integral form of the Taylors' expansion. This is left as an exercise. □

Below we review a simple piecewise linear interpolation procedure on a partition of an interval:

Vector space of piecewise linear functions on an interval. Given $I = [a, b]$, let $\mathcal{T}_h$: $a = x_0 < x_1 < x_2 < \cdots < x_{N-1} < x_N = b$ be a partition of I into subintervals $I_j = [x_{j-1}, x_j]$ of length $h_j = |I_j| := x_j - x_{j-1}$; $j = 1,2,\dots,N$. Let

$$V_h := \{v | v \text{ is a continuous, piecewise linear function on } \mathcal{T}_h\}, \tag{3.5.17}$$

then V_h is a vector space with the previously introduced *hat functions*: $\{\varphi_j\}_{j=0}^N$ as basis functions. Note that $\varphi_0(x)$ and $\varphi_N(x)$ are left and right *half-hat* functions, respectively. We now show that every function in V_h is a linear combination of φ_j:s.

Lemma 3.1 *We have that*

$$\forall v \in V_h; \qquad v(x) = \sum_{j=0}^{N} v(x_j)\varphi_j(x), \quad \Longrightarrow (\dim V_h = N+1). \tag{3.5.18}$$

Proof: Both the left- and right-hand sides are continuous piecewise linear functions. Thus, it suffices to show that they have the same nodal values: let $x = x_j$, then since $\varphi_i(x_j) = \delta_{ij}$,

$$\begin{aligned} \text{RHS}|_{x_j} &= v(x_0)\varphi_0(x_j) + v(x_1)\varphi_1(x_j) + \cdots + v(x_{j-1})\varphi_{j-1}(x_j) \\ &\quad + v(x_j)\varphi_j(x_j) + v(x_{j+1})\varphi_{j+1}(x_j) + \cdots + v(x_N)\varphi_N(x_j) \\ &= v(x_j) = \text{LHS}|_{x_j}. \end{aligned} \tag{3.5.19}$$

□

Definition 3.4 For a partition $\mathcal{T}_h$: $a = x_0 < x_1 < x_2 < \cdots < x_N = b$ of the interval $[a, b]$, we define the mesh function $h(x)$ as the piecewise constant function $h(x) := h_j = x_j - x_{j-1}$ for $x \in I_j = (x_{j-1}, x_j)$, $j = 1,2,\dots,N$.

Definition 3.5 Assume that f is a continuous function on $[a,b]$. Then, in the partition $\mathcal{T}_h$ above, the continuous piecewise linear interpolant of f is defined by

$$\pi_h f(x) = \sum_{j=0}^{N} f(x_j)\varphi_j(x), \quad x \in [a,b].$$

Here, the subindex h refers to the mesh function $h(x)$. Hence,

$$\pi_h f(x_j) = f(x_j), \quad j = 0, 1, \ldots, N. \tag{3.5.20}$$

Remark 3.7 Note that we denote the linear interpolant, defined for a single interval $[a,b]$, by $\pi_1 f$ which is a polynomial of degree 1, whereas the piecewise linear interpolant $\pi_h f$ is defined for a partition $\mathcal{T}_h$ of $[a,b]$ and is a *continuous, piecewise linear* function (see Figure 3.11). For the piecewise polynomial interpolants of (higher) degree q, we shall use the notation for *Cardinal functions* of Lagrange interpolation below.

Note that for each interval I_j, $j = 1, \ldots, N$, we have that

(i) $\pi_h f(x)$ is linear on $I_j \Longrightarrow \pi_h f(x) = c_0 + c_1 x$ for $x \in I_j$.
(ii) $\pi_h f(x_{j-1}) = f(x_{j-1})$ and $\pi_h f(x_j) = f(x_j)$.

Combining (i) and (ii), we get

$$\begin{cases} \pi_h f(x_{j-1}) = c_0 + c_1 x_{j-1} = f(x_{j-1}) \\ \pi_h f(x_j) = c_0 + c_1 x_j = f(x_j) \end{cases} \Longrightarrow \begin{cases} c_1 = \frac{f(x_j)-f(x_{j-1})}{x_j - x_{j-1}} \\ c_0 = \frac{-x_{j-1}f(x_j)+x_j f(x_{j-1})}{x_j - x_{j-1}}, \end{cases}$$

Thus, we may write

$$\begin{cases} c_0 = f(x_{j-1})\frac{x_j}{x_j - x_{j-1}} + f(x_j)\frac{-x_{j-1}}{x_j - x_{j-1}} \\ c_1 x = f(x_{j-1})\frac{-x}{x_j - x_{j-1}} + f(x_j)\frac{x}{x_j - x_{j-1}}. \end{cases} \tag{3.5.21}$$

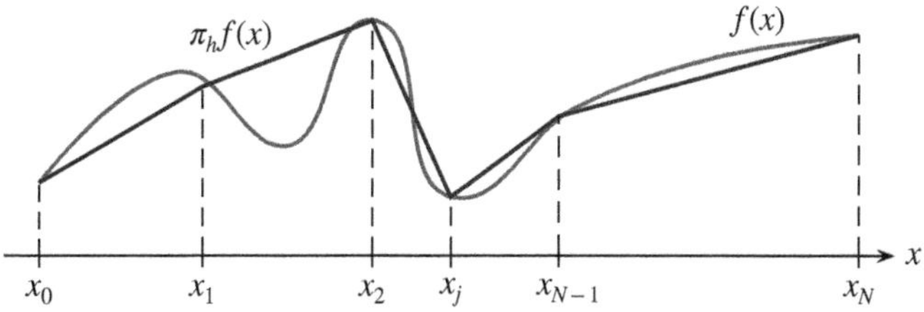

Figure 3.11 Piecewise linear interpolant $\pi_h f(x)$ of $f(x)$.

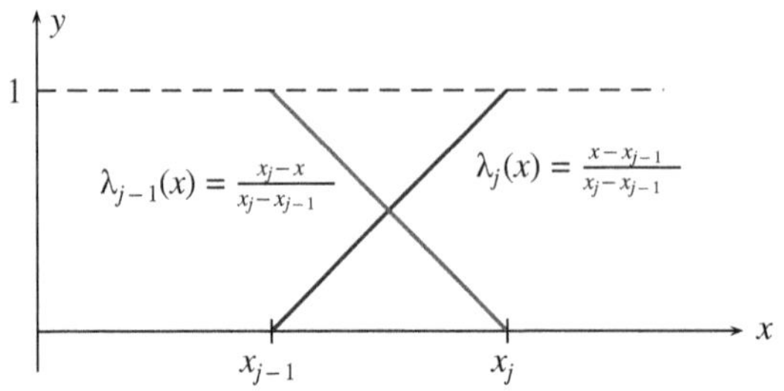

Figure 3.12 Linear Lagrange basis functions for $q = 1$ on subinterval I_j.

For $x_{j-1} \le x \le x_j$, $\quad j = 1, 2, \dots, N$, adding up the equations in (3.5.21) yields

$$\begin{aligned}\pi_h f(x) = c_0 + c_1 x &= f(x_{j-1})\frac{x_j - x}{x_j - x_{j-1}} + f(x_j)\frac{x - x_{j-1}}{x_j - x_{j-1}} \\ &= f(x_{j-1})\lambda_{j-1}(x) + f(x_j)\lambda_j(x),\end{aligned}$$

where, as seen in Figure 3.12, $\lambda_{j-1}(x)$ and $\lambda_j(x)$ are the restrictions of the piecewise linear basis functions $\varphi_{j-1}(x)$ and $\varphi_j(x)$ to I_j.

In Section 3.5.1, we generalize the above procedure and introduce Lagrange interpolation basis functions.

The main result of this section can be stated as follows:

Theorem 3.3 *Let $\pi_h v(x)$ be the piecewise linear interpolant of the function $v(x)$ on a partition $\mathcal{T}_h$ of $[0, T]$, as above. Then, assuming that v is sufficiently regular ($v \in C^2(a, b)$), there are interpolation constants c_i, $i = 1, 2, 3$, such that for $p = 1, 2, \infty$,*

$$\|\pi_h v - v\|_{L_p(a,b)} \le c_1 \|h^2 v''\|_{L_p(a,b)}, \tag{3.5.22}$$

$$\|(\pi_h v)' - v'\|_{L_p(a,b)} \le c_2 \|h v''\|_{L_p(a,b)}, \tag{3.5.23}$$

$$\|\pi_h v - v\|_{L_p(a,b)} \le c_3 \|h v'\|_{L_p(a,b)}. \tag{3.5.24}$$

Proof: Recalling the definition of the partition $\mathcal{T}_h$, we may write

$$\|\pi_h v - v\|^p_{L_p(a,b)} = \sum_{j=0}^{N+1} \|\pi_h v - v\|^p_{L_p(I_j)} \le \sum_{j=0}^{N+1} c_1^p \|h_j^2 v''\|^p_{L_p(I_j)} \le c_1^p \|h^2 v''\|^p_{L_p(a,b)}, \tag{3.5.25}$$

where in the first inequality, we apply Theorem 3.2 to an arbitrary partition interval I_j and then sum over j. The other two estimates are proved similarly. □

3.5.1 Lagrange Interpolation

Consider $\mathcal{P}^{(q)}(a, b)$; the vector space of all polynomials of degree $\le q$ on the interval (a, b), with the monomial basis functions $1, x, x^2, \dots, x^q$. In general, the monomials build up nonorthogonal basis (with respect to the scalar product (3.5.5) with, e.g. $a = 0$ and $b = 1$) that often leads to ill-conditioned coefficient matrices. We will now introduce a new set of basis functions, which have some useful properties.

Definition 3.6 (Cardinal functions) Lagrange basis is the set of polynomials $\{\lambda_i\}_{i=0}^{q} \subset P^{(q)}(a,b)$ associated with the $(q+1)$ distinct points, $a = x_0 < x_1 < \cdots < x_q = b$ in $[a,b]$ and determined by the requirement that: at the nodes, $\lambda_i(x_j) = 1$ for $i = j$, and 0; otherwise, ($\lambda_i(x_j) = 0$ for $i \neq j$), i.e. for $x \in [a,b]$,

$$\lambda_i(x) = \frac{(x-x_0)(x-x_1)\cdots(x-x_{i-1})\downarrow(x-x_{i+1})\cdots(x-x_q)}{(x_i-x_0)(x_i-x_1)\cdots(x_i-x_{i-1})\uparrow(x_i-x_{i+1})\cdots(x_i-x_q)}. \tag{3.5.26}$$

By the arrows $\downarrow, \uparrow$ in (3.5.26) we emphasize that $\lambda_i(x) = \prod_{j=1(\neq i)}^{q}\left(\frac{x-x_j}{x_i-x_j}\right)$ does not contain the singular factor $\frac{x-x_i}{x_i-x_i}$. Hence, for $i = 0, 1, \ldots, q$,

$$\lambda_i(x_j) = \frac{(x_j-x_0)(x_j-x_1)\cdots(x_j-x_{i-1})(x_j-x_{i+1})\cdots(x_j-x_q)}{(x_i-x_0)(x_i-x_1)\cdots(x_i-x_{i-1})(x_i-x_{i+1})\cdots(x_i-x_q)}.$$

Thus, $\lambda_i(x)$ is a polynomial of degree q on (a,b) with

$$\lambda_i(x_j) = \delta_{ij} = \begin{cases} 1 & i = j \\ 0 & i \neq j. \end{cases} \tag{3.5.27}$$

Example 3.5 Let $q = 2$, then, we have $a = x_0 < x_1 < x_2 = b$, where

$$i = 1, j = 2 \implies \delta_{12} = \lambda_1(x_2) = \frac{(x_2-x_0)(x_2-x_2)}{(x_1-x_0)(x_1-x_2)} = 0$$

$$i = j = 1 \implies \delta_{11} = \lambda_1(x_1) = \frac{(x_1-x_0)(x_1-x_2)}{(x_1-x_0)(x_1-x_2)} = 1.$$

A polynomial $P(x) \in \mathcal{P}^{(q)}(a,b)$ with the values $p_i = P(x_i)$ at the nodes x_i, $i = 0, 1, \ldots, q$, can be expressed in terms of the above Lagrange basis as

$$P(x) = p_0\lambda_0(x) + p_1\lambda_1(x) + \cdots + p_q\lambda_q(x). \tag{3.5.28}$$

Using (3.5.27), $P(x_i) = p_0\lambda_0(x_i) + p_1\lambda_1(x_i) + \cdots + p_i\lambda_i(x_i) + \cdots + p_q\lambda_q(x_i) = p_i$.

Recalling Definition 3.6, if we choose $a \leq \xi_0 < \xi_1 < \cdots < \xi_q \leq b$, as $q+1$ distinct interpolation nodes on $[a,b]$, then, the interpolation polynomial $\pi_q f \in P^{(q)}(a,b)$ satisfies

$$\pi_q f(\xi_i) = f(\xi_i), \quad i = 0, 1, \ldots, q \tag{3.5.29}$$

and the Lagrange formula (3.5.28) for $\pi_q f(x)$ reads as

$$\pi_q f(x) = f(\xi_0)\lambda_0(x) + f(\xi_1)\lambda_1(x) + \cdots + f(\xi_q)\lambda_q(x), \quad a \le x \le b.$$

Example 3.6 For $q = 1$, we have only the nodes a and b. Recall that $\lambda_a(x) = \frac{b-x}{b-a}$ and $\lambda_b(x) = \frac{x-a}{b-a}$, thus as in the introduction to this section

$$\pi_1 f(x) = f(a)\lambda_a(x) + f(b)\lambda_b(x). \tag{3.5.30}$$

Example 3.7 To interpolate $f(x) = x^3 + 1$ by piecewise polynomials of degree 2, in the partition $x_0 = 0,\ x_1 = 1,\ x_2 = 2$ of the interval $[0, 2]$, we have

$$\pi_2 f(x) = f(0)\lambda_0(x) + f(1)\lambda_1(x) + f(2)\lambda_2(x),$$

where $f(0) = 1,\ f(1) = 2,\ f(2) = 9$, and we may compute Lagrange basis as

$$\lambda_0(x) = \frac{1}{2}(x-1)(x-2), \quad \lambda_1(x) = -x(x-2), \quad \lambda_2(x) = \frac{1}{2}x(x-1).$$

This yields

$$\pi_2 f(x) = 1 \cdot \frac{1}{2}(x-1)(x-2) - 2 \cdot x(x-2) + 9 \cdot \frac{1}{2}x(x-1) = 3x^2 - 2x + 1.$$

Below we want to compare the Lagrange polynomial of degree q with another well-known polynomial: namely the *Taylor's polynomial* of degree q.

Definition 3.7 *(Taylor's theorem)* Suppose that the function f is $q+1$-times continuously differentiable at the point $x_0 \in (a, b)$. Then, f can be expressed by a Taylor expansion about x_0 as

$$f(x) = T_q f(x) + R_q f(x), \tag{3.5.31}$$

where

$$T_q f(x) = f(x_0) + f'(x_0)(x - x_0) + \frac{1}{2}f''(x_0)(x - x_0)^2 + \cdots + \frac{1}{q!}f^{(q)}(x_0)(x - x_0)^q,$$

is the *Taylor polynomial* of degree q, approximating f and

$$R_q f(x) = \frac{1}{(q+1)!}f^{(q+1)}(\xi)(x - x_0)^{q+1}, \tag{3.5.32}$$

is the *remainder* term, where ξ is a point between x_0 and x.

Theorem 3.4 *We have the following error estimates:*

- *For the Taylor polynomial expansion $T_q f(x)$ of $f(x)$,*

$$|f(x) - T_q f(x)| = |R_q(f)| \le \frac{1}{(q+1)!}|x - x_0|^{q+1} \cdot \max_{x\in[a,b]} |f^{(q+1)}(x)|,$$

which is accurate of order $q+1$ near $x = x_0$.

- *Whereas for the Lagrange interpolant $\pi_q f(x)$ of $f(x)$,*

$$|f(x) - \pi_q f(x)| \le \frac{1}{(q+1)!} \prod_{i=0}^{q} |x - x_i| \cdot \max_{x\in[a,b]} |f^{(q+1)}(x)|,$$

which is accurate of order 1 (the convergence rate is of order 1) at each interpolation node $x_0, x_1, \dots, x_q$.

Proof: The *Taylor polynomial error* follows immediately from (3.5.32). As for the *Lagrange* interpolation error, we note that at the node points x_i we have $f(x_i) - \pi_q f(x_i) = 0$, for $i = 0, 1, \dots, q$. Thus, $f(x) - \pi_q f(x)$ has $q + 1$ zeros in $[a, b]$ at the interpolation nodes x_i and hence, there is a function $g(x)$ defined on $[a, b]$ such that

$$f(x) - \pi_q f(x) = (x - x_0)(x - x_1) \cdots (x - x_q) g(x). \tag{3.5.33}$$

To determine $g(x)$, we define an auxiliary function φ by

$$\varphi(t) := f(t) - \pi_q f(t) - (t - x_0)(t - x_1) \cdots (t - x_q) g(x). \tag{3.5.34}$$

Note that $g(x)$ is independent of t. The function $\varphi(t)$ vanishes at the nodes x_i, $i = 0, \dots, q$ as well as for $t = x$, i.e. $\varphi(x_0) = \varphi(x_1) = \cdots = \varphi(x_q) = \varphi(x) = 0$. Thus $\varphi(t)$ has $(q + 2)$ roots in the interval $[a, b]$. Now by the *Generalized Rolle's theorem* (see below), there exists a point $\xi \in (a, b)$ such that $\varphi^{(q+1)}(\xi) = 0$. Taking the $(q + 1)$th derivative of the function $\varphi(t)$, with respect to t, we get using the fact that $\deg(\pi_q f(x)) = q$,

$$\varphi^{(q+1)}(t) = f^{(q+1)}(t) - 0 - (q + 1)! g(x), \tag{3.5.35}$$

where the last term is due to differentiating $(t - x_0) \cdots (t - x_q) = t^{q+1} + \alpha t^q + \cdots$, (for some constant α), and $g(x)$ is independent of t. Thus,

$$0 = \varphi^{(q+1)}(\xi) = f^{(q+1)}(\xi) - (q + 1)! g(x), \tag{3.5.36}$$

which yields

$$g(x) = \frac{f^{(q+1)}(\xi)}{(q+1)!}. \tag{3.5.37}$$

Inserting $g(x)$ in (3.5.33) we get the Lagrange interpolation error

$$E(x) = f(x) - \pi_q f(x) = \frac{f^{(q+1)}(\xi)}{(q+1)!} \prod_{i=0}^{q} (x - x_i), \tag{3.5.38}$$

and the proof is complete. □

Theorem 3.5 (Generalized Rolle's theorem) *If a function $u(x) \in C^{q+1}(a, b)$ has $(q + 2)$ roots, $x_0, x_1, \dots, x_q, x$, in a closed interval $[a, b]$, then there is a point $\xi \in (a, b)$, generated by $x_0, x_1, \dots, x_q, x$, such that $u^{(q+1)}(\xi) = 0$.*

3.6 Orthogonal- and L_2-Projection

Before presenting the quadrature rule for numerical integration, we introduce yet another concept: the L_2-projection, which is a kind of analytic approximation procedure that often serves as an intermediate step in the finite element approximation. Below we give a simple description:

Let $\mathbf{u} = (u_1, u_2, u_3) \in \mathbb{R}^3$ and assume that for some reasons, we only have u_1 and u_2 available. Letting $\mathbf{x} = (x_1, x_2, x_3) \in \mathbb{R}^3$, the objective then is to find $\mathbf{U} \in \{\mathbf{x} : x_3 = 0\}$, such that $(\mathbf{u} - \mathbf{U})$ is as small as possible. Obviously, in this case $\mathbf{U} = (u_1, u_2, 0)$ and we have $(\mathbf{u} - \mathbf{U}) \perp \mathbf{U}$, $\forall \mathbf{U}$ in the x_1x_2-plane, see Figure 3.13.

Note that, if $m < n$, and $\mathbf{u}_m$ is the projection of $\mathbf{u} = (u_1, u_2, \dots, u_{n-1}, u_n)$ on $\mathbb{R}^m$, then $\mathbf{u}_m = (u_1, u_2, \dots, u_m, u_{m+1} = 0, \dots, u_n = 0)$, and the Euclidean distance: $|\mathbf{u} - \mathbf{u}_m| = \sqrt{u_{m+1}^2 + u_{m+2}^2 + \cdots + u_n^2} \to 0$ as $m \to n$. This means the obvious fact that the accuracy of the orthogonal projection will improve by raising the dimension of the projection space.

3.6.1 The L_2-Projection onto the Space of Polynomials

A polynomial πf interpolating a given function $f(x)$ on an interval (a, b) agrees with point values of f at a certain discrete set of points $x_i \in (a, b)$: $\pi f(x_i) = f(x_i)$, $i = 1, \dots, n$, for some integer n. This concept can be generalized to determine a polynomial πf so that certain averages (a kind of weighting) agree. These could include the usual average of f over $[a, b]$ (as in Figure 3.14) defined by,

$$\frac{1}{b-a}\int_a^b f(x)\, dx,$$

or a *generalized average* of f with respect to a *weight function* w defined by

$$(f, w) = \int_a^b f(x)w(x)\, dx.$$

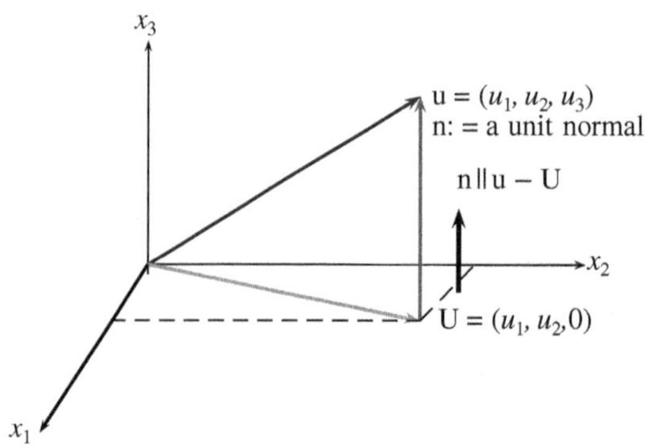

Figure 3.13 Example of a projection onto $\mathbb{R}^2$.

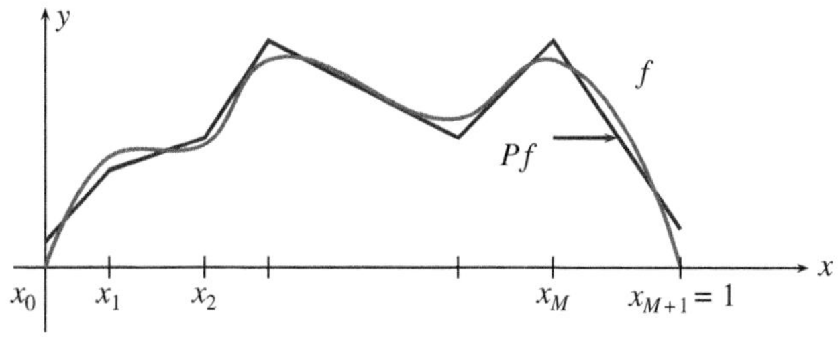

Figure 3.14 An example of a function f and its L_2 projection Pf in $[0, 1]$.

Definition 3.8 The orthogonal projection, or L_2-projection, of the function f onto $\mathcal{P}^{(q)}(a, b)$ is the polynomial $Pf \in \mathcal{P}^{(q)}(a, b)$ such that

$$(f, w) = (Pf, w) \iff (f - Pf, w) = 0 \quad \text{for all } w \in \mathcal{P}^{(q)}(a, b). \tag{3.6.1}$$

We see that Pf is defined so that certain weighted average values of Pf are the same as those of f. By the construction (3.6.1) is equivalent to a $(q + 1) \times (q + 1)$ system of equations.

We want to show that:

Lemma 3.2 *(i) Pf is uniquely defined by (3.6.1).*
(ii) Pf is the best approximation of f in $\mathcal{P}^{(q)}(a, b)$ in the $L_2(a, b)$-norm, i.e.

$$||f - Pf||_{L_2(a,b)} \le ||f - v||_{L_2(a,b)}, \quad \textit{for all } v \in \mathcal{P}^{q}(a, b). \tag{3.6.2}$$

Proof: (i) Suppose that P_1f and P_2f are two polynomials in $\mathcal{P}^{(q)}(a, b)$ such that

$$(f - P_1f, w) = 0 \quad \text{and} \quad (f - P_2f, w) = 0 \qquad \text{for all } w \in \mathcal{P}^{(q)}(a, b).$$

Subtracting the two relations, we conclude that

$$(P_2f - P_1f, w) = 0, \quad \text{for all } w \in \mathcal{P}^{(q)}(a, b).$$

Now choosing $w = P_2f - P_1f$, we get

$$\int_a^b |P_2f - P_1f|^2 \, dx = 0,$$

which yields $P_1f = P_2f$ since $|P_2f - P_1f|$ is a nonnegative continuous functions.

(ii) Using Cauchy–Schwarz' inequality, it follows that for all $v \in \mathcal{P}^{(q)}(a, b)$, since $(v - Pf) \in \mathcal{P}^{(q)}(a, b)$ and $(f - Pf) \perp \mathcal{P}^{(q)}(a, b)$,

$$\begin{aligned} ||f - Pf||^2_{L_2(a,b)} &= (f - Pf, f - Pf) = (f - Pf, f - v + v - Pf) \\ &= (f - Pf, f - v) + (f - Pf, v - Pf) \\ &= (f - Pf, f - v) \le ||f - Pf||_{L_2(a,b)} ||f - v||_{L_2(a,b)}, \end{aligned}$$

which gives the desired result. □

3.7 Numerical Integration, Quadrature Rule

In the finite element approximation procedure of solving differential equations, with a given source term (data) $f(x)$, we need to evaluate integrals of the form $\int f(x)\varphi_i(x)dx$, with $\varphi_i(x)$ being a finite element basis function. Such integrals are not easily computable for higher-order approximations (e.g. with φ_j being Lagrange basis of higher order) and more involved data. Further, we encounter matrices with entries being the integrals of products of these, higher order, basis functions, and their derivatives. Except for some special cases (see calculations for $\mathbf{A}$ and $\mathbf{A}_{\text{unif}}$ in Section 3.3), such integrations are usually performed approximately by using numerical methods. Below, we briefly review some of these numerical integration techniques.

We approximate the integral $I = \int_a^b f(x)\, dx$ using a partition of the interval $[a, b]$ into subintervals, where on each subinterval f is approximated by a polynomial of a certain degree d. We shall denote the approximate value of the integral I by I_d. To proceed we assume, without loss of generality, that $f(x) > 0$ on $[a, b]$ and that f is continuous on (a, b). Then, the integral $I = \int_a^b f(x)\, dx$ is interpreted as the area of the domain under the curve $y = f(x)$, limited by the x-axis and the lines $x = a$ and $x = b$. Here, we shall approximate this area using the values of f at certain points as follows.

We start by approximating the integral over a single interval $[a, b]$. These rules are referred to as *simple rules*.

(i) *Simple midpoint rule* uses the value of f at the midpoint $\bar{x} := \frac{a+b}{2}$ of $[a, b]$, i.e. $f\left(\frac{a+b}{2}\right)$. This means that f is approximated by the constant function (polynomial of degree 0) $P_0(x) = f\left(\frac{a+b}{2}\right)$, as in Figure 3.15, and the area under the curve $y = f(x)$ by (the area of the rectangle)

$$I = \int_a^b f(x)\, dx \approx (b-a)f\left(\frac{a+b}{2}\right). \tag{3.7.1}$$

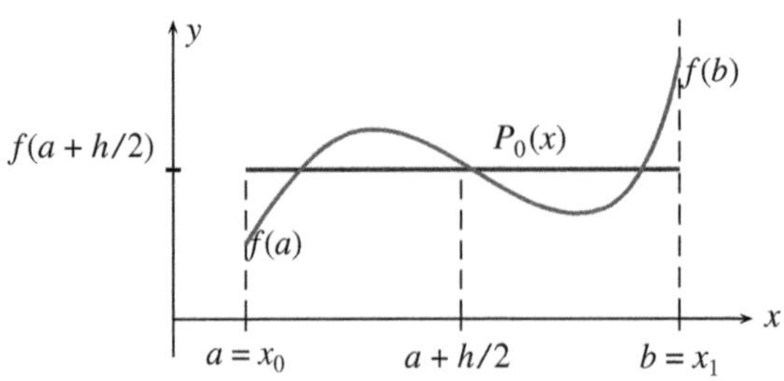

Figure 3.15 Midpoint approximation I_0 of the integral $I = \int_{x_0}^{x_1} f(x)\, dx$.

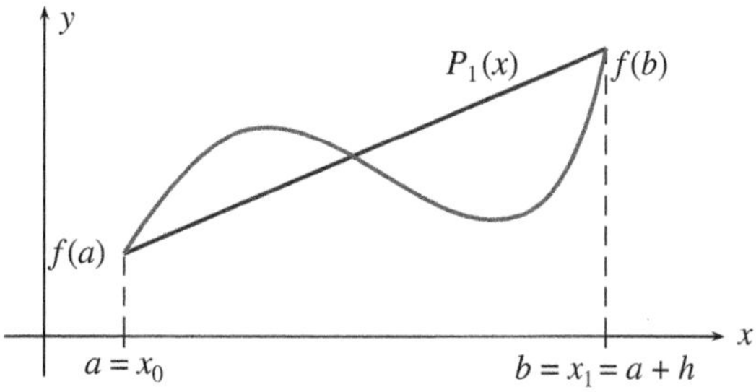

Figure 3.16 Trapezoidal approximation I_1 of the integral $I = \int_{x_0}^{x_1} f(x)\,dx$.

To prepare for generalizations, if we let $x_0 = a$ and $x_1 = b$ and assume that the length of the interval is h, then

$$I \approx I_0 = hf\left(a + \frac{h}{2}\right) = hf(\overline{x}) \tag{3.7.2}$$

(ii) *Simple trapezoidal rule* uses the values of f at two endpoints a and b, i.e. $f(a)$ and $f(b)$. Here f is approximated by the linear function (polynomial of degree 1) $P_1(x)$ passing through the two points $(a, f(a))$ and $(b, f(b))$, as in Figure 3.16. Consequently, the area under the curve $y = f(x)$ is approximated as

$$I = \int_a^b f(x)\,dx \approx (b - a)\frac{f(a) + f(b)}{2}. \tag{3.7.3}$$

This is the area of the trapezoidal between the lines $y = 0$, $x = a$, and $x = b$ and under the graph of $P_1(x)$, and therefore is referred to as the *simple trapezoidal rule*. Once again, for the purpose of generalization, we let $x_0 = a, x_1 = b$ and assume that the length of the interval is h, then (3.7.3) can be written as

$$I \approx I_1 = hf(a) + \frac{h[f(a+h) - f(a)]}{2} = h\frac{f(a) + f(a+h)}{2} = \frac{h}{2}[f(x_0) + f(x_1)]. \tag{3.7.4}$$

(iii) *Simple Simpson's rule* uses the values of f at the two endpoints a and b, and the midpoint $\frac{a+b}{2}$ of the interval $[a, b]$, i.e. $f(a), f(b)$, and $f\left(\frac{a+b}{2}\right)$. In this case the area under $y = f(x)$ is approximated by the area under the graph of the second degree polynomial $P_2(x)$; with $P_2(a) = f(a)$, $P_2\left(\frac{a+b}{2}\right) = f\left(\frac{a+b}{2}\right)$, and $P_2(b) = f(b)$ (see Figure 3.17). To determine $P_2(x)$, we may use Lagrange interpolation for $q = 2$: let $x_0 = a,\ x_1 = (a + b)/2$, and $x_2 = b$, then

$$P_2(x) = f(x_0)\lambda_0(x) + f(x_1)\lambda_1(x) + f(x_2)\lambda_2(x), \tag{3.7.5}$$

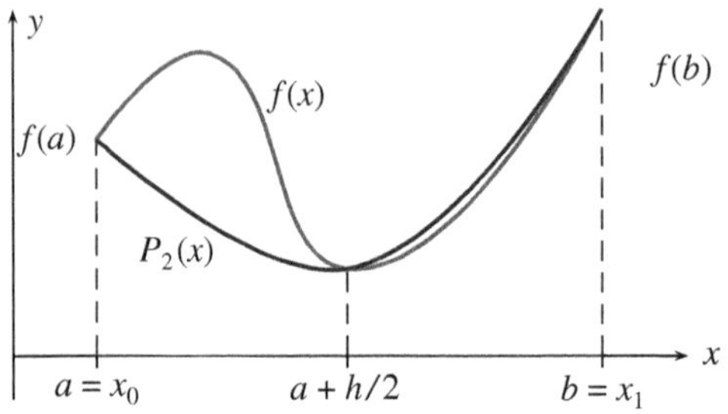

Figure 3.17 Simpson's rule approximation I_2 of the integral $I = \int_{x_0}^{x_1} f(x)\,dx$.

where

$$\begin{cases} \lambda_0(x) = \dfrac{(x-x_1)(x-x_2)}{(x_0-x_1)(x_0-x_2)}, \\ \lambda_1(x) = \dfrac{(x-x_0)(x-x_2)}{(x_1-x_0)(x_1-x_2)}, \\ \lambda_2(x) = \dfrac{(x-x_0)(x-x_1)}{(x_2-x_0)(x_2-x_1)}. \end{cases} \tag{3.7.6}$$

Thus

$$I = \int_a^b f(x)\,dx \approx \int_a^b P_2(x)\,dx = \sum_{i=0}^{2} f(x_i) \int_a^b \lambda_i(x)\,dx. \tag{3.7.7}$$

Now, we can easily compute the integrals

$$\int_a^b \lambda_0(x)\,dx = \int_a^b \lambda_2(x)\,dx = \frac{b-a}{6}, \quad \int_a^b \lambda_1(x)\,dx = \frac{4(b-a)}{6}. \tag{3.7.8}$$

Hence,

$$I = \int_a^b f(x)\,dx \approx \frac{b-a}{6}[f(x_0) + 4f(x_1) + f(x_2)], \tag{3.7.9}$$

and similar to the two previous cases, with $h = b - a, x_1 = x_0 + h/2$, and $x_2 = x_0 + h$, we thus have

$$I \approx I_2 = \frac{h}{6}[f(x_0) + 4f(x_0 + h/2) + f(x_1 + h)].$$

Obviously, these approximations are less accurate for large intervals, $[a, b]$ and/or oscillatory functions f. Following Riemann's idea, we can use these rules, instead of on the whole interval $[a, b]$, for the subintervals in an appropriate partition of $[a, b]$. Then, we get the generalized versions.

3.7.1 Composite Rules for Uniform Partitions

We shall use the following *General algorithm* to approximate the integral

$$I = \int_a^b f(x)\,dx.$$

(1) Divide the interval $[a, b]$, uniformly, into N subintervals

$$a = x_0 < x_1 < x_2 < \cdots < x_{N-1} < x_N = b. \tag{3.7.10}$$

(2) Write the integral as

$$\int_a^b f(x)\,dx = \int_{x_0}^{x_1} f(x)\,dx + \cdots + \int_{x_{N-1}}^{x_N} f(x)\,dx = \sum_{k=1}^{N} \int_{x_{k-1}}^{x_k} f(x)\,dx. \tag{3.7.11}$$

(3) For each subinterval $I_k := [x_{k-1}, x_k]$, $k = 1, 2, \ldots, N$, apply the *same* integration rule (i)–(iii). Then, we get the following generalizations.

(M) *Composite midpoint rule:* approximates f by constants (the values of f at the midpoint of the subinterval) on each subinterval (piecewise constant approximation): let

$$h = |I_k| = \frac{b-a}{N}, \quad \text{and} \quad \overline{x}_k = \frac{x_{k-1} + x_k}{2}, \qquad k = 1, 2, \ldots, N.$$

Then, using the simple midpoint rule for the interval $I_k := [x_{k-1}, x_k]$,

$$\int_{x_{k-1}}^{x_k} f(x)\,dx \approx \int_{x_{k-1}}^{x_k} f(\overline{x}_k)\,dx = hf(\overline{x}_k). \tag{3.7.12}$$

Summing over k, we get the composite midpoint rule as

$$\int_a^b f(x)\,dx \approx \sum_{k=1}^{N} hf(\overline{x}_k) = h[f(\overline{x}_1) + \cdots + f(\overline{x}_N)] := M_N. \tag{3.7.13}$$

(T) *Composite trapezoidal rule*: approximates f by simple trapezoidal rule on each subinterval I_k (piecewise linear approximation),

$$\int_{x_{k-1}}^{x_k} f(x)\,dx \approx \frac{h}{2}[f(x_{k-1}) + f(x_k)]. \tag{3.7.14}$$

Summing over k yields the composite trapezoidal rule

$$\begin{aligned}\int_a^b f(x)\,dx &\approx \sum_{k=1}^{N} \frac{h}{2}[f(x_{k-1}) + f(x_k)] \\ &= \frac{h}{2}[f(x_0) + 2f(x_1) + \cdots + 2f(x_{k-1}) + f(x_k)] := T_N.\end{aligned} \tag{3.7.15}$$

(S) *Composite Simpson's rule*: approximates f by simple Simpson's rule on each subinterval I_k (piecewise quadratic approximation),

$$\int_{x_{k-1}}^{x_k} f(x)\,dx \approx \frac{h}{6}\left[f(x_{k-1}) + 4f\left(\frac{x_{k-1} + x_k}{2}\right) + f(x_k)\right]. \tag{3.7.16}$$

To simplify, we introduce the following identification (see Figure 3.18) on each I_k:

$$z_{2k-2} = x_{k-1}, \quad z_{2k-1} = \frac{x_{k-1} + x_k}{2} := \overline{x}_k, \quad z_{2k} = x_k, \quad h_z = \frac{h}{2}. \tag{3.7.17}$$

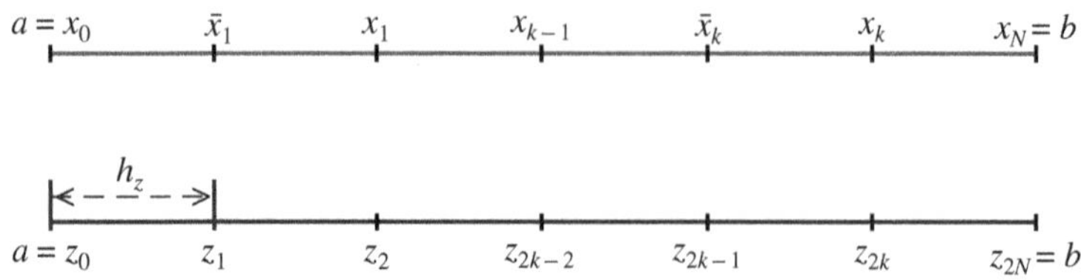

Figure 3.18 Identification of subintervals for composite Simpson's rule.

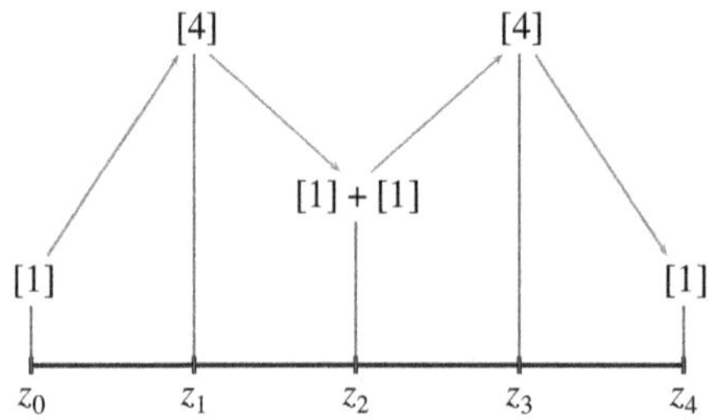

Figure 3.19 Coefficients for composite Simpson's rule.

Then, summing (3.7.16) over k and using the above identification, we obtain the composite Simpson's rule (see Figure 3.19 for the emerge of coefficients),

$$\int_a^b f(x)\,dx \approx \sum_{k=1}^{N} \frac{h}{6}\left[f(x_{k-1}) + 4f\left(\frac{x_{k-1}+x_k}{2}\right) + f(x_k)\right]$$
$$= \sum_{k=1}^{N} \frac{h_z}{3}[f(z_{2k-2}) + 4f(z_{2k-1}) + f(z_{2k})] \tag{3.7.18}$$
$$= \frac{h_z}{3}[f(z_0) + 4f(z_1) + 2f(z_2) + 4f(z_3) + 2f(z_4)$$
$$+ \cdots + 2f(z_{2N-2}) + 4f(z_{2N-1}) + f(z_{2N})] := S_N.$$

Figure 3.19 illustrates the starting procedure for the composite Simpson's rule. The numbers in the brackets indicate the actual coefficient on each subinterval. For instance the end of the first interval, $x_1 = z_2$ coincides with the start of the second interval, resulting in the add-up [1] + [1] = 2 as the coefficient of $f(z_2)$. Note that this is the case for each interior node x_k which are z_{2k} values; but *not!* for the midpoints: $z_{2k-1} = \bar{x}_k,\ k = 1, \dots, N-1$.

Remark 3.8 One can verify that the errors of these integration rules are depending on the regularity of the function and the size of interval (in simple rules) and the mesh size (in the composite rules). They read as follows:

Error in simple Midpoint rule

$$\left|\int_{x_{k-1}}^{x_k} f(x)\,dx - hf(\bar{x}_k)\right| = \frac{h^3}{24}|f''(\eta)|, \quad \eta \in (x_{k-1}, x_k).$$

Error in composite Midpoint rule

$$\left|\int_a^b f(x)\,dx - M_N\right| = \frac{h^2(b-a)}{24}|f''(\xi)|, \quad \xi \in (a,b).$$

Error in simple trapezoidal rule

$$\left|\int_{x_{k-1}}^{x_k} f(x)\,dx - \frac{h}{2}[f(\overline{x}_{k-1} + f(x_k)]\right| = \frac{h^3}{12}|f''(\eta)|, \quad \eta \in (x_{k-1}, x_k).$$

Error in composite trapezoidal rule

$$\left|\int_a^b f(x)\,dx - T_N\right| = \frac{h^2(b-a)}{12}|f''(\xi)|, \quad \xi \in (a,b),\ h = (b-a)/N.$$

Error in simple Simpson's rule,

$$\left|\int_a^b f(x)\,dx - \frac{b-a}{6}[f(a) + 4f((a+b)/2) + f(b)]\right|$$
$$= \frac{1}{90}\left(\frac{b-a}{2}\right)^5 |f^{(4)}(\eta)|, \quad \eta \in (a,b)$$

Error in composite Simpson's rule

$$\left|\int_a^b f(x)\,dx - S_N\right| = \frac{h^4(b-a)}{180} \max_{\xi\in[a,b]} |f^4(\xi)|, \quad h = (b-a)/N.$$

Remark 3.9 The rules (M), (T), and (S) use values of the function at *equally spaced* points. These are not always the best approximation methods. Below, we introduce a general and more optimal approach.

3.7.2 Gauss Quadrature Rule

The Gauss quadrature is an approximate integration rule aimed to choose the points of evaluation of an integrand f in an *optimal* manner, not necessarily at equally spaced points. Here, first, we illustrate this rule by an example and then construct a general approach and prove a corresponding theorem verifying its optimality.

Problem: Choose the nodes $x_i \in [a,b]$, and coefficients c_i, $1 \le i \le n$ such that, for the integral of an arbitrary function f, the following error is minimal:

$$\int_a^b f(x)\,dx - \sum_{i=1}^n c_i f(x_i). \tag{3.7.19}$$

Solution: Relation (3.7.19) contains $2n$ unknowns consisting of n nodes x_i and n coefficients c_i. Therefore, we need $2n$ equations. Thus, if we replace f by a polynomial, then an optimal choice of these $2n$ parameters yields a quadrature rule (3.7.19) which is *exact* for *polynomials*, f, of *degree* $\leq 2n-1$.

Example 3.8 Let $n = 2$ and $[a, b] = [-1, 1]$. Then, the coefficients are c_1 and c_2 and the nodes are x_1 and x_2. Thus, optimal choice of these four parameters should yield that the approximation

$$\int_{-1}^{1} f(x)\,dx \approx c_1 f(x_1) + c_2 f(x_2), \tag{3.7.20}$$

is indeed exact for $f(x)$ replaced by any polynomial of degree ≤ 3. So we replace f by a polynomial of the form $f(x) = Ax^3 + Bx^2 + Cx + D$ and require equality in (3.7.20). Thus, to determine the coefficients c_1, c_2 and the nodes x_1, x_2, in an optimal way, it suffices to change the above approximation to equality when f is replaced by the basis functions for polynomials of degree ≤ 3, i.e. $1, x, x^2$, and x^3. Consequently, we get the equation system

$$\begin{aligned}
&\int_{-1}^{1} 1\,dx = c_1 + c_2 \text{ and we get } [x]_{-1}^{1} = 2 = c_1 + c_2\\
&\int_{-1}^{1} x\,dx = c_1 \cdot x_1 + c_2 \cdot x_2 \text{ and } \left[\frac{x^2}{2}\right]_{-1}^{1} = 0 = c_1 \cdot x_1 + c_2 \cdot x_2\\
&\int_{-1}^{1} x^2\,dx = c_1 \cdot x_1^2 + c_2 \cdot x_2^2 \text{ and } \left[\frac{x^3}{3}\right]_{-1}^{1} = \frac{2}{3} = c_1 \cdot x_1^2 + c_2 \cdot x_2^2\\
&\int_{-1}^{1} x^3\,dx = c_1 \cdot x_1^3 + c_2 \cdot x_2^3 \text{ and } \left[\frac{x^4}{4}\right]_{-1}^{1} = 0 = c_1 \cdot x_1^3 + c_2 \cdot x_2^3,
\end{aligned} \tag{3.7.21}$$

which, although nonlinear, has the unique solution presented below:

$$\begin{cases} c_1 + c_2 = 2\\ c_1x_1 + c_2x_2 = 0\\ c_1x_1^2 + c_2x_2^2 = \frac{2}{3}\\ c_1x_1^3 + c_2x_2^3 = 0 \end{cases} \Longrightarrow \begin{cases} c_1 = 1\\ c_2 = 1\\ x_1 = -\frac{\sqrt{3}}{3}\\ x_2 = \frac{\sqrt{3}}{3}. \end{cases} \tag{3.7.22}$$

Hence, the approximation

$$\int_{-1}^{1} f(x)\,dx \approx c_1 f(x_1) + c_2 f(x_2) = f\left(-\frac{\sqrt{3}}{3}\right) + f\left(\frac{\sqrt{3}}{3}\right), \tag{3.7.23}$$

is exact for all polynomials of degree ≤ 3.

Example 3.9 Let $f(x) = 3x^2 + 2x + 1$. Then $\int_{-1}^{1}(3x^2 + 2x + 1)\,dx = [x^3 + x^2 + x]_{-1}^{1} = 4$, and we can easily check that $f(-\sqrt{3}/3) + f(\sqrt{3}/3) = 4$.

Higher-order Gauss quadrature. To generalize the Gauss quadrature rule, on an interval and to $n > 2$, Legendre polynomials are used. To illustrate, we choose $\{P_n\}_{n=0}^{\infty}$ such that

(1) For each n, P_n is a polynomial of degree n.
(2) $P_n \perp P_m$ if $m \neq n \Longleftrightarrow \int_{-1}^{1} P_n(x)P_m(x)\,dx = 0$.

The *Legendre polynomials*, on $[-1, 1]$, are obtained through the formula

$$P_n(x) = \frac{1}{2^n n!}\frac{d^n}{dx^n}((x^2-1)^n).$$

Here are the first few Legendre polynomials:

$$P_0(x) = 1, \quad P_1(x) = x, \quad P_2(x) = \frac{3}{2}x^2 - \frac{1}{2}, \quad P_3(x) = \frac{5}{2}x^3 - \frac{3}{2}x, \ldots,$$

The roots of Legendre polynomials are *distinct*, *symmetric*, with respect to $x = 0$, and the *correct choices* as *quadrature* points, i.e. the roots of the Legendre polynomial P_n ($P_0 = 1$ is an exception) are our optimal quadrature points x_i, $1 \le i \le n$ (see Theorem 3.6).

Example 3.10 Roots of the Legendre polynomial as *quadrature* points:

$$P_1(x) = x = 0.$$

$$P_2(x) = \frac{3}{2}x^2 - \frac{1}{2} = 0, \quad \text{gives} \quad x_{1,2} = \pm\frac{\sqrt{3}}{3} \quad \text{(compare with Example 3.8).}$$

$$P_3(x) = \frac{5}{2}x^3 - \frac{3}{2}x = 0, \quad \text{gives} \quad x_1 = 0,\ x_{2,3} = \pm\sqrt{\frac{3}{5}}.$$

Theorem 3.6 *Suppose that x_i, $i = 1, 2, \ldots, n$, are the roots of the nth Legendre polynomial P_n and that*

$$c_i = \int_{-1}^{1} \lambda_i(x)\,dx,$$

where

$$\lambda_i(x) = \prod_{\substack{j=1\\ j\neq i}}^{n}\left(\frac{x - x_j}{x_i - x_j}\right),$$

are the Lagrange basis functions with interpolation nodes x_i (being the roots of Legendre polynomial of degree n). If we choose the above (x_i, c_i) as nodes and coefficients (weights) of a Gauss quadrature rule on $[-1, 1]$, and if $f(x)$ is a polynomial of degree $<2n$, then

$$\int_{-1}^{1} f(x)\,dx \equiv \sum_{i=1}^{n} c_i f(x_i),$$

i.e. this quadrature approximation is exact for polynomials of degree $<2n$.

Proof: Consider a polynomial $R(x)$ of degree $<n$. We can rewrite $R(x)$ as a $(n-1)$th degree Lagrange interpolation polynomial with its nodes at the roots of the nth Legendre polynomial P_n:

$$R(x) = \sum_{i=1}^{n} \lambda_i(x) R(x_i).$$

This representation of $R(x)$ is exact, since for the error

$$E(x) = \frac{1}{n!}(x-x_1)(x-x_2)\cdots(x-x_n)R^{(n)}(\xi), \qquad R^{(n)}(\xi) \equiv 0. \tag{3.7.24}$$

Integrating we get

$$\int_{-1}^{1} R(x)\,dx = \int_{-1}^{1} \left[\sum_{i=1}^{n} \lambda_i(x)R(x_i)\right] dx = \sum_{i=1}^{n} \left[\int_{-1}^{1} \lambda_i(x)R(x_i)\right] dx. \tag{3.7.25}$$

Hence,

$$\int_{-1}^{1} R(x)\,dx = \sum_{i=1}^{n} c_i R(x_i), \tag{3.7.26}$$

which shows the assertion of theorem for polynomials of degree $<n$.

Now consider a polynomial, $P(x)$, of degree $<2n$. Dividing $P(x)$ by the nth Legendre polynomial $P_n(x)$, we get

$$P(x) = Q(x) \times P_n(x) + R(x), \quad \deg\ Q(x) < n, \ \deg\ R(x) < n, \tag{3.7.27}$$

and

$$\int_{-1}^{1} P(x)\,dx = \int_{-1}^{1} Q(x)P_n(x)\,dx + \int_{-1}^{1} R(x)\,dx. \tag{3.7.28}$$

Since $P_n(x) \perp Q(x)$, $\forall Q(x)$ with $\deg Q < n$. Using (3.7.27) it follows that

$$\int_{-1}^{1} Q(x)P_n(x)\,dx = 0 \Longrightarrow \int_{-1}^{1} P(x)\,dx = \int_{-1}^{1} R(x)\,dx. \tag{3.7.29}$$

Recall that, the x_is are the roots of $P_n(x)$, i.e. $P_n(x_i) = 0$. Thus, by (3.7.27),

$$P(x_i) = Q(x_i)P_n(x_i) + R(x_i) = R(x_i). \tag{3.7.30}$$

Hence, by (3.7.29), (3.7.26), and (3.7.30),

$$\int_{-1}^{1} P(x)\,dx = \int_{-1}^{1} R(x)\,dx = \sum_{i=1}^{n} c_i R(x_i) = \sum_{i=1}^{n} c_i P(x_i), \tag{3.7.31}$$

and the proof is complete.

3.8 Exercises

Problem 3.9 Show that $(f - P_h f, v) = 0,\ \forall v \in V_h$, if and only if $(f - P_h f, \varphi_i) = 0,\ i = 0, \dots, N$; where $\{\varphi_i\}_{i=0}^N \subset V_h$ is the basis of hat-functions, where $P_h f$ is the L_2-projection of f onto V_h.

Problem 3.10 Use the expressions $\lambda_a(x) = \frac{b-x}{b-a}$ and $\lambda_b(x) = \frac{x-a}{b-a}$ to show that

$$\lambda_a(x) + \lambda_b(x) = 1, \quad a\lambda_a(x) + b\lambda_b(x) = x.$$

Give a geometric interpretation by plotting, $\lambda_a(x)$, $\lambda_b(x)$, $\lambda_a(x) + \lambda_b(x)$, $a\lambda_a(x)$, $b\lambda_b(x)$, and $a\lambda_a(x) + b\lambda_b(x)$.

Problem 3.11 Let f: $[0.1] \to \mathbb{R}$ be a Lipschitz continuous function. Determine the linear interpolant $\pi f \in \mathcal{P}(0, 1)$ and plot f and πf in the same figure, when
(a) $f(x) = x^2$,
(b) $f(x) = \sin(\pi x)$.

Problem 3.12 Determine the linear interpolation of the function

$$f(x) = \frac{1}{\pi^2}(x - \pi)^2 - \cos^2\left(x - \frac{\pi}{2}\right), -\pi \le x \le \pi.$$

where the interval $[-\pi, \pi]$ is divided into four equal subintervals.

Problem 3.13 Assume that $w' \in L_1(I)$. Let $x, \bar{x} \in I = [a, b]$ and $w(\bar{x}) = 0$. Show that

$$|w(x)| \le \int_I |w'|\, dx. \tag{3.8.1}$$

Problem 3.14 Assume that v interpolates φ, at the points a and b.

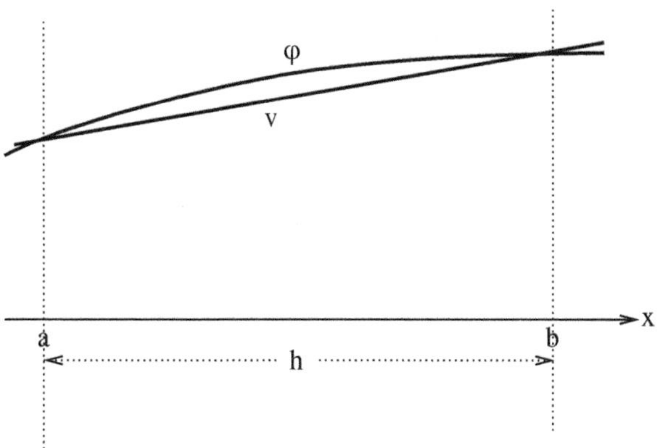

Show, using (3.8.1) that

(i) $|(\varphi - v)(x)| \leq \int_I |(\varphi - v)'|\, dx,$

(ii) $|(\varphi - v)'(x)| \leq \int_I |(\varphi - v)''|\, dx = \int_I |\varphi''|\, dx,$

(iii) $\max_I |\varphi - v| \leq \max_I |h^2\varphi''|,$

(iv) $\int_I |\varphi - v|\, dx \leq \int_I |h^2\varphi''|\, dx,$

(v) $\|\varphi - v\|_I \leq \|h^2\varphi''\|_I$ and $\|h^{-2}(\varphi - v)\|_I \leq \|\varphi''\|_I,$

where $\|w\|_I = \left(\int_I w^2\, dx\right)^{1/2}$ is the $L_2(I)$-norm.

Problem 3.15 Use, in the above problem

$$v' = \frac{\varphi(b) - \varphi(a)}{h} = \frac{1}{h}\int_a^b \varphi'\, dx \quad (\varphi' \text{ is constant on } I),$$

and show that

(vi) $|(\varphi - v)(x)| \leq 2\int_I |\varphi'|\, dx,$

(vii) $\int_I h^{-1}|\varphi - v|\, dx \leq 2\int_I |\varphi'|\, dx$ and $\|h^{-1}(\varphi - v)\| \leq 2\|\varphi'\|_I.$

Problem 3.16 Let now $v(t)$ be the constant interpolant of φ on I. Show that

$$\int_I h^{-1}|\varphi - v|\, dx \leq \int_I |\varphi'|\, dx. \tag{3.8.2}$$

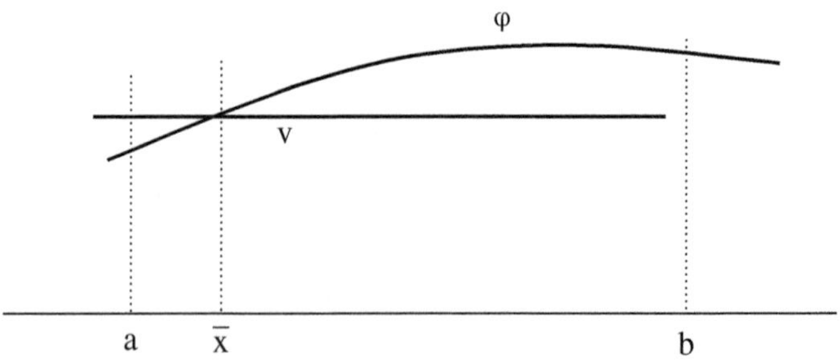

Problem 3.17 Show that

$$\mathcal{P}^q(a, b) := \{p(x)|p(x) \text{is a polynomial of degree } \leq q\},$$

is a vector space but

$$\mathcal{P}^q(a,b) := \{p(x)|p(x) \text{ is a polynomial of degree} = q\},$$

is not! a vector space.

Problem 3.18 Compute formulas for the linear interpolant of a continuous function f through the points a and $(b+a)/2$. Plot the corresponding Lagrange basis functions.

Problem 3.19 Prove the following interpolation error estimate:

$$||\pi_1 f - f||_{L_\infty(a,b)} \le \frac{1}{8}(b-a)^2||f''||_{L_\infty(a,b)}.$$

Hint: Using an scaling argument, it suffices to prove the inequality for $[a,b] \equiv [0,1]$.

Problem 3.20 Prove that any value of f on the subintervals, in a partition of (a,b), can be used to define $\pi_h f$ satisfying the error bound

$$||f - \pi_h f||_{L_\infty(a,b)} \le \max_{1\le i\le m+1} h_i||f'||_{L_\infty(I_i)} = ||hf'||_{L_\infty(a,b)}.$$

Prove that choosing the midpoint improves the bound by an extra factor $1/2$.

Problem 3.21 Compute and graph $\pi_4(e^{-8x^2})$ on $[-2,2]$, which interpolates e^{-8x^2} at five equally spaced points in $[-2,2]$.

Problem 3.22 Write down a basis for the set of piecewise quadratic polynomials $W_h^{(2)}$ on a partition $a = x_0 < x_1 < x_2 < \cdots < x_{m+1} = b$ of (a,b) into subintervals $I_i = (x_{i-1}, x_i)$, where

$$W_h^{(q)} = \{v\colon v|_{I_i} \in \mathcal{P}^q(I_i), i = 1,\dots,m+1\}.$$

Problem 3.23 Determine a set of basis functions for the space of continuous piecewise quadratic functions $V_h^{(2)}$ on $I = (a,b)$, where

$$V_h^{(q)} = \{v \in W_h^{(q)} \ :\ v \text{ is continuous on } I\}.$$

Problem 3.24 Prove that

$$\int_{x_0}^{x_1} f'\left(\frac{x_1+x_0}{2}\right)\left(x - \frac{x_1+x_0}{2}\right)\,dx = 0.$$

Problem 3.25 Prove that

$$\left| \int_{x_0}^{x_1} f(x)\,dx - f\left(\frac{x_1+x_0}{2}\right)(x_1-x_0) \right|$$
$$\leq \frac{1}{2} \max_{[x_0,x_1]} |f''| \int_{x_0}^{x_1} \left(x - \frac{x_1+x_0}{2}\right)^2 dx \leq \frac{1}{24}(x_1-x_0)^3 \max_{[x_0,x_1]} |f''|.$$

Hint: Use Taylor expansion of f about $x = \frac{x_1+x_0}{2}$.

4

Linear Systems of Equations

We have seen that the numerical solution of a differential equation, e.g. using the Galerkin finite element method, is an approximation of the exact solution with a function in a finite dimensional vector space. If the differential equation is linear, then the procedure ends up solving a linear system of equations of the form $\mathbf{Ax} = \mathbf{b}$, where the coefficient matrix $\mathbf{A}$ is a square matrix, roughly, of the same order as the dimension of the approximation space, and is related to the differential operator in the equation, $\mathbf{x}$ is a vector in the approximation space with entries consisting of certain nodal values of the approximate solution, and the vector $\mathbf{b}$ is related to the data and basis functions of the approximation space.

The criterion for the quality of a numerical method, to solve a certain problem, lies in its potential in speed (rate) of convergence of the approximate solution to the exact solution in an adequate measuring environment (norm). Quantitatively, this is expressed by "how fast" the approximate solution would converge to the exact solution by increasing the approximation degree (the dimension of the approximation space), which is a theoretical procedure. In computations, however, with an already justified convergence, it is important to have a numerical algorithm that solves the approximate problem reasonably fast (takes shorter time which may be achieved, e.g. by taking a fewer number of approximation points).

This chapter is devoted to the numerical solution of linear systems of equations of type $\mathbf{Ax} = \mathbf{b}$. *Throughout this chapter, we assume that* $\det \mathbf{A} \neq 0$, *i.e. the matrix* $\mathbf{A}$ *is invertible.* Then $\mathbf{Ax} = \mathbf{b} \Leftrightarrow \mathbf{x} = \mathbf{A}^{-1}\mathbf{b}$, but we wish to circumvent inverting the matrix $\mathbf{A}$. To this approach, we shall review the well-known direct method of Gauss elimination and then continue with some more efficient iterative approaches. A thorough study of this type is in the realm of numerical linear algebra where the solution algorithms for linear systems of equations is undoubtedly one of the most applied tools, see, e.g. Golub and Loan [67].

An Introduction to the Finite Element Method for Differential Equations, First Edition.
Mohammad Asadzadeh.

4.1 Direct Methods

Consider the general form of an $n \times n$ linear system of equations given by

$$\mathbf{Ax} = \mathbf{b} \Leftrightarrow \sum_{j=1}^{n} a_{ij}x_j = b_i, \quad i = 1, \ldots, n, \qquad \text{or equivalently}$$

$$\begin{cases} a_{11}x_1 + a_{12}x_2 + \cdots + a_{1n}x_n = & b_1 \\ a_{21}x_1 + a_{22}x_2 + \cdots + a_{2n}x_n = & b_2 \\ \vdots & \\ a_{n1}x_1 + a_{n2}x_2 + \cdots + a_{nn}x_n = & b_n. \end{cases} \tag{4.1.1}$$

We introduce the extended $n \times (n+1)$ matrix $\mathcal{E}$ enlarging the coefficient matrix **A** by putting the right-hand side **b** as the additional last column:

$$\mathcal{E} := \begin{pmatrix} a_{11} & a_{12} & \cdots & a_{1n} & b_1 \\ a_{21} & a_{22} & \cdots & a_{2n} & b_2 \\ \vdots & \vdots & \vdots & \vdots & \vdots \\ a_{n1} & a_{n2} & \cdots & a_{nn} & b_n \end{pmatrix}. \tag{4.1.2}$$

Note that to solve the equation system $\mathbf{Ax} = \mathbf{b}$, it is a bad idea to calculate $\mathbf{A}^{-1}$ and then multiply by **b**. However, if **A** is an upper (or lower) triangular matrix, i.e. if $a_{ij} = 0$ for $i > j$ (or $i < j$), and **A** is invertible, then we can solve **x** using the *back substitution method* that yields

$$\begin{cases} x_1 = \dfrac{1}{a_{11}}[b_1 - a_{12}x_2 - \cdots - a_{1n}x_n] \\ \qquad \vdots \quad \vdots \quad \vdots \\ x_{n-1} = \dfrac{1}{a_{n-1,n-1}}[b_{n-1} - a_{n-1,n}x_n] \\ \\ \qquad\qquad x_n = \dfrac{b_n}{a_{nn}}. \end{cases} \tag{4.1.3}$$

Number of operations. The speed of convergence depends on the number of operations performed during the computations. In the above procedure, additions and subtractions are not considered as time-consuming operations; therefore, we shall count only the number of multiplications and divisions.

- The number of multiplications to solve for x_n from (4.1.3) is 0 and the number of divisions is 1.
- To solve for x_{n-1}, we need one multiplication and one division.
- To solve for x_1, we need $(n-1)$ multiplication and one division.

Thus, to solve the triangular linear system of equations (4.1.3) we need

$$1 + 2 + \cdots + (n-1) = \frac{n(n-1)}{2} := \frac{n^2}{2} + \mathbf{Q}(n),$$

multiplications, where $\mathbf{Q}(n)$ is a remainder of order n, and n divisions, i.e. totally

$$\frac{n(n-1)}{2} + n = \frac{n(n+1)}{2},$$

operations.

Gaussian elimination method is based on the facts that a solution to a linear system of equations is not changed under *elementary row operations*:

(i) interchanging two equations
(ii) adding a multiple of one equation to another
(iii) multiplying an equation by a nonzero constant.

The Gauss elimination procedure is based on splitting the coefficient matrix **A** to two factors, an upper triangular matrix **U** and a lower triangular matrix **L**, known as **LU** factorization. Below we give a brief introduction of the Gauss *elimination procedure* (detailed description can be found in any elementary text in linear algebra). To this end, we start from the first row of the coefficient matrix and use elementary row operations to eliminate the elements $a_{i1}, i > 1$, under a_{11} (make $a_{i1} = 0$),

Remark 4.1 If $a_{11} = 0$, then we interchange equations, and replace the first equation by an equation in the system with $a_{i1} \neq 0$.

The equation system corresponding to this newly obtained matrix $\tilde{\mathbf{A}}$ with elements $\tilde{a}_{ij}$, and where $\tilde{a}_{i1} = 0, i > 1$, has the same solution as the original one. We repeat the same procedure of elementary row operations to eliminate the elements $\tilde{a}_{i2}, i > 2$, from the matrix $\tilde{\mathbf{A}}$. Continuing in this way, we obtain an upper triangular matrix **U** with a corresponding equation system which is *equivalent* to the original system (has the same solution). Below we shall illustrate this procedure through an *example*.

Example 4.1 Solve the linear system of equations

$$\begin{cases} 2x_1 + x_2 + x_3 &= 2 \\ 4x_1 - x_2 + 3x_3 &= 0 \\ 2x_1 + 6x_2 - 2x_3 &= 10. \end{cases} \tag{4.1.4}$$

Solution: In the extended coefficient matrix:

$$\mathcal{E} = \begin{pmatrix} 2 & 1 & 1 & | & 2 \\ 4 & -1 & 3 & | & 0 \\ 2 & 6 & -2 & | & 10 \end{pmatrix}, \tag{4.1.5}$$

we have that $a_{11} = 2$, $a_{21} = 4$, and $a_{31} = 2$. We introduce the multipliers m_{i1}, $i > 1$ by letting

$$m_{21} = \frac{a_{21}}{a_{11}} = \frac{4}{2} = 2 \qquad m_{31} = \frac{a_{31}}{a_{11}} = \frac{2}{2} = 1. \tag{4.1.6}$$

Now, we multiply the first row by m_{21} and then subtract it from row 2 and replace the result in row 2. Then, we multiply the first row by $m_{31} = 1$, subtract it from row 3, and replace the result in row 3:

$$\tilde{\mathcal{E}} := \left(\begin{array}{ccc|c} 2 & 1 & 1 & 2 \\ 0 & -3 & 1 & -4 \\ 0 & 5 & -3 & 8 \end{array}\right). \tag{4.1.7}$$

In this setting, we have $\tilde{a}_{22} = -3$ and $\tilde{a}_{32} = 5$. Now, we let $m_{32} = \tilde{a}_{32}/\tilde{a}_{22} = -5/3$, then multiplying the second row in $\tilde{\mathcal{E}}$ by m_{32} and subtracting the result from row 3 yields

$$\left(\begin{array}{ccc|c} 2 & 1 & 1 & 2 \\ 0 & -3 & 1 & -4 \\ 0 & 0 & -\frac{4}{3} & \frac{4}{3} \end{array}\right), \tag{4.1.8}$$

where we have obtained the upper triangular matrix and the resulting new equivalent equation system:

$$\mathbf{U} = \begin{pmatrix} 2 & 1 & 1 \\ 0 & -3 & 1 \\ 0 & 0 & -\frac{4}{3} \end{pmatrix}, \quad \text{and} \quad \begin{cases} 2x_1 + x_2 + x_3 = & 2 \\ \quad -3x_2 + x_3 = & -4 \\ \qquad -\frac{4}{3}x_3 = & \frac{4}{3} \end{cases}, \tag{4.1.9}$$

with the solution $x_1 = 1$, $x_2 = 1$, and $x_3 = -1$ which, as we can verify, is also the solution of the original equation system (4.1.4).

Remark 4.2 For an $n \times n$ matrix $\mathbf{A}$, the number of operation in the Gauss elimination procedure leading to an upper triangular matrix $\mathbf{U}$, is of order $\mathcal{O}(n^3)$. Recall that the number of operations to solve an $n \times n$ upper triangular system is $\mathcal{O}(n^2)$. Therefore, the total number of operations to solve an $n \times n$ linear system of equations using the Gauss elimination is of the order $\mathcal{O}(n^3)$.

Definition 4.1 We define the lower triangular matrices:

$$\mathbf{L}_1 = \begin{pmatrix} 1 & 0 & 0 \\ -m_{21} & 1 & 0 \\ -m_{31} & 0 & 1 \end{pmatrix}, \mathbf{L}_2 = \begin{pmatrix} 1 & 0 & 0 \\ 0 & 1 & 0 \\ 0 & -m_{32} & 1 \end{pmatrix}, \text{ and } \mathbf{L} = \begin{pmatrix} 1 & 0 & 0 \\ m_{21} & 1 & 0 \\ m_{31} & m_{32} & 1 \end{pmatrix},$$

where m_{21}, m_{31}, and m_{32} are the multipliers. The matrices $\mathbf{L}_1$, $\mathbf{L}_2$, and $\mathbf{L}$ are *unit* (ones on the diagonal) lower triangular 3×3-matrices satisfying

$$\mathbf{L} = (\mathbf{L}_2\mathbf{L}_1)^{-1} = \mathbf{L}_1^{-1}\mathbf{L}_2^{-1}, \qquad \text{and} \qquad \mathbf{A} = \mathbf{L}\mathbf{U}. \tag{4.1.10}$$

Example 4.2 Continuing the previous example, we have $m_{21} = 2, m_{31} = 1$, and $m_{32} = -5/3$, consequently

$$\mathbf{L}_1\mathbf{A} = \begin{pmatrix} 1 & 0 & 0 \\ -2 & 1 & 0 \\ -1 & 0 & 1 \end{pmatrix} \begin{pmatrix} 2 & 1 & 1 \\ 4 & -1 & 3 \\ 2 & 6 & -2 \end{pmatrix} = \begin{pmatrix} 2 & 1 & 1 \\ 0 & -3 & 1 \\ 0 & 5 & -3 \end{pmatrix} = \tilde{\mathbf{A}},$$

which corresponds to the first two elementary row operations in the Gaussian elimination. Further

$$\mathbf{L}_2\mathbf{L}_1\mathbf{A} = \begin{pmatrix} 1 & 0 & 0 \\ 0 & 1 & 0 \\ 0 & \frac{5}{3} & 1 \end{pmatrix} \begin{pmatrix} 2 & 1 & 1 \\ 0 & -3 & 1 \\ 0 & 5 & -3 \end{pmatrix} = \begin{pmatrix} 2 & 1 & 1 \\ 0 & -3 & 1 \\ 0 & 0 & -\frac{4}{3} \end{pmatrix} = \mathbf{U},$$

which corresponds to the last (third) elementary row operation performed in Example 4.1.

Note that, the last relation means also

$$\mathbf{A} = (\mathbf{L}_2\mathbf{L}_1)^{-1}\mathbf{U} = \mathbf{LU}.$$

In the general setting, we have the following result:

Proposition 4.1 *The $n \times n$ unit lower triangular matrix $\mathbf{L}$ is given by*

$$\mathbf{L} = (\mathbf{L}_{n-1}\mathbf{L}_{n-2} \dots \mathbf{L}_1)^{-1},$$

where $\mathbf{L}_i$, $i = 1, \dots, n-1$ are the corresponding $n \times n$ row-operation matrices, namely, Example 4.2

4.1.1 LU Factorization of an $n \times n$ Matrix A

To generalize the above procedure to an $n \times n$ linear system of equations, we use the factorization $\mathbf{A} = \mathbf{LU}$ of the coefficient matrix $\mathbf{A}$, where $\mathbf{L}$ is the unit lower triangular matrix and $\mathbf{U}$ is the upper triangular matrix obtained from $\mathbf{A}$ by Gaussian elimination. To solve the system $\mathbf{Ax} = \mathbf{b}$, assuming we have an $\mathbf{LU}$ factorization, we let $\mathbf{y} = \mathbf{Ux}$, and first solve $\mathbf{Ly} = \mathbf{b}$ by forward substitution (from the first row to the last) and obtain the vector $\mathbf{y}$, then using $\mathbf{y}$ as the known right-hand side finally we solve $\mathbf{Ux} = \mathbf{y}$ by backward substitution (from the last row to the first) and get the solution $\mathbf{x}$. Thus,

$$\mathbf{Ax} = \mathbf{b} \Longleftrightarrow (\mathbf{LU})\mathbf{x} = \mathbf{b} \Longleftrightarrow \mathbf{L}(\mathbf{Ux}) = \mathbf{b} \Longleftrightarrow (\mathbf{Ly} = \mathbf{b} \quad \wedge \quad \mathbf{Ux} = \mathbf{y}).$$

In the Gauss elimination procedure, $\mathbf{Ly} = \mathbf{b}$ is solved "by automatic," viz.

$$\mathbf{y} = \mathbf{L}^{-1}\mathbf{b}.$$

Then, we solve

$$\mathbf{Ux} = \mathbf{L}^{-1}\mathbf{b} = \mathbf{y},$$

by backward substitution.

Example 4.3 We return to Example 4.2, where we have that

- $\mathbf{Ly} = \mathbf{b}$ yields the system of equations

$$\begin{pmatrix} 1 & 0 & 0 \\ 2 & 1 & 0 \\ 1 & -\frac{5}{3} & 1 \end{pmatrix}\begin{pmatrix} y_1 \\ y_2 \\ y_3 \end{pmatrix} = \begin{pmatrix} 2 \\ 0 \\ 10 \end{pmatrix} \Longleftrightarrow \begin{cases} y_1 = 2 \\ 2y_1 + y_2 = 0 \\ y_1 - \frac{5}{3}y_2 + y_3 = 10. \end{cases}$$

  Using forward substitution, we get $y_1 = 2, \; y_2 = -4, \; y_3 = 4/3$. Thus
- $\mathbf{Ux} = \mathbf{y}$ yields

$$\begin{pmatrix} 2 & 1 & 1 \\ 0 & -3 & 1 \\ 0 & 0 & -\frac{4}{3} \end{pmatrix}\begin{pmatrix} x_1 \\ x_2 \\ x_3 \end{pmatrix} = \begin{pmatrix} 2 \\ -4 \\ \frac{4}{3} \end{pmatrix} \Longleftrightarrow \begin{cases} 2x_1 + x_2 + x_3 = 2 \\ -3x_2 + x_3 = -4 \\ -\frac{4}{3}x_3 = \frac{4}{3}. \end{cases}$$

  Using backward substitution, we get the solution: $x_1 = 1, \; x_2 = 1, \; x_3 = -1.$

Remark 4.3 In some cases, we can predict the solubility of an equation system: Let $\mathbf{A}$ be an $n \times n$ matrix. Then, we have the following properties:

- If $\det(\mathbf{A}) \neq 0$, then the system of equations $\mathbf{Ax} = \mathbf{b}$ has a unique solution, given by $\mathbf{x} = \mathbf{A}^{-1}\mathbf{b}$.
- $\mathbf{A}$ has an $\mathbf{LU}$ factorization if $\det(\Delta_k) \neq 0, \quad k = 1, \ldots, n-1$, where Δ_k is Sylvester's matrix:

$$\Delta_k = \begin{pmatrix} a_{11} & \cdots & a_{1k} \\ \vdots & \vdots & \vdots \\ a_{k1} & \cdots & a_{kk} \end{pmatrix}.$$

- If the $\mathbf{LU}$ factorization exists and $\det(\mathbf{A}) \neq 0$, then the $\mathbf{LU}$ factorization is unique and $\det(\mathbf{A}) = \mathbf{U}_{11} \cdot \cdots \cdot \mathbf{U}_{nn}$.

In particular for the "symmetric matrices," we have:

Theorem 4.1 Cholesky's method *Let* $\mathbf{A}$ *be a symmetric matrix,* $(a_{ij} = a_{ji})$*, then the following statements are equivalent:*

(i) **A** *is positive definite.*
(ii) *The eigenvalues of* **A** *are positive.*
(iii) Sylvester's criterion *yields:* $\det(\Delta_k) > 0$ *for* $k = 1, 2, \ldots, n$.
(iv) $\mathbf{A} = \mathbf{L}\mathbf{L}^T$ *where* **L** *is lower triangular and has positive diagonal elements (Cholesky factorization).*

We do not give a proof of this theorem. The interested reader is referred to literature in linear algebra and matrix theory, e.g. G. Golub and C.V. Loan [67].

4.2 Iterative Methods

To speed up the solution procedure, instead of solving $\mathbf{Ax} = \mathbf{b}$ directly, for the exact solution $\mathbf{x}$, we consider iterative solution methods based on computing a sequence of approximations $\mathbf{x}^{(k)}$, $k = 1, 2, \ldots$ such that

$$\lim_{k\to\infty} \mathbf{x}^{(k)} = \mathbf{x} \quad \text{or} \quad \lim_{k\to\infty} \|\mathbf{x}^{(k)} - \mathbf{x}\| = 0, \quad \text{for some matrix norm.}$$

Definition 4.2 For an $n \times n$ matrix $\mathbf{A}$, $\mathbf{L}_1$ and $\mathbf{L}_\infty$ matrix norms are defined as

$$||\mathbf{A}||_1 = \max_j \sum_{i=1}^{n} |a_{ij}|, \qquad \text{and} \quad ||\mathbf{A}||_\infty = \max_i \sum_{j=1}^{n} |a_{ij}|.$$

We consider the general $n \times n$ linear system of equations (4.1.1) $\mathbf{Ax} = \mathbf{b}$, where both the coefficient matrix $\mathbf{A}$ and the vector $\mathbf{b}$ have real entries. For this system we introduce the two main iterative methods.

4.2.1 Jacobi Iteration

Assume that $a_{ii} \neq 0$, then (4.1.1) can be rewritten as:

$$\begin{cases} x_1 = -\dfrac{1}{a_{11}}[a_{12}x_2 + a_{13}x_3 + \cdots + a_{1n}x_n - b_1] \\ x_2 = -\dfrac{1}{a_{22}}[a_{21}x_1 + a_{23}x_3 + \cdots + a_{2n}x_n - b_2] \\ \vdots \\ x_n = -\dfrac{1}{a_{nn}}[a_{n1}x_1 + a_{n2}x_2 + \cdots + a_{n,n-1}x_{n-1} - b_n]. \end{cases}$$

We start with a given initial approximation for the solution, viz.

$$\mathbf{x}^{(0)} = (x_1^{(0)} = c_1,\ x_2^{(0)} = c_2,\ \ldots,\ x_n^{(0)} = c_n),$$

and, based on the above system, define a successive iteration procedure, for $k = 0, 1, 2, \dots,$ as:

$$\begin{cases} x_1^{(k+1)} & = & -\dfrac{1}{a_{11}}[a_{12}x_2^{(k)} + a_{13}x_3^{(k)} + \cdots + a_{1n}x_n^{(k)} - b_1] \\ x_2^{(k+1)} & = & -\dfrac{1}{a_{22}}[a_{21}x_1^{(k)} + a_{23}x_3^{(k)} + \cdots + a_{2n}x_n^{(k)} - b_2] \\ \vdots & & \vdots \\ x_n^{(k+1)} & = & -\dfrac{1}{a_{nn}}[a_{n1}x_1^{(k)} + a_{n2}x_2^{(k)} + \cdots + a_{n,n-1}x_{n-1}^{(k)} - b_n], \end{cases}$$

or in compact form in *Jacobi coordinates*, by

$$\begin{cases} \sum_{j=1}^{n} a_{ij}x_j = b_i \Longleftrightarrow a_{ii}x_i = -\sum_{\substack{j=1 \\ j\neq i}}^{n} a_{ij}x_j + b_i, \\ a_{ii}x_i^{(k+1)} = -\sum_{\substack{j=1 \\ j\neq i}}^{n} a_{ij}x_j^{(k)} + b_i. \end{cases} \tag{4.2.1}$$

If n is large and the number of iterations are small ($< n$), then the Jacobi method requires less operations than the usual Gauss elimination method. Below we state a convergence criterion (the proof is left as an exercise).

4.2.2 Convergence Criterion

The Jacobi method gives convergence to the exact solution if the matrix $\mathbf{A}$ is *strictly diagonally dominant*, i.e.

$$|a_{ii}| > \sum_{\substack{j=1 \\ j\neq i}}^{n} |a_{ij}| \quad i = 1, 2, \dots, n. \tag{4.2.2}$$

Example 4.4 Solve $\mathbf{Ax} = \mathbf{b}$ where $\mathbf{A} = \begin{pmatrix} 2 & -1 \\ -1 & 2 \end{pmatrix}$ and $\mathbf{b} = \begin{pmatrix} 1 \\ 1 \end{pmatrix}$.

$\mathbf{A}$ is diagonally dominant and the matrix equation $\mathbf{Ax} = \mathbf{b}$ is equivalent to the linear equation system

$$\begin{cases} 2x_1 - x_2 = 1 \\ -x_1 + 2x_2 = 1. \end{cases} \tag{4.2.3}$$

We choose zero initial values for x_1 and x_2, i.e. $x_1^{(0)} = 0$ and $x_2^{(0)} = 0$ and use (4.2.3) to build the Jacobi iteration system

$$\begin{cases} 2x_1^{(k+1)} = x_2^{(k)} + 1 \\ 2x_2^{(k+1)} = x_1^{(k)} + 1, \end{cases} \tag{4.2.4}$$

where k is the iteration step. Then, we have

$$\begin{cases} 2x_1^{(1)} = x_2^{(0)} + 1 \\ 2x_2^{(1)} = x_1^{(0)} + 1 \end{cases} \quad \text{with the solution} \quad \begin{cases} x_1^{(1)} = 1/2 \\ x_2^{(1)} = 1/2. \end{cases} \tag{4.2.5}$$

In the next iteration step:

$$\begin{cases} 2x_1^{(2)} = x_2^{(1)} + 1 \\ 2x_2^{(2)} = x_1^{(1)} + 1 \end{cases} \Rightarrow \begin{cases} 2x_1^{(2)} = 1/2 + 1 \\ 2x_2^{(2)} = 1/2 + 1 \end{cases} \Rightarrow \begin{cases} x_1^{(2)} = 3/4 \\ x_2^{(2)} = 3/4. \end{cases} \tag{4.2.6}$$

Continuing we obviously have $\lim_{k\to\infty} x_i^{(k)} = x_i, \quad i = 1, 2,$ where $x_1 = x_2 = 1$. Below we show the first few iterations giving the corresponding $x_1^{(k)}$ and $x_2^{(k)}$ values

k	$x_1^{(k)}$	$x_2^{(k)}$
0	0	0
1	1/2	1/2
2	3/4	3/4
3	7/8	7/8

Now if we use the maximum norm: $\|\mathbf{e}_k\|_\infty := \max_{i=1,2} |x_i^{(k)} - x_i|$, then

$$\begin{aligned}
\|\mathbf{e}_0\|_\infty &= \max(|x_1^{(0)} - x_1|, |x_2^{(0)} - x_2|) = \max(|0-1|, |0-1|) = 1 \\
\|\mathbf{e}_1\|_\infty &= \max(|x_1^{(1)} - x_1|, |x_2^{(1)} - x_2|) = \max\left(\left|\frac{1}{2} - 1\right|, \left|\frac{1}{2} - 1\right|\right) = \frac{1}{2} \\
\|\mathbf{e}_2\|_\infty &= \max(|x_1^{(2)} - x_1|, |x_2^{(2)} - x_2|) = \max\left(\left|\frac{3}{4} - 1\right|, \left|\frac{3}{4} - 1\right|\right) = \frac{1}{4} \\
\|\mathbf{e}_3\|_\infty &= \max(|x_1^{(3)} - x_1|, |x_2^{(3)} - x_2|) = \max\left(\left|\frac{7}{8} - 1\right|, \left|\frac{7}{8} - 1\right|\right) = \frac{1}{8}.
\end{aligned}$$

In this way $\|\mathbf{e}_{k+1}\|_\infty = \frac{1}{2}\|\mathbf{e}_k\|_\infty$, where $\mathbf{e}_k$ is the error in step $k \geq 0$. Iterating, we see that for the k-th Jacobi iteration, the convergence rate is $\left(\frac{1}{2}\right)^k$:

$$\|\mathbf{e}_k\|_\infty = \frac{1}{2}\|\mathbf{e}_{k-1}\|_\infty = \left(\frac{1}{2}\right)^2 \|\mathbf{e}_{k-2}\|_\infty = \cdots = \left(\frac{1}{2}\right)^k \|\mathbf{e}_0\|_\infty = (1/2)^k.$$

4.2.3 Gauss–Seidel Iteration

We start again with an initial approximation of the solution of the form

$$\mathbf{x} \approx (x_1^{(0)}, x_2^{(0)}, \ldots, x_n^{(0)}).$$

Here, we use the fact that the first row in the k-th Jacobi iteration gives $x_1^{(k+1)}$ and in the $i+1$-th row $x_1^{(k+1)}, \ldots, x_i^{(k+1)}$ are already computed values given on the left-hand sides of the previous (first i) rows. Then, the idea in the Gauss–Seidel procedure is that, in the very same iteration step, one simultaneously inserts these previously computed values. More specifically, the Gauss–Seidel iteration steps are given by

$$\begin{cases} x_1^{(k+1)} = \dfrac{-1}{a_{11}}[a_{12}x_2^{(k)} + a_{13}x_3^{(k)} + \cdots + a_{1n}x_n^{(k)} - b_1] \\ x_2^{(k+1)} = \dfrac{-1}{a_{22}}[a_{21}x_1^{(k+1)} + a_{23}x_3^{(k)} + \cdots + a_{2n}x_n^{(k)} - b_2] \\ \vdots \\ x_{n-1}^{(k+1)} = \dfrac{-1}{a_{n-1,n-1}}[a_{n-1,1}x_1^{(k+1)} + \cdots + a_{n-1,n-2}x_{n-2}^{(k+1)} + a_{n-1,n}x_n^{(k)} - b_{n-1}] \\ x_n^{(k+1)} = \dfrac{-1}{a_{nn}}[a_{n1}x_1^{(k+1)} + a_{n2}x_2^{(k+1)} + \cdots + a_{n,n-1}x_{n-1}^{(k+1)} - b_n], \end{cases}$$

or in a compact form in *Gauss–Seidel coordinates*, as

$$\mathbf{Ax} = \mathbf{b} \Longleftrightarrow \sum_{j=1}^{n} a_{ij}x_j = b_i \Longleftrightarrow \sum_{j=1}^{i} a_{ij}x_j + \sum_{j=i+1}^{n} a_{ij}x_j = b_i. \tag{4.2.7}$$

Therefore, the iterative form for the Gauss–Seidel method is given by

$$\begin{cases} \sum_{j=1}^{i} a_{ij}x_j^{(k+1)} = -\sum_{j=i+1}^{n} a_{ij}x_j^{(k)} + b_i \Longleftrightarrow \\ a_{ii}x_i^{(k+1)} = -\sum_{j=1}^{i-1} a_{ij}x_j^{(k+1)} - \sum_{j=i+1}^{n} a_{ij}x_j^{(k)} + b_i. \end{cases} \tag{4.2.8}$$

Example 4.5 We consider the same example as above: $\mathbf{Ax} = \mathbf{b}$ with

$$\mathbf{A} = \begin{pmatrix} 2 & -1 \\ -1 & 2 \end{pmatrix}, \quad \mathbf{x} = \begin{pmatrix} x_1 \\ x_2 \end{pmatrix}, \quad \text{and} \quad \mathbf{b} = \begin{pmatrix} 1 \\ 1 \end{pmatrix}.$$

Recall the Jacobi iteration system

$$\begin{cases} 2x_1^{(k+1)} = x_2^{(k)} + 1, \\ 2x_2^{(k+1)} = x_1^{(k)} + 1. \end{cases} \tag{4.2.9}$$

The corresponding *Gauss–Seidel* iteration system reads as follows:

$$\begin{cases} 2x_1^{(k+1)} = x_2^{(k)} + 1, \\ 2x_2^{(k+1)} = x_1^{(k+1)} + 1. \end{cases} \tag{4.2.10}$$

We choose the same initial values for x_1 and x_2 as in the Jacobi iterations, i.e. $x_1^{(0)} = 0$, and $x_2^{(0)} = 0$. Now from the first equation in (4.2.10), we have

$$2x_1^{(1)} = x_2^{(0)} + 1 \Rightarrow x_1^{(1)} = \frac{1}{2}.$$

Inserting this value of $x_1^{(1)} = \frac{1}{2}$ into the second equation in (4.2.10) yields

$$2x_2^{(1)} = x_1^{(1)} + 1 \Rightarrow 2x_2^{(1)} = \frac{1}{2} + 1 \Rightarrow x_2^{(1)} = \frac{3}{4}.$$

Below we list the first few iteration steps for this Gauss–Seidel approach:

k	$x_1^{(k)}$	$x_2^{(k)}$
0	0	0
1	1/2	3/4
2	7/8	15/16
3	31/32	63/64

Obviously, $\lim_{k\to\infty} x_1^{(k)} = \lim_{k\to\infty} x_2^{(k)} = 1$. Now with $\|\mathbf{e}_k\|_\infty = \max_{i=1,2}|x_i^{(k)} - x_i|$, we get the successive iteration errors:

$$\|\mathbf{e}_1\|_\infty = \max(|x_1^{(1)} - x_1|, |x_2^{(1)} - x_2|) = \max\left(\left|\frac{1}{2} - 1\right|, \left|\frac{3}{4} - 1\right|\right) = \frac{1}{2},$$

$$\|\mathbf{e}_2\|_\infty = \max\left(\left|\frac{7}{8} - 1\right|, \left|\frac{15}{16} - 1\right|\right) = \frac{1}{8}, \quad \|\mathbf{e}_3\|_\infty = \max\left(\frac{1}{32}, \frac{1}{64}\right) = \frac{1}{32}.$$

Thus, for the Gauss–Seidel iteration $\|\mathbf{e}_{k+1}\|_\infty = \dfrac{1}{4}\|\mathbf{e}_k\|_\infty$, where $\mathbf{e}_k$ is the error for step k, and hence, we can conclude that the Gauss–Seidel method converges faster than the Jacobi method:

$$\|\mathbf{e}_k\|_\infty = \frac{1}{4}\|\mathbf{e}_{k-1}\|_\infty = \left(\frac{1}{4}\right)^2 \|\mathbf{e}_{k-2}\|_\infty = \cdots = \left(\frac{1}{4}\right)^k \|\mathbf{e}_0\|_\infty = \left(\frac{1}{4}\right)^k.$$

4.2.4 The Successive Over-Relaxation Method (S.O.R.)

The SOR method is a modified version of the Gauss–Seidel iteration. The iteration procedure is given by

$$x_i^{(k+1)} = (1-\omega)x_i^{(k)} + \frac{\omega}{a_{ii}}\left[b_i - \sum_{j=1}^{i-1} a_{ij}x_j^{(k+1)} - \sum_{j=i+1}^{n} a_{ij}x_j^{(k)}\right]. \tag{4.2.11}$$

For $\omega > 1$, the method is called an *over-relaxation method* and if $0 < \omega < 1$, it is referred to as an *under-relaxation method*. In the *S.O.R. coordinates*, we have

$$a_{ii}x_i^{(k+1)} = a_{ii}x_i^{(k)} - \omega\left(\sum_{j=1}^{i-1} a_{ij}x_j^{(k+1)} + \sum_{j=i}^{n} a_{ij}x_j^{(k)} - b_i\right). \tag{4.2.12}$$

Remark 4.4 Note that for $\omega = 1$, (4.2.12) gives the Gauss–Seidel method.

4.2.5 Abstraction of Iterative Methods

In our procedures, we have considered $\mathbf{Ax} = \mathbf{b}$ and $\mathbf{x} = \mathbf{Bx} + \mathbf{c}$ as equivalent linear systems of equations, where $\mathbf{B}$ is the iteration matrix and $\mathbf{x}_{k+1} = \mathbf{Bx}_k + \mathbf{c}$.

Some advantages of iterative methods over direct methods are

(i) Iterative methods are faster (depends on $\mathbf{B}$, accuracy is required)
(ii) Iterative methods require less memory (sparsity of $\mathbf{A}$ can be preserved).

4.2.5.1 Questions

For a qualitative study of an iteration procedure, the following questions are of vital interest:

(Q1) For a given $\mathbf{A}$, what is a good choice for $\mathbf{B}$?
(Q2) When does $\mathbf{x}^{(k)} \to \mathbf{x}$?
(Q3) What is the rate of convergence?

The error at step k is $\mathbf{e}_k = \mathbf{x}^{(k)} - \mathbf{x}$ and that of step $(k+1)$ is $\mathbf{e}_{k+1} = \mathbf{x}^{(k+1)} - \mathbf{x}$. Then, we have $\mathbf{e}_{k+1} = \mathbf{x}^{(k+1)} - \mathbf{x} = (\mathbf{Bx}^{(k)} + \mathbf{c}) - (\mathbf{Bx} + \mathbf{c}) = \mathbf{B}(\mathbf{x}^{(k)} - \mathbf{x}) = \mathbf{Be}_k$. Iterating, we have

$$\mathbf{e}_k = \mathbf{Be}_{k-1} = \mathbf{B} \cdot \mathbf{Be}_{k-2} = \mathbf{B}^3\mathbf{e}_{k-3} = \cdots = \mathbf{B}^k\mathbf{e}_{k-k} = \mathbf{B}^k\ \mathbf{e}_0.$$

Let now

$$\mathbf{L} = \begin{pmatrix} 0 & \cdots & \cdots & 0 \\ a_{21} & 0 & \cdots & 0 \\ \vdots & \vdots & \vdots & \vdots \\ a_{n1} & \cdots & a_{n,n-1} & 0 \end{pmatrix}, \qquad \mathbf{U} = \begin{pmatrix} 0 & a_{12} & \cdots & a_{1n} \\ \vdots & \vdots & \vdots & \vdots \\ \cdots & \cdots & 0 & a_{n-1,n} \\ 0 & \cdots & \cdots & 0 \end{pmatrix}$$

and define the diagonal matrix

$$\mathbf{D} = \text{diag}(a_{11}, a_{22}, \dots, a_{nn}).$$

Then, $\mathbf{A} = \mathbf{L} + \mathbf{D} + \mathbf{U}$. We rewrite $\mathbf{Ax} = \mathbf{b}$ as $(\mathbf{L} + \mathbf{D} + \mathbf{U})\mathbf{x} = \mathbf{b}$, then $\mathbf{Dx} = -(\mathbf{L} + \mathbf{U})\mathbf{x} + \mathbf{b}$, and we may reformulate the iterative methods, viz.

4.2.6 Jacobi's Method

$$\mathbf{Dx}^{(k+1)} = -(\mathbf{L} + \mathbf{U})\mathbf{x}^{(k)} + \mathbf{b} \Rightarrow \mathbf{B}_J = -\mathbf{D}^{-1}(\mathbf{L} + \mathbf{U}),$$

where $\mathbf{B}_J$ is the *Jacobi iteration matrix.*

Example 4.6 We consider Example 4.5 and write the linear system in matrix form as $\mathbf{x} = \mathbf{B}_J\mathbf{x} + \mathbf{c}$, i.e.

$$\begin{cases} 2x_1 - x_2 = 1 \\ -x_1 + 2x_2 = 1 \end{cases} \Rightarrow \begin{cases} x_1 = \dfrac{1}{2}x_2 + \dfrac{1}{2} \\ x_2 = \dfrac{1}{2}x_1 + \dfrac{1}{2} \end{cases} \quad \text{which in matrix form is}$$

$$\begin{pmatrix} x_1 \\ x_2 \end{pmatrix} = \begin{pmatrix} 0 & \frac{1}{2} \\ \frac{1}{2} & 0 \end{pmatrix} \begin{pmatrix} x_1 \\ x_2 \end{pmatrix} + \begin{pmatrix} \frac{1}{2} \\ \frac{1}{2} \end{pmatrix}, \quad \text{where}$$

$$\mathbf{x} = \begin{pmatrix} x_1 \\ x_2 \end{pmatrix}, \quad \mathbf{B}_J = \begin{pmatrix} 0 & \frac{1}{2} \\ \frac{1}{2} & 0 \end{pmatrix} \quad \text{and} \quad \mathbf{c} = \begin{pmatrix} \frac{1}{2} \\ \frac{1}{2} \end{pmatrix}.$$

Example 4.7 Determine the same matrix $\mathbf{B}_J$ by $\mathbf{B}_J = -\mathbf{D}^{-1}(\mathbf{L} + \mathbf{U})$,

$$\mathbf{A} = \begin{pmatrix} 2 & -1 \\ -1 & 2 \end{pmatrix}, \mathbf{L} = \begin{pmatrix} 0 & 0 \\ -1 & 0 \end{pmatrix}, \mathbf{U} = \begin{pmatrix} 0 & -1 \\ 0 & 0 \end{pmatrix}, \mathbf{D} = \begin{pmatrix} 2 & 0 \\ 0 & 2 \end{pmatrix}.$$

We can easily see that

$$\mathbf{D}^{-1} = \frac{1}{2} \begin{pmatrix} 1 & 0 \\ 0 & 1 \end{pmatrix},$$

and thus

$$\mathbf{B}_J = -\mathbf{D}^{-1}(\mathbf{L} + \mathbf{U}) = -\frac{1}{2} \begin{pmatrix} 1 & 0 \\ 0 & 1 \end{pmatrix} \begin{pmatrix} 0 & -1 \\ -1 & 0 \end{pmatrix} = \frac{1}{2} \begin{pmatrix} 0 & 1 \\ 1 & 0 \end{pmatrix}.$$

4.2.7 Gauss–Seidel's Method

As in the Jacobi case, we may write $\mathbf{Ax} = \mathbf{b}$ as $(\mathbf{L} + \mathbf{D} + \mathbf{U})\mathbf{x} = \mathbf{b}$, but now we choose $(\mathbf{L} + \mathbf{D})\mathbf{x} = -\mathbf{Ux} + \mathbf{b}$. Similar to the previous procedure, we then have $(\mathbf{L} + \mathbf{D})\mathbf{x}^{(k+1)} = -\mathbf{Ux}^{(k)} + \mathbf{b}$, and hence $\mathbf{B}_{GS} = -(\mathbf{L} + \mathbf{D})^{-1}\mathbf{U}$, where $\mathbf{B}_{GS}$ stands for the *Gauss–Seidel iteration matrix*.

4.2.7.1 Relaxation

Gauss–Seidel corresponds to $(\mathbf{L} + \mathbf{D})\mathbf{x} = -\mathbf{Ux} + \mathbf{b}$, thus the iteration procedure is

$$\mathbf{Dx}^{(k+1)} = \mathbf{Dx}^{(k)} - [\mathbf{Lx}^{(k+1)} + (\mathbf{D} + \mathbf{U})\mathbf{x}^{(k)} - \mathbf{b}].$$

Now, we define the general relaxation procedure

$$\mathbf{x}^{(k+1)} = (1 - \omega)\mathbf{x}^{(k)} + \omega\mathbf{D}^{-1}(\mathbf{b} - \mathbf{Lx}^{(k+1)} - \mathbf{Ux}^{(k)}), \tag{4.2.13}$$

where the coefficient of ω, in the second term, is just the right-hand side of Gauss–Seidel's method: the term in the bracket in (4.2.11). We may rewrite (4.2.13) as

$$(\omega\mathbf{L} + \mathbf{D})\mathbf{x}^{(k+1)} = [(1 - \omega)\mathbf{D} - \omega\mathbf{U}]\mathbf{x}^{(k)} + \omega\mathbf{b},$$

where ω is the relaxation parameter, ($\omega = 1$ gives the Gauss–Seidel iteration). Thus, the *relaxation iteration matrix* is

$$\mathbf{B}_\omega = (\omega\mathbf{L} + \mathbf{D})^{-1}[(1 - \omega)\mathbf{D} - \omega\mathbf{U}].$$

4.3 Exercises

Problem 4.1 Show that $\mathbf{A} = \begin{pmatrix} 4 & 2 & 1 \\ 1 & 5 & 1 \\ 0 & 1 & 3 \end{pmatrix}$ is diagonally dominant.

Problem 4.2 Illustrate the **LU** factorization for the matrix

$$\mathbf{A} = \begin{bmatrix} 1 & 3 & 2 \\ -2 & -6 & 1 \\ 2 & 5 & 7 \end{bmatrix}.$$

Problem 4.3 Solve $\mathbf{A}^4 x = b$ for

$$\mathbf{A} = \begin{bmatrix} -1 & 2 \\ 2 & -3 \end{bmatrix} \qquad b = \begin{bmatrix} 144 \\ -233 \end{bmatrix}$$

Problem 4.4 Find the unique $\mathbf{LD\tilde{U}}$ factorization for the matrix

$$\mathbf{A} = \begin{bmatrix} 1 & 1 & -3 \\ 0 & 1 & 1 \\ 3 & -1 & 1 \end{bmatrix}.$$

Hint: First find the **LU** factorization of **A**. Then, **D** is a diagonal matrix with its diagonal elements being those of **U**. $\tilde{\mathbf{U}}$ is an upper triangular matrix obtained from **U** by replacing the diagonal elements of **U** by ones.

Problem 4.5 Show that every orthogonal 2×2 matrix is of the form

$$\mathbf{A}_1 = \begin{bmatrix} c & s \\ -s & c \end{bmatrix} \qquad \text{or} \qquad \mathbf{A}_2 = \begin{bmatrix} c & s \\ s & -c \end{bmatrix},$$

where $c^2 + s^2 = 1$

Problem 4.6 Find the **LU** factorization for the matrix **A**

$$\mathbf{A} = \begin{bmatrix} 1 & 1 & 0 & 0 \\ 1 & 3 & 1 & 0 \\ 0 & 2 & 4 & 1 \\ 0 & 0 & 3 & 5 \end{bmatrix}.$$

Problem 4.7 Solve the following system of equations

$$\begin{bmatrix} 4 & -1 \\ -1 & 4 \end{bmatrix} \begin{bmatrix} u_1 \\ u_2 \end{bmatrix} = \begin{bmatrix} 1 \\ -3 \end{bmatrix}$$

using the following iteration methods and a starting value of $u^0 = (0, 0)^T$.

(a) Jacobi Method.
(b) Gauss–Seidel Method.
(c) Optimal successive over relaxation (SOR) (you must compute the optimal value of $\omega = \omega_0$ first).

Problem 4.8 Prove strict diagonal dominance (4.2.2) implies convergence of Jacobi and Gauss–Seidel methods.
Hint: One may use *fix-point* iterations.

Problem 4.9 $\mathbf{A}$ is an orthogonal matrix if $\mathbf{A}^T\mathbf{A} = \mathbf{I}$ ($\mathbf{I}$ being the identity matrix). Show that for any vector $\mathbf{v}$ with $||\mathbf{v}||_2 = 1$, the matrix $\mathbf{I} - \mathbf{v}\mathbf{v}^T$ is orthogonal.

Problem 4.10 Suppose that $\det(a_{ij}) = 0$, but $a_{ii} \neq 0$ for all i. Prove that the Jacobi's method does not converge.

Problem 4.11 Consider the $\mathbf{N} \times \mathbf{N}$ matrix $\mathbf{A} = (a_{ij})$ defined by

$$a_{ij} = \begin{cases} -1 & |i-j| = 1 \\ 1 & i = j = 1 \\ 1 & i = j = \mathbf{N} \\ 2 & 1 < i = j < \mathbf{N} \end{cases}$$

(a) Show that $\mathbf{A}$ is positive semidefinite, i.e. $\mathbf{x}^T\mathbf{A}\mathbf{x} \geq 0$ for all $\mathbf{x} \in \mathbf{R}^N$.
(b) Show that $\mathbf{x}^T\mathbf{A}\mathbf{x} = 0$ if and only if $\mathbf{x}$ is a constant multiple of the one vector $(1, 1, \dots, 1)^T$.

Problem 4.12 We define the $\mathbf{L}_p$-norm ($p = 1, 2, \infty$) of a vector $\mathbf{v}$ as

$$\begin{aligned} ||\mathbf{v}||_1 &= \sum_{i=1}^{n} |v_i| \\ ||\mathbf{v}||_2 &= \left(\sum_{i=1}^{n} |v_i|^2\right)^{1/2} \\ ||\mathbf{v}||_\infty &= \max\{|v_i|;\ i = 1, \dots, n\}. \end{aligned}$$

Show that

$$||\mathbf{v}||_\infty \leq ||\mathbf{v}||_2 \leq ||\mathbf{v}||_1.$$

Problem 4.13 Assume that the matrix $\mathbf{B}$ is invertible. Let $\mathbf{A} = \mathbf{B}\mathbf{B}^T$ and show that $\mathbf{A}$ is symmetric and positive definite.

Problem 4.14 For an invertible matrix **A**, we define the condition number of **A** as

$$\kappa(\mathbf{A}) = \|\mathbf{A}^{-1}\|\|\mathbf{A}\|.$$

Let now $a > 0$ and

$$\mathbf{A}(a) = \begin{pmatrix} 1 & 1 & 1 \\ 2 & 1 & 1 \\ 0 & 1+a & 1 \end{pmatrix}.$$

(a) Determine the inverse and condition number (in ∞-norm) of **A** as a function of a.
(b) Write a Matlab code (use the command "*norm(***A** *,inf)*" to determine the condition number for $\mathbf{A}(a)$ for the values $a = 10^{-3},\ 0.1,\ 1,\ 2,\ 10,\ 10^3$.
(c) Which a-value (above) gives the most well-conditioned matrix.

5

Two-Point Boundary Value Problems

In this chapter, we focus on finite element approximation procedure for two-point *boundary value problems* (BVPs). For each problem, we formulate a corresponding *finite element method* (FEM): a discrete version of the *variational formulation* (VF). The well-posedness of the discrete (finite dimensional) problems is built up in the construction of the finite element strategies which yields *well-conditioned* coefficient matrices for the corresponding linear system of equations. We derive stability estimates and perform convergence analysis, proving both *a priori* and *a posteriori* error estimates in adequate energy norms.

5.1 The Finite Element Method (FEM)

Below we describe the finite element procedure for BVPs in a bounded interval I. For simplicity, we let $I := [0,1]$ and consider a partition $\mathcal{T}_h = \{0 = x_0 < x_1 < \cdots < x_M < x_{M+1} = 1\}$ of I into subintervals $I_k = [x_{k-1}, x_k]$ and set $h_k = x_k - x_{k-1}$, $k = 1, 2, \dots, M+1$, see Figure 5.1.

For this partition, we define the piecewise constant *mesh function* $h(x) := x_k - x_{k-1} = h_k$ for $x \in I_k$. Let now

$$C(I, P_1(I_k)) := \{f : f \text{ is continuous in } I \text{ and linear on each } I_k\},$$

denote the set of all continuous piecewise linear functions on the partition $\mathcal{T}_h$, and define

$$V_h^{(0)} = \{v : v \in C(I, P_1(I_k)), \quad v(0) = v(1) = 0\}. \tag{5.1.1}$$

Note that $V_h^{(0)}$ is a finite dimensional ($\dim V_h^{(0)} = M$) subspace of

$$H_0^1 = \left\{v : \int_0^1 v'(x)^2 \, dx < \infty, \quad \text{and} \quad v(0) = v(1) = 0\right\}, \tag{5.1.2}$$

An Introduction to the Finite Element Method for Differential Equations, First Edition.
Mohammad Asadzadeh.

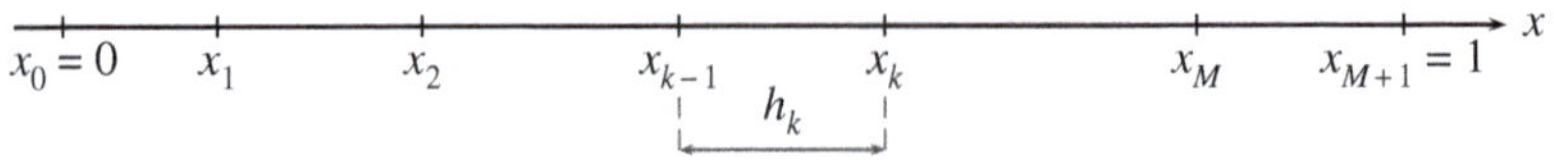

Figure 5.1 A partition of $[0, 1]$ into $M + 1$ nonuniform subintervals.

where the basis functions for $V_h^{(0)}$ are the hat functions φ_k on x_k, $k = 1, \dots, M$, due to the fact that the value at $x_0 = 0$ and $x_{M+1} = 1$ are known.

Continuous Galerkin of degree 1, cG(1). In Chapter 2, we considered the Dirichlet boundary value problem $(\text{BVP})_1$, (with $a(x) > 0$),

$$(\text{BVP})_1 \quad -(a(x)u'(x))' = f(x), \quad 0 < x < 1 \qquad u(0) = u(1) = 0, \tag{5.1.3}$$

and introduced its corresponding variational formulation $(\text{VF})_1$: find the continuous (exact) solution $u \in H_0^1$ such that

$$(\text{VF})_1 \quad \int_0^1 a(x)u'(x)v'(x)\,dx = \int_0^1 f(x)v(x)\,dx, \quad \forall v \in H_0^1. \tag{5.1.4}$$

A finite element formulation for problem (5.1.3) is given by the following: find the discrete solution $u_h \in V_h^{(0)}$ such that the following discrete version of the variational formulation holds true

$$(\text{FEM}) \quad \int_0^1 a(x)u_h'(x)v'(x)\,dx = \int_0^1 f(x)v(x)\,dx, \quad \forall v \in V_h^{(0)}. \tag{5.1.5}$$

The FEM is a finite dimensional version of the variational formulation $(\text{VF})_1$, where the test (also trial) functions are in a finite dimensional subspace $V_h^{(0)}$, of H_0^1, spanned by the hat-functions, $\varphi_j(x)$, $j = 1, \dots, M$.. Thus, if in $(\text{VF})_1$, we restrict v to $V_h^{(0)}$ (rather than H_0^1) and subtract FEM from it, we get the *Galerkin orthogonality* relation:

$$\int_0^1 a(x)(u'(x) - u_h'(x))v'(x)\,dx = 0, \quad \forall v \in V_h^{(0)}. \tag{5.1.6}$$

Now the purpose is to *estimate the error* arising in approximating the solution for $(\text{BVP})_1$ (a similar procedure is applied for $(\text{BVP})_2$) by functions in $V_h^{(0)}$. To this end, we need to introduce the *energy norm*: a weighted version of the L_2-norm for the first derivative, as

Weighted L_2-norm $\qquad \|v\|_a := \left(\int_0^1 a(x)|v(x)|^2\,dx\right)^{1/2}, \quad a(x) > 0$

Energy-norm $\qquad \|v\|_E := \left(\int_0^1 a(x)|v'(x)|^2\,dx\right)^{1/2},$

Note that $\qquad \|v\|_E = \|v'\|_a.$

Here $\|v\|_E$ describes the *internal elastic energy* of the elastic bar modeled by our Dirichlet boundary value problem $(\text{BVP})_1$.

5.2 Error Estimates in the Energy Norm

We shall study two types of error estimates:

(i) An *a priori error estimate*, where a certain norm of the error is estimated by some norm of the *exact solution u*. Here, the error analysis gives information about the certain "size" of the error, depending on an adequate "size" of (unknown) exact solution u, before any computational steps.
(ii) An a *posteriori error estimate*, where a certain norm of the error is estimated by some norm of the *residual* of the approximate solution. Recall that the residual is the difference between the left- and right-hand side of the differential equation when the exact solution $u(x)$ is replaced by its approximation $u_h(x)$. Hence, a posteriori error estimates give quantitative information about the size of the error after that the approximate solution $u_h(x)$ has been computed.

Below, in the a priori part, first, we shall prove a qualitative result which shows that the finite element solution is the best approximate solution, in the underlying finite element space, to the Dirichlet problem in the energy norm. Then, we continue to quantify "how best is this approximation" in the sense that what is the *optimal convergence* rate due to the maximal available regularity of the exact solution?

Theorem 5.1 *Let $u(x)$ be the solution to the Dirichlet BVP (5.1.3) and $u_h(x)$ its finite element approximation given by (5.1.5), then*

$$\|u - u_h\|_E \le \|u - v\|_E, \quad \forall v(x) \in V_h^{(0)}. \tag{5.2.1}$$

This means that the finite element solution $u_h \in V_h^{(0)}$ is the best approximation of the solution u, in the energy norm, by functions in $V_h^{(0)}$.

Proof: We take an arbitrary $v \in V_h^{(0)}$, then using the energy norm, we have

$$\begin{aligned}
\|u - u_h\|_E^2 &= \int_0^1 a(x)(u'(x) - u_h'(x))^2\, dx \\
&= \int_0^1 a(x)(u'(x) - u_h'(x))(u'(x) - v'(x) + v'(x) - u_h'(x))\, dx \\
&= \int_0^1 a(x)(u'(x) - u_h'(x))(u'(x) - v'(x))\, dx \\
&\quad + \int_0^1 a(x)(u'(x) - u_h'(x))(v'(x) - u_h'(x))\, dx.
\end{aligned} \tag{5.2.2}$$

Since $v - u_h \in V_h^{(0)} \subset H_0^1$, by Galerkin orthogonality (5.1.6), the last integral is zero. Thus, since $a(x) > 0$,

$$\begin{aligned}
\|u - u_h\|_E^2 &= \int_0^1 a(x)(u'(x) - u_h'(x))(u'(x) - v'(x))\, dx \\
&= \int_0^1 a(x)^{\frac{1}{2}}(u'(x) - u_h'(x))a(x)^{\frac{1}{2}}(u'(x) - v'(x))dx \\
&\leq \left(\int_0^1 a(x)(u'(x) - u_h'(x))^2\, dx\right)^{\frac{1}{2}} \left(\int_0^1 a(x)(u'(x) - v'(x))^2\, dx\right)^{\frac{1}{2}} \\
&= \|u - u_h\|_E \cdot \|u - v\|_E,
\end{aligned}$$

where, in the last estimate, we used Cauchy–Schwarz' inequality. Hence,

$$\|u - u_h\|_E \leq \|u - v\|_E, \quad \forall v \in V_h^{(0)}, \tag{5.2.3}$$

and the proof is complete. □

The next step is to show that there exists a function $v \in V_h^{(0)}$ such that $\|u - v\|_E$ is not *too large*. The function that we have in mind is $\pi_h u(x)$: the *piecewise linear interpolant of* $u(x)$, introduced in Chapter 3.

Theorem 5.2 (An a priori error estimate) *Let u and u_h be the solutions of the Dirichlet problem $(BVP)_1$ and the FEM, respectively. Then, there exists an interpolation constant C_i, depending only on $a(x)$, such that*

$$\|u - u_h\|_E \leq C_i \|hu''\|_a. \tag{5.2.4}$$

Proof: Since $\pi_h u(x) \in V_h^{(0)}$, we may take $v = \pi_h u(x)$ in (5.2.1) and use, e.g. the second estimate in the interpolation Theorem 3.3 (slightly generalized to the weighted norm $\|\cdot\|_a$, see Remark 5.1) to get

$$\begin{aligned}
\|u - u_h\|_E &\leq \|u - \pi_h u\|_E = \|u' - (\pi_h u)'\|_a \\
&\leq C_i \|hu''\|_a = C_i \left(\int_0^1 a(x)h^2(x)u''(x)^2\, dx\right)^{1/2},
\end{aligned} \tag{5.2.5}$$

which is the desired result and the proof is complete. □

Remark 5.1 The interpolation theorem is not in the weighted norm. The $a(x)$ dependence of the interpolation constant C_i can be shown as follows:

$$\begin{aligned}
\|u' - (\pi_h u)'\|_a &= \left(\int_0^1 a(x)(u'(x) - (\pi_h u)'(x))^2\, dx\right)^{1/2} \\
&\leq \left(\max_{x\in[0,1]} a(x)^{1/2}\right) \cdot \|u' - (\pi_h u)'\|_{L_2} \leq c_i \left(\max_{x\in[0,1]} a(x)^{1/2}\right) \|hu''\|_{L_2}
\end{aligned}$$

$$= c_i \left(\max_{x\in[0,1]} a(x)^{1/2} \right) \left(\int_0^1 h(x)^2 u''(x)^2 \, dx \right)^{1/2}$$

$$\leq c_i \frac{\left(\max_{x\in[0,1]} a(x)^{1/2} \right)}{\left(\min_{x\in[0,1]} a(x)^{1/2} \right)} \cdot \left(\int_0^1 a(x) h(x)^2 u''(x)^2 \, dx \right)^{1/2}.$$

Thus,

$$C_i = c_i \frac{(\max_{x\in[0,1]} a(x)^{1/2})}{(\min_{x\in[0,1]} a(x)^{1/2})}, \tag{5.2.6}$$

where $c_i = c_2$ is the interpolation constant in the second estimate in Theorem 3.3.

Remark 5.2 If the objective is to divide $[0, 1]$ into a finite number of subintervals, then one can use the result of Theorem 5.2 to obtain an optimal partition of $[0, 1]$, where whenever $a(x)u''(x)^2$ gets large, we compensate by making $h(x)$ smaller. This, however, "requires that the *exact solution $u(x)$ is known*." [1] Now, we shall study *a posteriori error analysis*, which instead of the unknown solution $u(x)$, uses the residual of the computed solution $u_h(x)$.

Theorem 5.3 An a posteriori error estimate. *There is an interpolation constant c_i depending only on $a(x)$ such that the error in the finite element approximation of the Dirichlet BVP (5.1.3), satisfies*

$$\|e\|_E \leq C_i \left(\int_0^1 \frac{1}{a(x)} h^2(x) R^2(u_h(x)) \, dx \right)^{1/2}, \tag{5.2.7}$$

where $R(u_h(x)) = f + (a(x)u_h'(x))'$ is the residual, and $e := u - u_h \in H_0^1$.

Proof: By the definition of the *energy norm*, we have

$$\|e\|_E^2 = \int_0^1 a(x)(e'(x))^2 \, dx = \int_0^1 a(x)(u'(x) - u_h'(x))e'(x) \, dx$$

$$= \int_0^1 a(x)u'(x)e'(x) \, dx - \int_0^1 a(x)u_h'(x)e'(x) \, dx. \tag{5.2.8}$$

Since $e \in H_0^1$ the variational formulation $(VF)_1$ gives that

$$\int_0^1 a(x)u'(x)e'(x) \, dx = \int_0^1 f(x)e(x) \, dx. \tag{5.2.9}$$

---

1 Note that when a is a given constant then, $-u''(x) = (1/a)f(x)$ is known.

Hence, we can write

$$\|e\|_E^2 = \int_0^1 f(x)e(x)\,dx - \int_0^1 a(x)u_h'(x)e'(x)\,dx. \tag{5.2.10}$$

Adding and subtracting the interpolant $\pi_h e(x)$ and its derivative $(\pi_h e)'(x)$ to e and e' in the integrands above yields

$$\|e\|_E^2 = \int_0^1 f(x)(e(x) - \pi_h e(x))\,dx + \underbrace{\int_0^1 f(x)\pi_h e(x)\,dx}_{\text{(i)}}$$
$$- \int_0^1 a(x)u_h'(x)(e'(x) - (\pi_h e)'(x))\,dx - \underbrace{\int_0^1 a(x)u_h'(x)(\pi_h e)'(x)\,dx.}_{\text{(ii)}}$$

Since $u_h(x)$ is the solution of the (FEM) given by (5.1.5) and $\pi_h e(x) \in V_h^{(0)}$ we have that (i)–(ii)= 0. Hence,

$$\|e\|_E^2 = \int_0^1 f(x)(e(x) - \pi_h e(x))\,dx - \int_0^1 a(x)u_h'(x)(e'(x) - (\pi_h e)'(x))\,dx$$
$$= \int_0^1 f(x)(e(x) - \pi_h e(x))\,dx - \sum_{k=1}^{M+1} \int_{x_{k-1}}^{x_k} a(x)u_h'(x)(e'(x) - (\pi_h e)'(x))\,dx.$$

To continue, we integrate by parts in the integrals in the summation above:

$$- \int_{x_{k-1}}^{x_k} a(x)u_h'(x)(e'(x) - (\pi_h e)'(x))\,dx$$
$$= -[a(x)u_h'(x)(e(x) - \pi_h e(x))]_{x_{k-1}}^{x_k} + \int_{x_{k-1}}^{x_k} (a(x)u_h'(x))'(e(x) - \pi_h e(x))\,dx.$$

Now, using $e(x_k) = \pi_h e(x_k),\ \ k = 0, 1, \ldots, M + 1$, where the x_k's are the interpolation nodes, the boundary terms vanish and thus, we end up with

$$- \int_{x_{k-1}}^{x_k} a(x)u_h'(x)(e'(x) - (\pi_h e)'(x))\,dx = \int_{x_{k-1}}^{x_k} (a(x)u_h'(x))'(e(x) - \pi_h e(x))\,dx.$$

Thus, summing over k, we have

$$- \int_0^1 a(x)u_h'(x)(e'(x) - (\pi_h e)'(x))\,dx = \int_0^1 (a(x)u_h'(x))'(e(x) - \pi_h e(x))\,dx,$$

where $(a(x)u_h'(x))'$ should be interpreted locally on each subinterval $[x_{k-1}, x_k]$. (Since $u_h'(x)$ in general is discontinuous, $u_h''(x)$ does not exist globally on $[0, 1]$. In the case of linear approximation, it is just $\sum_{j=1}^{M} d_j a(x)|_{I_j}$, with piecewise constants $d_j := u_h'|_{I_j}$.)

Therefore,

$$\|e\|_E^2 = \int_0^1 f(x)(e(x) - \pi_h e(x))\,dx + \int_0^1 (a(x)u_h'(x))'(e(x) - \pi_h e(x))\,dx$$
$$= \int_0^1 \{f(x) + (a(x)u_h'(x))'\}(e(x) - \pi_h e(x))\,dx.$$

Now, let $R(u_h(x)) = f(x) + (a(x)u_h'(x))'$, i.e. $R(u_h(x))$ is the *residual error*, which is a well-defined function except in the set $\{x_k\}$, $k = 1, \dots, M$; where $(a(x_k)u_h'(x_k))'$ is not defined. Then, using Cauchy–Schwarz' inequality we get the following estimate

$$\begin{aligned}
\|e\|_E^2 &= \int_0^1 R(u_h(x))(e(x) - \pi_h e(x))\, dx \\
&= \int_0^1 \frac{1}{\sqrt{a(x)}} h(x) R(u_h(x)) \cdot \sqrt{a(x)} \left(\frac{e(x) - \pi_h e(x)}{h(x)} \right) dx \\
&\le \left(\int_0^1 \frac{1}{a(x)} h^2(x) R^2(u_h(x))\, dx \right)^{1/2} \left(\int_0^1 a(x) \left(\frac{e(x) - \pi_h e(x)}{h(x)} \right)^2 dx \right)^{1/2}.
\end{aligned}$$

Further, by the definition of the weighted L_2-norm we have,

$$\left\| \frac{e - \pi_h e}{h} \right\|_a = \left(\int_0^1 a(x) \left(\frac{e(x) - \pi_h e(x)}{h(x)} \right)^2 dx \right)^{1/2}. \tag{5.2.11}$$

To estimate (5.2.11), we can use the third interpolation estimate (in Theorem 3.3) for $e(x)$ in each subinterval and get

$$\left\| \frac{e - \pi_h e}{h} \right\|_a \le C_i \|e'\|_a = C_i \|e\|_E, \tag{5.2.12}$$

where C_i, as above, depends on $a(x)$. Thus,

$$\|e\|_E^2 \le \left(\int_0^1 \frac{1}{a(x)} h^2(x) R^2(u_h(x))\, dx \right)^{1/2} \cdot C_i \|e(x)\|_E, \tag{5.2.13}$$

and the proof is complete. □

Remark 5.3 The detailed derivation of (5.2.12) is similar to that of the a priori case in Remark 5.1. We include it for the convenience of the reader:

$$\begin{aligned}
\left\| \frac{e - \pi_h e}{h} \right\|_a &= \left(\sum_{j=1}^{M+1} \int_{I_j} a(x) \left(\frac{e(x) - \pi_h e(x)}{h_j} \right)^2 dx \right)^{1/2} \\
&\le \max_{x \in [0,1]} a(x)^{1/2} \left(\sum_{j=1}^{M+1} \frac{1}{h_j^2} \int_{I_j} (e(x) - \pi_h e(x))^2\, dx \right)^{1/2} \\
&\le \max_{x \in [0,1]} a(x)^{1/2} \left(\sum_{j=1}^{M+1} \frac{1}{h_j^2} \cdot c_i^2 \cdot \int_{I_j} (h_j e'(x))^2\, dx \right)^{1/2} \\
&\le c_i \frac{\max_{x \in [0,1]} a(x)^{1/2}}{\min_{x \in [0,1]} a(x)^{1/2}} \|e'\|_a.
\end{aligned}$$

5.2.1 Adaptivity

Below we briefly outline the adaptivity procedure based on the a posteriori error estimate which uses the *approximate* solution and which can be used for mesh-refinements. Loosely speaking, the estimate (5.2.7) predicts local mesh refinement, i.e. indicates the regions (subintervals) which should be subdivided further. More specifically, the idea is as follows: assume that one seeks an error less than a given error tolerance TOL > 0:

$$||e||_E \leq \text{TOL}. \tag{5.2.14}$$

Then, one may use the following steps as a mesh refinement strategy:

(i) Make an initial partition of the interval with the mesh function h.
(ii) Compute the corresponding FEM solution $u_h(x)$ and residual $R(u_h(x))$.
(iii) If $||e||_E > \text{TOL}$, refine the mesh in the subdomains (subintervals), where $\frac{1}{a(x)}R^2(u_h(x))$ is large and perform the steps (ii) and (iii) with this new mesh (note that $h = h(x) = |I_k|$ on each subinterval $I_k = (x_{k-1}, x_k)$, $k = 1, \ldots, M+1$).

5.3 FEM for Convection–Diffusion–Absorption BVPs

We now return to the Galerkin approximation of a solution to the BVPs and give a framework for the cG(1) (*continuous Galerkin of degree 1*) finite element procedure leading to a linear system of equations of the form $\mathbf{A}\xi = \mathbf{b}$. More specifically, we shall extend our previous approach for the stationary heat equation to cases involving *absorption* and/or *convection* terms. We also consider nonhomogeneous Dirichlet boundary conditions. We illustrate this procedure through the Examples 5.1 and 5.2.

Example 5.1 Determine the coefficient matrix and load vector for the cG(1) finite element approximation of the BVP

$$-u''(x) + 4u(x) = 0, \quad 0 < x < 1; \qquad u(0) = \alpha \neq 0, \quad u(1) = \beta \neq 0,$$

on a uniform partition $\mathcal{T}_h$ of the interval [0, 1] into $n+1$ subintervals.

Solution: The problem is to construct an approximate solution u_h in a finite dimensional space spanned by the continuous, piecewise linear basis functions (hat-functions) $\varphi_j(x)$, $j = 0, 1, \ldots, n+1$ on the partition $\mathcal{T}_h$. This results in a discrete problem represented by a linear system of equations $\mathbf{A}\xi = \mathbf{b}$, for the unknown $\xi = \{c_j\}_{j=1}^n$. ($c_0 = \alpha$ and $c_{n+1} = \beta$ are given as boundary data.)

The continuous solution is assumed to be in the Hilbert space

$$H^1 = \left\{ w : \int_0^1 (w(x)^2 + w'(x)^2)\, dx < \infty \right\}.$$

Since both $u(0) = \alpha$ och $u(1) = \beta$ are given, we need to take the trial functions in

$$V := \{w \in H^1 : \quad w(0) = \alpha, \quad w(1) = \beta\},$$

and the test functions in

$$V^0 := H_0^1 = \{w \in H^1 : \quad w(0) = w(1) = 0\}.$$

We multiply the partial differential equation (PDE) by a test function $v \in V^0$ and integrate over $(0, 1)$. Integrating by parts, we get

$$-u'(1)v(1) + u'(0)v(0) + \int_0^1 u'v'\, dx + 4\int_0^1 uv\, dx = 0 \iff$$

$$\text{(VF)}: \quad \text{Find} \quad u \in V \quad \text{so that} \quad \int_0^1 u'v'\, dx + 4\int_0^1 uv\, dx = 0, \quad \forall v \in V^0.$$

The partition $\mathcal{T}_h$, of $[0, 1]$ into $n + 1$ uniform subintervals $I_1 = [0, h]$, $I_2 = [h, 2h]$, …, and $I_{n+1} = [nh, (n+1)h]$, is also described by the nodes $x_0 = 0, x_1 = h, \dots, x_n = nh$ and $x_{n+1} = (n+1)h = 1$. The corresponding discrete function spaces are (varying with h and hence, with n),

$$V_h := \{w_h : w_h \text{ is piecewise linear, continuous on } \mathcal{T}_h,\ w_h(0) = \alpha,\ w_h(1) = \beta\},$$

and

$$V_h^0 := \{v_h : v_h \text{ is piecewise linear and continuous on } \mathcal{T}_h, \quad v_h(0) = v_h(1) = 0\}.$$

Note that here, the basis functions needed to represent functions in V_h are the hat-functions φ_j, $j = 0, \dots, n+1$, as in Figure 5.2. (including the two *half-hat-functions* φ_0 and φ_{n+1}, Thus, $\dim V_h = n + 2$), whereas the basis functions describing V_h^0 are φ_i :s for $i = 1, \dots, n$, i.e. all *full-hat-functions* but not φ_0 and φ_{n+1} (hence, $\dim V_h^0 = n$). Thus, the test and trial function spaces are different. This is due to the fact that the values $u(0) = \alpha$ and $u(1) = \beta$ are given and therefore, we do not need to determine those two nodal values approximately, rather transfer their contributions to the right hand side as data.

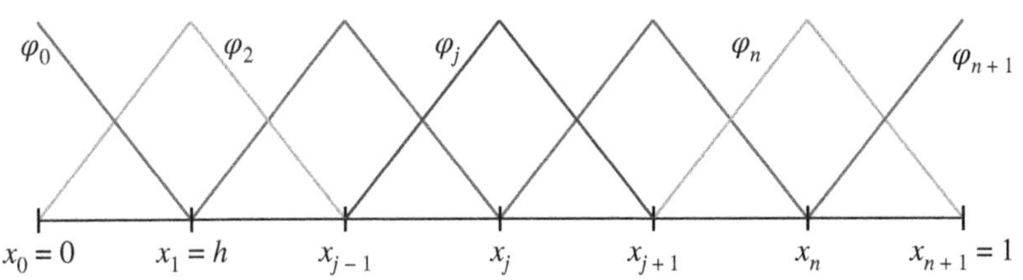

Figure 5.2 The basis functions $\varphi_0, \dots, \varphi_{n+1}$.

Now, the finite element formulation is: find $u_h \in V_h$ such that

$$\text{(FEM)} \quad \int_0^1 u_h' v' \, dx + 4 \int_0^1 u_h v \, dx = 0, \quad \forall v \in V_h^0.$$

We have that $u_h(x) = c_0\varphi_0(x) + \sum_{j=1}^n c_j\varphi_j(x) + c_{n+1}\varphi_{n+1}(x)$, where $c_0 = \alpha$, $c_{n+1} = \beta$ and

$$\varphi_0(x) = \frac{1}{h}\begin{cases} h - x & 0 \le x \le h \\ 0, & \text{else} \end{cases}, \quad \varphi_j(x) = \frac{1}{h}\begin{cases} x - x_{j-1}, & x_{j-1} \le x \le x_j \\ x_{j+1} - x & x_j \le x \le x_{j+1} \\ 0 & x \notin [x_{j-1}, x_{j+1}]. \end{cases}$$

and

$$\varphi_{n+1}(x) = \frac{1}{h}\begin{cases} x - x_n & nh \le x \le (n+1)h \\ 0, & \text{else}. \end{cases}$$

Inserting u_h into (FEM), and choosing $v = \varphi_i(x)$, $i = 1, \dots, n$, we end up with

$$\begin{aligned}
&\sum_{j=1}^{n} \left(\int_0^1 \varphi_j'(x)\varphi_i'(x) \, dx + 4 \int_0^1 \varphi_j(x)\varphi_i(x) \, dx \right) c_j \\
&= -\left(\int_0^1 \varphi_0'(x)\varphi_i'(x) \, dx + 4 \int_0^1 \varphi_0(x)\varphi_i(x) \, dx \right) c_0 \\
&\quad - \left(\int_0^1 \varphi_{n+1}'(x)\varphi_i'(x) \, dx + 4 \int_0^1 \varphi_{n+1}(x)\varphi_i(x) \, dx \right) c_{n+1}.
\end{aligned}$$

In matrix form, this corresponds to $\mathbf{A}\xi = \mathbf{b}$ with $\mathbf{A} = \mathbf{S} + 4\mathbf{M}$, where $\mathbf{S} = \mathbf{A}_{\text{unif}}$ is the, previously computed, stiffness matrix:

$$\mathbf{S} = \frac{1}{h}\begin{bmatrix} 2 & -1 & 0 & 0 & \cdots & 0 \\ -1 & 2 & -1 & 0 & \cdots & 0 \\ 0 & -1 & 2 & -1 & \cdots & 0 \\ \cdots & \cdots & \cdots & \cdots & \cdots & \cdots \\ 0 & \cdots & \cdots & -1 & 2 & -1 \\ 0 & \cdots & \cdots & \cdots & -1 & 2 \end{bmatrix}, \tag{5.3.1}$$

and $\mathbf{M}$ is the mass-matrix: given by

$$\mathbf{M} = \begin{bmatrix} \int_0^1 \varphi_1\varphi_1 & \int_0^1 \varphi_2\varphi_1 & \cdots & \int_0^1 \varphi_n\varphi_1 \\ \int_0^1 \varphi_1\varphi_2 & \int_0^1 \varphi_2\varphi_2 & \cdots & \int_0^1 \varphi_n\varphi_2 \\ \vdots & \vdots & \cdots & \vdots \\ \int_0^1 \varphi_1\varphi_n & \int_0^1 \varphi_2\varphi_n & \cdots & \int_0^1 \varphi_n\varphi_n \end{bmatrix}. \tag{5.3.2}$$

Note the index locations in the matrices $\mathbf{S}$ and $\mathbf{M}$:

$$s_{ij} = \int_0^1 \varphi_j'(x)\varphi_i'(x) \, dx, \qquad m_{ij} = \int_0^1 \varphi_j(x)\varphi_i(x) \, dx.$$

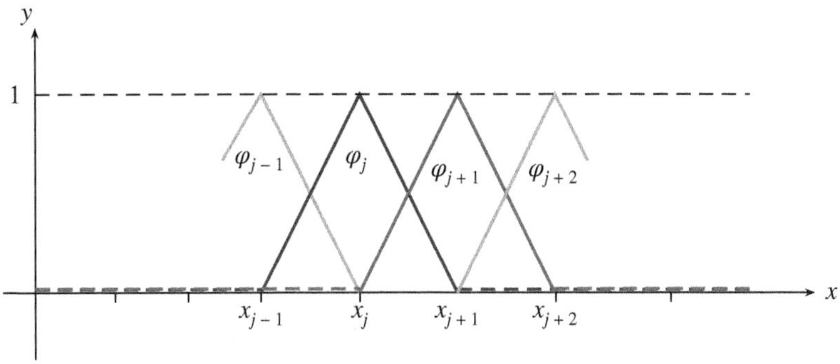

Figure 5.3 φ_{j-1}, φ_j, φ_{j+1}, and φ_{j+2}.

This, however, does not make any difference in the current example, since as seen, both **S** and **M** are symmetric. To compute the entries of **M**, we follow the same procedure as in Chapter 3 and notice that, as **S**, also **M** is symmetric and its elements m_{ij} are (see Figure 5.3),

$$m_{ij} = m_{ji} = \begin{cases} \int_0^1 \varphi_i \varphi_j \, dx = 0, \quad \forall i,j & \text{with } |i-j| > 1 \\ \int_0^1 \varphi_j^2(x) \, dx, & \text{for } i = j \\ \int_0^1 \varphi_j(x)\varphi_{j+1}(x) \, dx, & \text{for } i = j+1. \end{cases} \tag{5.3.3}$$

The diagonal elements are

$$\begin{aligned} m_{jj} &= \int_0^1 \varphi_j(x)^2 \, dx = \frac{1}{h^2}\left(\int_{x_{j-1}}^{x_j} (x - x_{j-1})^2 \, dx + \int_{x_j}^{x_{j+1}} (x_{j+1} - x)^2 \right) dx \\ &= \frac{1}{h^2}\left[\frac{(x - x_{j-1})^3}{3} \right]_{x_{j-1}}^{x_j} - \frac{1}{h^2}\left[\frac{(x_{j+1} - x)^3}{3} \right]_{x_j}^{x_{j+1}} \\ &= \frac{1}{h^2} \cdot \frac{h^3}{3} + \frac{1}{h^2} \cdot \frac{h^3}{3} = \frac{2}{3}h, \quad j = 1, \dots, n, \end{aligned} \tag{5.3.4}$$

and the two super- and subdiagonals can be computed as

$$\begin{aligned} m_{j,j+1} = m_{j+1,j} &= \int_0^1 \varphi_j \varphi_{j+1} \, dx = \frac{1}{h^2} \int_{x_j}^{x_{j+1}} (x_{j+1} - x)(x - x_j) \, dx = [\mathrm{PI}] \\ &= \frac{1}{h^2}\left[(x_{j+1} - x)\frac{(x - x_j)^2}{2} \right]_{x_j}^{x_{j+1}} - \frac{1}{h^2} \int_{x_j}^{x_{j+1}} -\frac{(x - x_j)^2}{2} \, dx \\ &= \frac{1}{h^2}\left[\frac{(x - x_j)^3}{6} \right]_{x_j}^{x_{j+1}} = \frac{1}{6}h, \quad j = 1, \dots, n-1. \end{aligned}$$

Thus, the mass matrix for this case is

$$\mathbf{M} = h\begin{bmatrix} \frac{2}{3} & \frac{1}{6} & 0 & 0 \cdots & 0 \\ \frac{1}{6} & \frac{2}{3} & \frac{1}{6} & 0 \cdots & 0 \\ 0 & \frac{1}{6} & \frac{2}{3} & \frac{1}{6} \cdots & 0 \\ \cdots & \cdots & \cdots & \cdots & \cdots \\ 0 \cdots\cdots & & \frac{1}{6} & \frac{2}{3} & \frac{1}{6} \\ 0 \cdots\cdots & & \cdots & \frac{1}{6} & \frac{2}{3} \end{bmatrix} = \frac{h}{6}\begin{bmatrix} 4 & 1 & 0 & 0 \cdots & 0 \\ 1 & 4 & 1 & 0 \cdots & 0 \\ 0 & 1 & 4 & 1 \cdots & 0 \\ \cdots & \cdots & \cdots & \cdots & \cdots \\ 0 \cdots\cdots & & 1 & 4 & 1 \\ 0 \cdots\cdots & & \cdots & 1 & 4 \end{bmatrix}.$$

Hence, for $i, j = 1, \dots, n$, the coefficient matrix $\mathbf{A} = \mathbf{S} + 4\mathbf{M}$ is given by

$$[\mathbf{A}]_{ij} = \int_0^1 \varphi_i' \varphi_j' \, dx + 4\int_0^1 \varphi_i \varphi_j(x) \, dx = \begin{cases} \frac{2}{h} + \frac{8h}{3}, & i = j, \\ -\frac{1}{h} + \frac{2h}{3}, & |i - j| = 1, \\ 0 & \text{else.} \end{cases}$$

Finally, with $c_0 = \alpha$ och $c_{n+1} = \beta$, we get entries of the load vector as

$$b_1 = -\left(-\frac{1}{h} + \frac{2h}{3}\right) c_0 = \alpha \left(\frac{1}{h} - \frac{2h}{3}\right),$$

$$b_2 = \cdots = b_{n-1} = 0,$$

$$b_n = -\left(-\frac{1}{h} + \frac{2h}{3}\right) c_{n+1} = \beta \left(\frac{1}{h} - \frac{2h}{3}\right).$$

Now, for each particular choice of h (i.e. n), α and β we may solve $\mathbf{A}\xi = \mathbf{b}$ to obtain the nodal values of the approximate solution u_h at the inner nodes x_j, $j = 1, \dots, n$. That is, $\xi = (c_1, \dots, c_n)^T := (u_h(x_1), \dots, u_h(x_n))^T$. Connecting the points $(x_j, u_h(x_j))$, $j = 0, \dots, n+1$ by straight lines, we obtain the desired, continuous, piecewise linear approximation of the solution u.

Remark 5.4 An easier way to compute the above integrals $m_{j,j+1}$ (as well as m_{jj}) is through Simpson's rule, which is exact for polynomials of degree ≤ 2. Since $\varphi_j(x)\varphi_{j+1}(x) = 0$ at $x = x_j$ and $x = x_{j+1}$, we need to evaluate only the midterm of the Simpson's formula, i.e.

$$\int_0^1 \varphi_j \varphi_{j+1} \, dx = 4 \cdot \frac{h}{6} \cdot \varphi_j\left(\frac{x_j + x_{j+1}}{2}\right) \cdot \varphi_{j+1}\left(\frac{x_j + x_{j+1}}{2}\right) = 4 \cdot \frac{h}{6} \cdot \frac{1}{2} \cdot \frac{1}{2} = \frac{h}{6}.$$

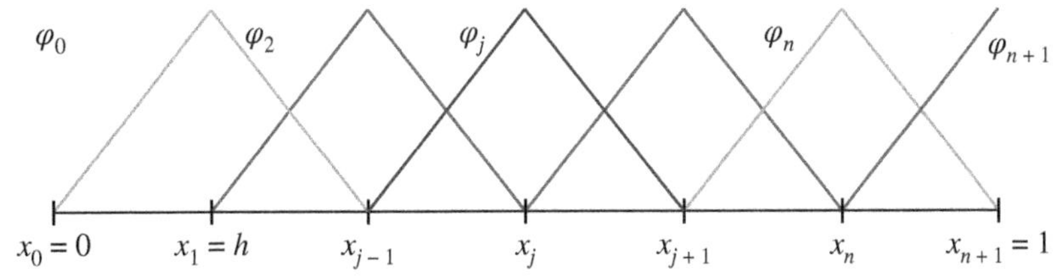

Figure 5.4 The basis test and trial functions for this problem.

For a uniform partition, one may use $\varphi_0 = 1 - x/h$ and $\varphi_1 = x/h$ on $(0, h)$:

$$\int_0^1 \varphi_0\varphi_1\, dx = \int_0^h \left(1 - \frac{x}{h}\right)\frac{x}{h}\, dx = \left[\left(1 - \frac{x}{h}\right)\frac{x^2}{2h}\right]_0^h - \int_0^h \frac{(-1)}{h}\cdot\frac{x^2}{2h}\, dx = \frac{h}{6}.$$

Example 5.2 Below we consider a *convection–diffusion problem*:

$$-\varepsilon u''(x) + pu'(x) = r, \quad 0 < x < 1; \qquad u(0) = 0, \quad u'(1) = \beta \neq 0,$$

where ε and p are positive real numbers and $r \in \mathbf{R}$. Here $-\varepsilon u''$ is the diffusion term, pu' corresponds to convection, and r is a given (here for simplicity a constant) source ($r > 0$) or sink ($r < 0$). We would like to answer the same question as in Example 5.1.

Solution: This time with $c_0 = u(0) = 0$. Then, the test function at $x = 0$; φ_0 will not be necessary. But since $u(1)$ is not given, we shall need the test function at $x = 1$: φ_{n+1}. Then the *trial and test function spaces are the same* (as in Figure 5.4):

$$V := \left\{ w : \int_0^1 (w(x)^2 + w'(x)^2)\, dx < \infty, \quad \text{and} \quad w(0) = 0 \right\}.$$

We multiply the PDE by a test function $v \in V$ and integrate over $(0, 1)$. Then, integration by parts yields

$$-\varepsilon u'(1)v(1) + \varepsilon u'(0)v(0) + \varepsilon \int_0^1 u'v'\, dx + p\int_0^1 u'v\, dx = r\int_0^1 v\, dx.$$

Hence, we end up with the variational formulation: find $u \in V$ such that

$$\text{(VF)} \quad \varepsilon \int_0^1 u'v'\, dx + p\int_0^1 u'v\, dx = r\int_0^1 v\, dx + \varepsilon\beta v(1), \quad \forall v \in V.$$

The corresponding discrete test and trial function space is

$$V_h^0 := \{w_h : w_h \text{ is piecewise linear and continuous on } \mathcal{T}_h, \text{ and } w_h(0) = 0\}.$$

Thus, the basis functions for V_h^0 are the hat-functions $\varphi_j, j = 1, \dots, n+1$ (including the *half-hat-function* φ_{n+1}), and hence, $\dim(V_h^0) = n + 1$.

Now, the finite element formulation reads as follows: find $u_h \in V_h^0$ such that

$$\text{(FEM)} \quad \varepsilon \int_0^1 u_h' v' \, dx + p \int_0^1 u_h' v \, dx = r \int_0^1 v \, dx + \varepsilon \beta v(1), \quad \forall v \in V_h^0.$$

Inserting the ansatz $u_h(x) = \sum_{j=1}^{n+1} \xi_j \varphi_j(x)$ into (FEM), and choosing the test functions as $v = \varphi_i(x)$, $i = 1, \dots, n+1$, we get

$$\sum_{j=1}^{n+1} \left(\varepsilon \int_0^1 \varphi_j'(x)\varphi_i'(x) \, dx + p \int_0^1 \varphi_j'(x)\varphi_i(x) \, dx \right) \xi_j = r \int_0^1 \varphi_i(x) \, dx + \varepsilon \beta \varphi_i(1),$$

In matrix form, this corresponds to the linear system of equations $\mathbf{A}\xi = \mathbf{b}$ with $\mathbf{A} = \varepsilon\, \tilde{\mathbf{S}} + p\, \mathbf{C}$, where $\tilde{\mathbf{S}}$ is computed as $\mathbf{A}_{unif}$ and is the $(n+1) \times (n+1)$-stiffness matrix *with its last diagonal element*, corresponding to the half-hat-function φ_{n+1}, i.e. as $\tilde{s}_{n+1,n+1} = \int_0^1 \varphi_{n+1}'(x)\varphi_{n+1}'(x) \, dx = 1/h$, and $\mathbf{C}$ is the *convection matrix* with the elements

$$c_{ij} = \int_0^1 \varphi_j'(x)\varphi_i(x) \, dx.$$

Evidently, we have that

$$\tilde{\mathbf{S}} = \frac{1}{h} \begin{bmatrix} 2 & -1 & 0 & 0 & \cdots & 0 \\ -1 & 2 & -1 & 0 & \cdots & 0 \\ 0 & -1 & 2 & -1 & \cdots & 0 \\ \cdots & \cdots & \cdots & \cdots & \cdots & \cdots \\ 0 & \cdots & \cdots & -1 & 2 & -1 \\ 0 & \cdots & \cdots & \cdots & -1 & 1 \end{bmatrix}.$$

To compute the entries for $\mathbf{C}$, we note that like, $\mathbf{S}$, $\mathbf{M}$, and $\tilde{\mathbf{S}}$, also $\mathbf{C}$ is a tridiagonal matrix. But $\mathbf{C}$ is *anti-symmetric*. Its entries are (compute!)

$$\begin{cases} c_{ij} = 0, & \text{for } |i-j| > 1 \\ c_{ii} = \int_0^1 \varphi_i(x)\varphi_i'(x) \, dx = 0, & \text{for } i = 1, \dots, n \\ c_{n+1,n+1} = \int_0^1 \varphi_{n+1}(x)\varphi_{n+1}'(x) \, dx = 1/2, & \\ c_{i,i+1} = \int_0^1 \varphi_i(x)\varphi_{i+1}'(x) \, dx = 1/2, & \text{for } i = 1, \dots, n \\ c_{i+1,i} = \int_0^1 \varphi_{i+1}(x)\varphi_i'(x) \, dx = -1/2, & \text{for } i = 1, \dots, n. \end{cases} \tag{5.3.5}$$

Finally, we have the entries b_i of the load vector $\mathbf{b}$ as

$$b_1 = \cdots = b_n = rh, \qquad b_{n+1} = rh/2 + \varepsilon\beta.$$

Thus,

$$\mathbf{C} = \frac{1}{2}\begin{bmatrix} 0 & 1 & 0 & 0 \cdots & 0 \\ -1 & 0 & 1 & 0 \cdots & 0 \\ 0 & -1 & 0 & 1 \cdots & 0 \\ \cdots & \cdots & \cdots & \cdots & \cdots \\ 0 & \cdots \cdots & -1 & 0 & 1 \\ 0 & \cdots \cdots & \cdots & -1 & 1 \end{bmatrix}, \quad \mathbf{b} = rh\begin{bmatrix} 1 \\ 1 \\ 1 \\ \cdots \\ 1 \\ 1/2 \end{bmatrix} + \varepsilon\beta\begin{bmatrix} 0 \\ 0 \\ 0 \\ \cdots \\ 0 \\ 1 \end{bmatrix}.$$

Remark 5.5 Note that in Example 5.2, since $u(0) = 0$, the *half-hat-function* φ_0 is not needed. Whereas if $u(0) = a \neq 0$, then the trial space should contain φ_0 in order to count for the *nonzero contribution* $u(0) = a$. Hence, V_h^{trial} would have one-dimension more than the test function space:

$$V_h^{\text{trial}} = \{w_h : w_h \text{ is continuous, piecewise linear on } \mathcal{T}_h,\ w_h(0) = a\}$$

while the test function space remains as V_h^0 above. Further, since $u(1)$ is not given (the data at $x = 1$ is $u'(1) = \beta$), both test and trial spaces should contain the *half-hat-function* φ_{n+1} at $x = 1$.

Remark 5.6 In the convection-dominated case $\frac{\varepsilon}{p} \ll 1$, this standard FEM will not work. Spurious oscillations in the approximate solution will appear. The standard FEM has to be modified in this case. Roughly speaking, this is done by a clever choice of modified test functions with stabilizing properties in the sense that the new weak formulation can be interpreted as a variational formulation for a problem with additional diffusion than the original PDE that we start with. This may be achieved using the, so-called, *streamline diffusion* method, briefly addressed in Chapter 7.

Last we would like to go backward (examine reversibility) and determine the underlying BVP from a linear system of equations that arise as a final result of a finite element approximation.

Example 5.3 Consider a uniform partition $0 = x_0 < x_1 < \cdots < x_N = 1$ of the interval $[0, 1]$ and let $\{\varphi_i\}_{i=0}^N$ be a set of piecewise linear continuous basis functions: $\varphi_i(x_j) = 1$ for $i = j$ and $\varphi_i(x_j) = 0$ if $i \neq j$. Given a FEM in the form of a linear system of equations $(\mathbf{S} + \mathbf{C})\xi = \mathbf{b} + \mathbf{d}$, with $\mathbf{S}$ and $\mathbf{C}$ N-by-N matrices, and ξ, $\mathbf{b}$ and $\mathbf{d}$ vectors of length N, where for $i, j = 0, \ldots, N-1$, $\mathbf{S}_{ij} = (\varphi_i', \varphi_j')$, $\mathbf{C}_{ij} = (\varphi_i, \varphi_j')$, and $\mathbf{b}_i = (\varphi_i, f)$ with f a given function. $d_0 = \alpha$ is the only nonzero element of $\mathbf{d}$.

(a) Derive the variational formulation and the strong formulation for the PDE from the above data.

(b) Let now both $u(0) = 0$ and $\alpha = 0$ and derive the continuous stability estimate $||u_x|| \le ||f||$.

Solution: (a) It is clear that the homogeneous Dirichlet condition is used at $x = 1$ since the basis function φ_N is not present in the matrices and there are no modifications corresponding to the last element of the load vector. Consider now the solution space $V^0 = \{w : ||w|| + ||w'|| < \infty, \;\; w(1) = 0\}$ where $||\cdot||$ is the usual L_2-norm over I. Then, the variational formulation reads as follows: *find* $u \in V^0$ such that

$$(v_x, u_x) + (v, u_x) = (v, f) + \alpha v(0), \quad \forall \; v \in V^0. \tag{5.3.6}$$

For the basis functions given, $\varphi_0(0) = 1$, which explains the first element of the vector **d**.

Backward integration by parts, together with the Dirichlet data on v yields

$$(v, f) + \alpha v(0) = (v, -u_{xx} + u_x) + v(0)u_x(0). \tag{5.3.7}$$

Thus, the strong formulation (PDE) is:

$$-u_{xx} + u_x = f \quad 0 < x < 1 \qquad u_x(0) = \alpha, \quad u(1) = 0. \tag{5.3.8}$$

(b) Let in (5.3.6) $\alpha = 0$ and $v = u$, then

$$(u_x, u_x) + (u, u_x) = (u, f).$$

Using integration by parts and $u(0) = 0$,

$$(u, u_x) = u^2(x)|_{x=0}^{x=1} - (u_x, u) \implies 2(u, u_x) = u^2(1) - u^2(0) = 0.$$

Hence, using Cauchy–Schwarz and Poincare inequalities:

$$||u_x||^2 = (u, f) \le ||u||\,||f|| \le ||u_x||\,||f||,$$

we get the desired result.

5.4 Exercises

Problem 5.1 Consider the two-point boundary value problem

$$-u'' = f, \quad 0 < x < 1; \qquad u(0) = u(1) = 0. \tag{5.4.1}$$

Let $V = \{v : ||v|| + ||v'|| < \infty, \;\; v(0) = v(1) = 0\}$, $||\cdot||$ denotes the L_2-norm.

a. Use V to derive a variational formulation of (5.4.1).
b. Discuss why V is valid as a vector space of test functions.
c. Classify which of the following functions are admissible test functions

$$\sin \pi x, \quad x^2, \quad x \ln x, \quad e^x - 1, \quad x(1 - x).$$

Problem 5.2 Assume that $u(0) = u(1) = 0$, and that u satisfies

$$\int_0^1 u'v'\,dx = \int_0^1 fv\,dx,$$

for all $v \in V = \{v : \|v\| + \|v'\| < \infty, \quad v(0) = v(1) = 0\}$.

a. Show that u minimizes the functional

$$F(v) = \frac{1}{2}\int_0^1 (v')^2\,dx - \int_0^1 fv\,dx. \tag{5.4.2}$$

Hint: $F(v) = F(u + w) = F(u) + \cdots \geq F(u)$.

b. Prove that the above minimization problem is equivalent to

$$-u'' = f, \quad 0 < x < 1; \qquad u(0) = u(1) = 0.$$

Problem 5.3 Consider the two-point boundary value problem

$$-u'' = 1, \quad 0 < x < 1; \qquad u(0) = u(1) = 0. \tag{5.4.3}$$

Let $\mathcal{T}_h : x_j = \frac{j}{4}$, $j = 0, 1, \ldots, 4$, denote a partition of the interval $0 < x < 1$ into four subintervals of equal length $h = 1/4$ and let V_h be the corresponding space of continuous piecewise linear functions vanishing at $x = 0$ and $x = 1$.

a. Compute a finite element approximation $U \in V_h$ to (5.4.3).
b. Prove that $U \in V_h$ is unique.

Problem 5.4 Consider once again the two-point boundary value problem

$$-u'' = f, \quad 0 < x < 1; \qquad u(0) = u(1) = 0.$$

a. Prove that the finite element approximation $U \in V_h$ to u satisfies

$$\|(u - U)'\| \leq \|(u - v)'\|, \quad \forall\, v \in V_h.$$

b. Use this result to deduce that

$$\|(u - \pi_h u)'\| \leq C\|hu''\|, \tag{5.4.4}$$

where C is a constant and $\pi_h u$ a piecewise linear interpolant to u.

Problem 5.5 Consider the two-point boundary value problem

$$\begin{aligned} -(au')' &= f, \qquad && x \in I := (0, 1), \\ u(0) &= 0, && a(1)u'(1) = g_1, \end{aligned} \tag{5.4.5}$$

where $a > 0$ is a positive function and g_1 is a constant.

a. Derive the variational formulation of (5.4.5).
b. Discuss how the boundary conditions are implemented.

Problem 5.6 Consider the two-point boundary value problem

$$-u'' = 0, \quad x \in I := (0,1); \qquad u(0) = 0, \quad u'(1) = 7. \tag{5.4.6}$$

Divide the interval I into two subintervals of length $h = \frac{1}{2}$ and let V_h be the corresponding space of continuous piecewise linear functions vanishing at $x = 0$.
a. Formulate a finite element method for (5.4.6).
b. Calculate by hand the finite element approximation $U \in V_h$ to (5.4.6). Study how the boundary condition at $x = 1$ is approximated.

Problem 5.7 Consider the two-point boundary value problem

$$-u'' = 0, \quad 0 < x < 1; \qquad u'(0) = 5, \quad u(1) = 0. \tag{5.4.7}$$

Let $\mathcal{T}_h : x_j = jh,\ j = 0, 1, \dots, N,\ h = 1/N$ be a uniform partition of the interval $0 < x < 1$ into N subintervals and let V_h be the corresponding space of continuous piecewise linear functions.
a. Use V_h, with $N = 3$, and formulate a finite element method for (5.4.7).
b. Compute the finite element approximation $U \in V_h$ assuming $N = 3$.

Problem 5.8 Consider the problem of finding a solution approximation to

$$-u'' = 1, \quad 0 < x < 1; \qquad u'(0) = u'(1) = 0. \tag{5.4.8}$$

Let $\mathcal{T}_h$ be a partition of the interval $0 < x < 1$ into two subintervals of equal length $h = \frac{1}{2}$ and let V_h be the corresponding space of continuous piecewise linear functions.
a. Find the exact solution to (5.4.8) by integrating twice.
b. Compute a finite element approximation $U \in V_h$ to u if possible.

Problem 5.9 Consider the two-point boundary value problem

$$-((1 + x)u')' = 0, \quad 0 < x < 1; \qquad u(0) = 0, \quad u'(1) = 1. \tag{5.4.9}$$

Divide the interval $0 < x < 1$ into three subintervals of equal length $h = \frac{1}{3}$ and let V_h be the corresponding space of continuous piecewise linear functions vanishing at $x = 0$.
a. Use V_h to formulate a finite element method for (5.4.9).
b. Verify that the stiffness matrix $\mathbf{A}$ and the load vector $\mathbf{b}$ are given by

$$\mathbf{A} = \frac{1}{2}\begin{bmatrix} 16 & -9 & 0 \\ -9 & 20 & -11 \\ 0 & -11 & 11 \end{bmatrix}, \qquad \mathbf{b} = \begin{bmatrix} 0 \\ 0 \\ 1 \end{bmatrix}.$$

c. Show that $\mathbf{A}$ is symmetric tridiagonal, and positive definite.

d. Derive a simple way to compute the energy norm $||U||_E^2$, defined by

$$||U||_E^2 = \int_0^1 (1+x)U'(x)^2\,dx,$$

where $U \in V_h$ is the finite element solution approximation.

Problem 5.10 Consider the two-point boundary value problem

$$-u'' = 0, \quad 0 < x < 1; \qquad u(0) = 0, \quad u'(1) = k(u(1) - 1), \quad k \neq 1. \tag{5.4.10}$$

Let $\mathcal{T}_h : 0 = x_0 < x_1 < x_2 < x_3 = 1$, where $x_1 = \frac{1}{3}$ and $x_2 = \frac{2}{3}$ be a partition of the interval $0 \leq x \leq 1$ and let V_h be the corresponding space of continuous piecewise linear functions, which vanish at $x = 0$.

a. Discuss the solubility of (5.4.10) assuming $k = 1$.
b. Discuss how the parameter k influence the boundary condition at $x = 1$. In particular when $k \to \infty$ and $k \to 0$.

Problem 5.11 Consider the finite element method applied to

$$-u'' = 0, \quad 0 < x < 1; \qquad u(0) = \alpha, \quad u'(1) = \beta,$$

where α and β are given constants. Assume that the interval $0 \leq x \leq 1$ is divided into three subintervals of equal length $h = 1/3$ and that $\{\varphi_j\}_0^3$ is a nodal basis of V_h, the corresponding space of continuous piecewise linear functions.

a. Verify that the ansatz

$$U(x) = \alpha\varphi_0(x) + \xi_1\varphi_1(x) + \xi_2\varphi_2(x) + \xi_3\varphi_3(x),$$

yields the following system of equations

$$\frac{1}{h}\begin{bmatrix} -1 & 2 & -1 & 0 \\ 0 & -1 & 2 & -1 \\ 0 & 0 & -1 & 1 \end{bmatrix}\begin{bmatrix} \alpha \\ \xi_1 \\ \xi_2 \\ \xi_3 \end{bmatrix} = \begin{bmatrix} 0 \\ 0 \\ \beta \end{bmatrix}. \tag{5.4.11}$$

b. If $\alpha = 2$ and $\beta = 3$ show that (5.4.11) can be reduced to

$$\frac{1}{h}\begin{bmatrix} 2 & -1 & 0 \\ -1 & 2 & -1 \\ 0 & -1 & 1 \end{bmatrix}\begin{bmatrix} \xi_1 \\ \xi_2 \\ \xi_3 \end{bmatrix} = \begin{bmatrix} 2h^{-1} \\ 0 \\ 3 \end{bmatrix}.$$

c. Solve the above system of equations to find $U(x)$.

Problem 5.12 Compute a finite element solution approximation to

$$-u'' + u = 1; \quad 0 \leq x \leq 1, \qquad u(0) = u(1) = 0, \tag{5.4.12}$$

using the continuous piecewise linear ansatz $U = \xi_1\varphi_1(x) + \xi_2\varphi_2(x)$ where

$$\varphi_1(x) = \begin{cases} 3x, & 0 < x < \frac{1}{3} \\ 2-3x, & \frac{1}{3} < x < \frac{2}{3}, \\ 0, & \frac{2}{3} < x < 1 \end{cases} \qquad \varphi_2(x) = \begin{cases} 0, & 0 < x < \frac{1}{3} \\ 3x-1, & \frac{1}{3} < x < \frac{2}{3}. \\ 3-3x, & \frac{2}{3} < x < 1 \end{cases}$$

Problem 5.13 Consider the following eigenvalue problem

$$-au'' + bu = 0; \quad 0 \le x \le 1, \qquad u(0) = u'(1) = 0, \tag{5.4.13}$$

where $a,\ b > 0$ are constants. Let $\mathcal{T}_h : 0 = x_0 < x_1 < \cdots < x_N = 1$, be a nonuniform partition of the interval $0 \le x \le 1$ into N intervals of length $h_i = x_i - x_{i-1}$, $i = 1, 2, \ldots, N$ and let V_h be the corresponding space of continuous piecewise linear functions. Compute the stiffness and mass matrices.

Problem 5.14 Show that the FEM with the mesh size h for the problem:

$$\begin{cases} -u'' = 1 & 0 < x < 1 \\ u(0) = 7u'(1) = 0, \end{cases} \tag{5.4.14}$$

with

$$U(x) = 7\varphi_0(x) + U_1\varphi_1(x) + \cdots + U_m\varphi_m(x). \tag{5.4.15}$$

leads to the linear system of equations: $\tilde{\mathbf{A}} \cdot \tilde{\mathbf{U}} = \tilde{\mathbf{b}}$, where

$$\tilde{\mathbf{A}} = \frac{1}{h}\underbrace{\begin{bmatrix} -1 & 2 & -1 & 0 \\ 0 & -1 & 2 & -1 \\ \cdots & \cdots & \cdots & \cdots \\ 0 & \cdots & 0 & \cdots \end{bmatrix}}_{m\times(m+1)}, \quad \tilde{\mathbf{U}} = \underbrace{\begin{bmatrix} 7 \\ U_1 \\ \vdots \\ U_m \end{bmatrix}}_{(m+1)\times 1}, \quad \tilde{\mathbf{b}} = \underbrace{\begin{bmatrix} h \\ \cdots \\ h \\ h/2 \end{bmatrix}}_{m\times 1}.$$

which is reduced to $\mathbf{AU} = \mathbf{b}$, with

$$\mathbf{A} = \frac{1}{h}\begin{bmatrix} 2 & -1 & 0 & \cdots & 0 \\ -1 & 2 & -1 & 0 & \cdots \\ \cdots & \cdots & \cdots & \cdots & \cdots \\ \cdots & 0 & -1 & 2 & -1 \\ 0 & 0 & 0 & -1 & 2 \end{bmatrix}, \quad \mathbf{U} = \begin{bmatrix} U_1 \\ U_2 \\ \vdots \\ U_m \end{bmatrix}, \quad \mathbf{b} = \begin{bmatrix} h + \frac{7}{h} \\ h \\ \cdots \\ h \\ h/2 \end{bmatrix}.$$

Problem 5.15 Prove an a priori and an a posteriori error estimate for a finite element method (for example cG(1)) for the problem

$$-u'' + \alpha u = f, \quad \text{in } I = (0,1), \qquad u(0) = u(1) = 0,$$

where the coefficient $\alpha = \alpha(x)$ is a bounded positive function on I, $(0 \le \alpha(x) \le K,\ \ x \in I)$.
Hint: Consider $\alpha = 1$ and use the energy norm $||u||_E^2 = ||u||^2 + ||u'||^2$.

Problem 5.16

a. Formulate a cG(1) method for the problem

$$\begin{cases} (a(x)u'(x))' = 0, & 0 < x < 1, \\ a(0)u'(0) = u_0, & u(1) = 0. \end{cases}$$

and give an a posteriori error estimate.

b. Let $u_0 = 3$ and compute the approximate solution in (a) for a uniform partition of $I = [0, 1]$ into four intervals and

$$a(x) = \begin{cases} 1/4, & x < 1/2, \\ 1/2, & x > 1/2. \end{cases}$$

c. Show that, with these special choices, the computed solution is equal to the exact one, i.e. the error is equal to 0.

Problem 5.17 Prove an a priori error estimate for the finite element method for the problem

$$-u''(x) + u'(x) = f(x), \quad 0 < x < 1, \qquad u(0) = u(1) = 0.$$

Problem 5.18

a. Prove an *a priori* error estimate for the cG(1) approximation of the boundary value problem

$$-u'' + cu' + u = f \quad \text{in} \quad I = (0, 1), \qquad u(0) = u(1) = 0,$$

where $c \ge 0$ is constant.

b. For which value of c is the *a priori* error estimate optimal?

Problem 5.19 Let U be the piecewise linear finite element approximation for

$$-u''(x) + 2xu'(x) + 2u(x) = f(x), \quad x \in (0, 1), \qquad u(0) = u(1) = 0,$$

in a partition $\mathcal{T}_h$ of the interval $[0, 1]$. Set $e = u - U$ and derive a priori and a posteriori error estimates in the energy-norm:

$$||e||_E^2 = ||e'||^2 + ||e||^2, \quad \text{where } ||w||^2 = \int_0^1 w(x)^2\, dx.$$

6

Scalar Initial Value Problems

This chapter is devoted to the finite element methods for time discretizations. Here, we shall consider problems depending on the time variable, only. The approximation techniques developed in this chapter, combined with those of Chapter 5 for boundary value problems, can be used for the numerical study of the initial boundary value problems, such as the heat and the wave equations.

The model problem that we consider here is the classical example of the population dynamics described by the following ordinary differential equation (ODE)

$$\begin{array}{l} \text{(DE)} \\ \text{(IV)} \end{array} \begin{cases} \dot{u}(t) + a(t)u(t) = f(t), & 0 < t \leq T, \\ \qquad\qquad\quad u(0) = u_0, \end{cases} \tag{6.0.1}$$

where $f(t)$ is the source term and $\dot{u}(t) = \dfrac{du}{dt}$. The coefficient $a(t)$ is a bounded function. If $a(t) \geq 0$, Problem (6.0.1) is called *parabolic*; while $a(t) \geq \alpha > 0$ yields a *dissipative problem*, in the sense that, with increasing t, perturbations of solutions to (6.0.1), e.g. introduced by numerical discretization, will decay. In general, in numerical approximations for (6.0.1), the error accumulates when advancing in time, i.e. the error of previous time steps adds up to the error of the present time step. The different types of error accumulation/perturbation growth are referred to as stability properties of the initial value problem (IVP).

6.1 Solution Formula and Stability

Theorem 6.1 *The solution of problem (6.0.1) is given by*

$$u(t) = u_0\, e^{-A(t)} + \int_0^t e^{-(A(t)-A(s))} f(s)\, ds, \tag{6.1.1}$$

where $A(t) = \int_0^t a(s)\, ds$ *and* $e^{A(t)}$ *is the integrating factor.*

An Introduction to the Finite Element Method for Differential Equations, First Edition.
Mohammad Asadzadeh.

Proof: Multiplying the (DE) by the integrating factor $e^{A(t)}$, we have

$$\dot{u}(t)e^{A(t)} + \dot{A}(t)e^{A(t)}u(t) = e^{A(t)}f(t), \tag{6.1.2}$$

where we assume that $a(t) = \dot{A}(t)$. Equation (6.1.2) can be rewritten as

$$\frac{d}{dt}\left(u(t)e^{A(t)}\right) = e^{A(t)}f(t).$$

We relabel the variable by s and integrate from 0 to t to get

$$\int_0^t \frac{d}{ds}\left(u(s)e^{A(s)}\right)\,ds = \int_0^t e^{A(s)}f(s)\,ds.$$

Thus,

$$u(t)e^{A(t)} - u(0)e^{A(0)} = \int_0^t e^{A(s)}f(s)\,ds.$$

Now, inserting $A(0) = 0$ and $u(0) = u_0$, we get the desired result (6.1.1). □

This representation for $u(t)$ is also referred to as the *Variation of constants.*

Theorem 6.2 (Stability estimates) *Using the solution formula* (6.1.1), *we can derive the following stability estimates:*

(i) If $a(t) \geq \alpha > 0$, *then* $|u(t)| \leq e^{-\alpha t}|u_0| + \frac{1}{\alpha}(1 - e^{-\alpha t})\max_{0\leq s\leq t}|f(s)|$,
(ii) If $a(t) \geq 0$ *(i.e.* $\alpha = 0$*; the parabolic case), then*

$$|u(t)| \leq |u_0| + \int_0^t |f(s)|\,ds, \quad \text{i.e.} \quad |u(t)| \leq |u_0| + \|f\|_{L_1(0,t)}. \tag{6.1.3}$$

Proof:

(i) For $a(t) \geq \alpha > 0$, we have that $A(t) = \int_0^t a(s)\,ds$ is an increasing function of t, $A(t) \geq \alpha t$ and

$$A(t) - A(s) = \int_0^t a(r)\,dr - \int_0^s a(r)\,dr = \int_s^t a(r)\,dr \geq \alpha(t - s). \tag{6.1.4}$$

Thus, $e^{-A(t)} \leq e^{-\alpha t}$ and $e^{-(A(t)-A(s))} \leq e^{-\alpha(t-s)}$. Hence, using (6.1.1), we get

$$|u(t)| \leq |u_0|e^{-\alpha t} + \int_0^t e^{-\alpha(t-s)}|f(s)|\,ds, \tag{6.1.5}$$

which yields

$$|u(t)| \leq e^{-\alpha t}|u_0| + \max_{0\leq s\leq t}|f(s)|\left[\frac{1}{\alpha}e^{-\alpha(t-s)}\right]_{s=0}^{s=t}, \qquad \text{i.e.}$$

$$|u(t)| \leq e^{-\alpha t}|u_0| + \frac{1}{\alpha}(1 - e^{-\alpha t})\max_{0\leq s\leq t}|f(s)|.$$

(ii) Let $\alpha = 0$ in (6.1.5) (which is true also in this case), then we get (ii). □

Remark 6.1 (i) Expresses that the effect of the initial data u_0 decays exponentially with time, and that the effect of the source term f on the right hand side does not depend on the length of the time interval, only on the maximum value of f and on the value of α. In case (ii), the influence of u_0 remains bounded in time, and the integral of f indicates an accumulation in time.

6.2 Finite Difference Methods for IVP

Recall the forward Euler method discussed in Section 3.2.1, discretizing

$$\dot{u}(t) = \lambda u(t), \quad 0 < t \le T, \quad u(0) = u_0. \tag{6.2.1}$$

Recall, further, that the analytic solution $u(t) = e^{\lambda t}$ would blow up, for $\lambda > 0$, as $t \to \infty$ (corresponds to $a(t) < 0$). Therefore, the physical problem setting suggests $\lambda < 0$. The definition of the derivative as a limit of difference quotients suggests a method of discretization as

$$\frac{du}{dt}(t) \approx \frac{u(t + \Delta t) - u(t)}{\Delta t}, \tag{6.2.2}$$

where Δt is a small positive parameter. Now, let us consider a uniform partition of the interval $[0, T]$ into N subintervals $I_k = (t_{k-1}, t_k]$, $k = 1, \dots, N$ of length Δt and on each I_k approximate $u(t)$, $t \in I_k$ by its values at either end point of I_k or as its mean value at the end points of I_k. This yields iterative algorithms generating a sequence of values, viz.

$$\frac{u(t_k) - u(t_{k-1})}{\Delta t} = \lambda \begin{cases} u(t_{k-1}), \\ u(t_k), \quad \text{or} \\ \dfrac{1}{2}(u(t_{k-1}) + u(t_k)). \end{cases} \tag{6.2.3}$$

Now, assuming that the value of $u(t)$ is known at the start point: t_{k-1} of I_k, the algorithm above gives three different approximation methods for $u(t_k)$:

$$\begin{cases} u(t_k) = (1 + \lambda\Delta t)u(t_{k-1}), & \text{Forward (explicit) Euler} \\ u(t_k) = (1 - \lambda\Delta t)^{-1}u(t_{k-1}), & \text{Backward (implicit) Euler} \\ u(t_k) = \dfrac{1 + \lambda\Delta t/2}{1 - \lambda\Delta t/2}u(t_{k-1}), & \text{Crank–Nicolson.} \end{cases} \tag{6.2.4}$$

Recall that $\lambda < 0$ so that all coefficients in (6.2.4) are finite. Iterating the above algorithms and inserting the given initial value $u(0) = u_0$, we have for $k = 0, 1, \ldots, N$; that

$$\begin{cases} u(t_k) = (1 + \lambda\Delta t)^k u_0, & \text{For Forward (explicit) Euler} \\ u(t_k) = (1 - \lambda\Delta t)^{-k} u_0, & \text{For Backward (implicit) Euler} \\ u(t_k) = \left(\dfrac{1 + \lambda\Delta t/2}{1 - \lambda\Delta t/2}\right)^k u_0, & \text{For Crank–Nicolson.} \end{cases} \tag{6.2.5}$$

The generalization of (6.2.4) to an IVP of the form

$$\dot{u}(t) = f(t, u(t)), \quad 0 < t \leq T, \quad u(0) = u_0, \tag{6.2.6}$$

is straightforward and yields the algorithms

$$\begin{cases} u(t_k) = u(t_{k-1}) + \Delta t f(t_{k-1}, u(t_{k-1})), & \text{Forward Euler} \\ u(t_k) = u(t_{k-1}) + \Delta t f(t_k, u(t_k)), & \text{Backward Euler} \\ u(t_k) = u(t_{k-1}) + \dfrac{\Delta t}{2}(f(t_{k-1}, u(t_{k-1})) + f(t_k, u(t_k))), & \text{Crank–Nicolson.} \end{cases} \tag{6.2.7}$$

Below we illustrate some features of these finite difference algorithms.

Example 6.1 Consider the IVP:

$$\dot{u}(t) + au(t) = 0, \quad t > 0, \quad u(0) = 1.$$

(a) Let $a = 40$, and the time step $k = 0.1$. Draw the graph of $U_n := U(nk)$, $k = 1, 2, \ldots$, approximating u using (i) explicit Euler, (ii) implicit Euler, and (iii) Crank–Nicolson methods.

(b) Consider the case $a = i$, $(i^2 = -1)$, having the complex solution $u(t) = e^{-it}$ with $|u(t)| = 1$ for all t. Show that this property is preserved in Crank–Nicolson approximation, i.e. $|U_n| = 1$, but not in any of the Euler methods.

Solution:

(a) With $a = 40$ and $k = 0.1$, we get the explicit Euler (Figure 6.1):

$$\begin{cases} U_n - U_{n-1} + 40 \times (0.1)U_{n-1} = 0, \\ U_0 = 1. \end{cases} \Longrightarrow \begin{cases} U_n = -3U_{n-1}, & n = 1,\ 2\ \ldots \\ U_0 = 1. & \text{(unstable)} \end{cases}$$

Implicit Euler (Figure 6.1):

$$\begin{cases} U_n = \dfrac{1}{1 + 40 \times (0.1)} U_{n-1} = \dfrac{1}{5} U_{n-1}, & n = 1,\ 2,\ \ldots \\ U_0 = 1. & \text{(strongly stable)} \end{cases}$$

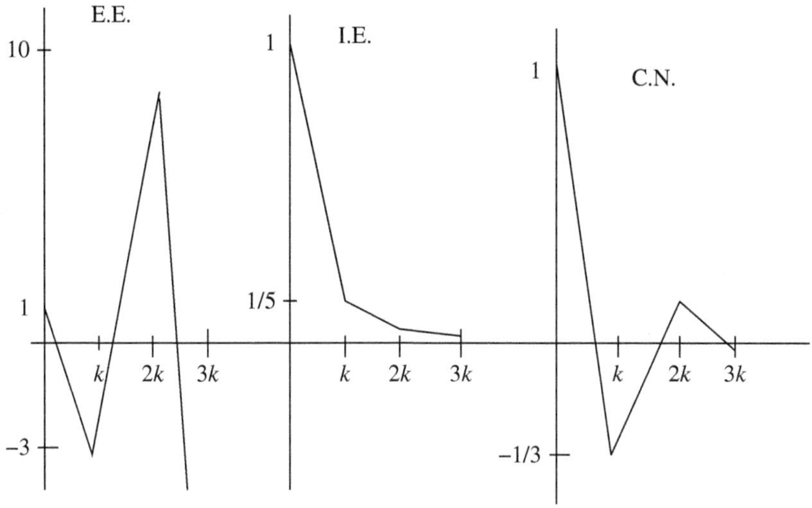

Figure 6.1 Example of finite difference schemes for population model. E.E., explicit Euler; I.E., implicit Euler; C.N., Crank Nicolson

Crank–Nicolson (Figure 6.1):

$$\begin{cases} U_n = \dfrac{1 - \frac{1}{2} \times 40 \times (0.1)}{1 + \frac{1}{2} \times 40 \times (0.1)} U_{n-1} = -\dfrac{1}{3} U_{n-1}, & n = 1,\ 2,\ \ldots \\ U_0 = 1. & \text{(stable, however, oscillatory)} \end{cases}$$

(b) With $a = i$, we get

Explicit Euler

$$|U_n| = |1 - (0.1) \times i||U_{n-1}| = \sqrt{1 + 0.01}|U_{n-1}| \Longrightarrow |U_n| \geq |U_{n-1}|.$$

Implicit Euler

$$|U_n| = \left|\frac{1}{1 + (0.1) \times i}\right| |U_{n-1}| = \frac{1}{\sqrt{1 + 0.01}} |U_{n-1}| \leq |U_{n-1}|.$$

Crank–Nicolson

$$|U_n| = \left|\frac{1 - \frac{1}{2}(0.1) \times i}{1 + \frac{1}{2}(0.1) \times i}\right| |U_{n-1}| = |U_{n-1}|.$$

6.3 Galerkin Finite Element Methods for IVP

Recall that we refer to the space of functions where we seek the approximate solution as the *trial space* and the space of functions used for the orthogonality condition (multipliers), as the *test space*. The polynomial approximation procedure introduced in Chapter 3 along with (3.2.10)–(3.2.18), for the initial value Problem (3.2.1), and consequently also for (6.0.1), being over the whole time interval $(0, T)$, is referred to as the *global Galerkin method*. In this section, first we introduce two versions of the global Galerkin method and then extend them to partitions of the interval $(0, T)$ using piecewise polynomial test (multiplier) and trial (solution) functions of certain orders. In most practical applications, it suffices to consider two simple, low degree polynomial, approximations: namely *piecewise continuous Galerkin of degree 1*, referred as cG(1) and *piecewise discontinuous Galerkin of degree 0*, referred as dG(0). We shall prove stability results and derive both a priori and a posteriori error estimates for these two approximation procedures.

- *The continuous Galerkin method of degree 1; cG(1).* In this case, the trial functions are piecewise linear and continuous, while the test functions are piecewise constant and discontinuous, i.e. the trial and the test functions belong to different polynomial spaces (this may also be interpreted as the effect of the initial data, reducing one degree of freedom).
- *The discontinuous Galerkin method of degree 0; dG(0).* Here, both the trial and the test functions are chosen to be piecewise constant and discontinuous.

Below first, we introduce the global versions formulated for approximations with polynomials of arbitrary degree, for the whole time interval $(0, T)$ and then extend them to the local versions for partitions of $(0, T)$.

6.3.1 The Continuous Galerkin Method

Recall the global Galerkin method of degree q; (3.2.13) for the IVP (6.0.1): find $U \in \mathcal{P}^{(q)}(0, T)$, with $U(0) = u_0$ such that

$$\int_0^T (\dot{U} + aU)v\, dt = \int_0^T fv\, dt, \quad \forall v \in \mathcal{P}^{(q)}(0, T), \quad \text{with} \quad v(0) = 0. \tag{6.3.1}$$

We formulate an alternative version for the above global method:
find $U \in \mathcal{P}^{(q)}(0, T)$ with $U(0) = u_0$ such that

$$\int_0^T (\dot{U} + aU)v\, dt = \int_0^T fv\, dt, \quad \forall v \in \mathcal{P}^{(q-1)}(0, T), \tag{6.3.2}$$

Note that in (6.3.1) we have that $v \in \text{span}\{t, t^2, \dots, t^q\}$, whereas in (6.3.2) $v \in \text{span}\{1, t, t^2, \dots, t^{q-1}\}$. Hence, the difference between these two formulation lies in the choice of their test function spaces. We shall focus on (6.3.2), due to the fact that it yields a more accurate approximation than the original method (6.3.1). The following example illustrates this phenomenon.

Example 6.2 Consider the IVP

$$\dot{u}(t) + u(t) = 0 \quad 0 \leq t \leq 1, \qquad u(0) = 1. \tag{6.3.3}$$

The exact solution is given by $u(t) = e^{-t}$. The continuous piecewise linear approximation, with the ansatz $U(t) = 1 + \xi_1 t$ in,

$$\int_0^1 (\dot{U}(t) + U(t))v(t)\, dt = 0, \tag{6.3.4}$$

and $v(t) = t$, hence $v(0) = 0$, (i.e. (6.3.1)) yields

$$\int_0^1 (\xi_1 + 1 + \xi_1 t)t\, dt = 0 \Longrightarrow \left[(\xi_1 + 1)\frac{t^2}{2} + \xi_1 \frac{t^3}{3}\right]_0^1 = 0 \Longrightarrow \xi_1 = -3/5.$$

In this case, the approximate solution that we denote by U_1 is given by $U_1(t) = 1 - (3/5)t$; whereas (6.3.2) for this problem uses $v(t) = 1$ and gives

$$\int_0^1 (\xi_1 + 1 + \xi_1 t)\, dt = 0 \Longrightarrow \left[(\xi_1 + 1)t + \xi_1 \frac{t^2}{2}\right]_0^1 = 0 \Longrightarrow \xi_1 = -2/3.$$

Now, the approximate solution denoted by U_2 is $U_2(t) = 1 - (2/3)t$ and, as we can see in Figure 6.2 below U_2, is a better approximation for e^{-t} than U_1.

Before generalizing (6.3.2) to approximation by *piecewise* polynomials of degree q, which is the cG(q) method, we consider an example.

Example 6.3 Consider (6.3.2) with $q = 1$, then we may choose $v \equiv 1$, thus

$$\int_0^T (\dot{U} + aU)v\, dt = \int_0^T (\dot{U} + aU)\, dt = U(T) - U(0) + \int_0^T aU(t)\, dt \tag{6.3.5}$$

But $U(t)$ being a linear function on $[0, T]$ is given by

$$U(t) = U(0)\frac{T - t}{T} + U(T)\frac{t}{T}. \tag{6.3.6}$$

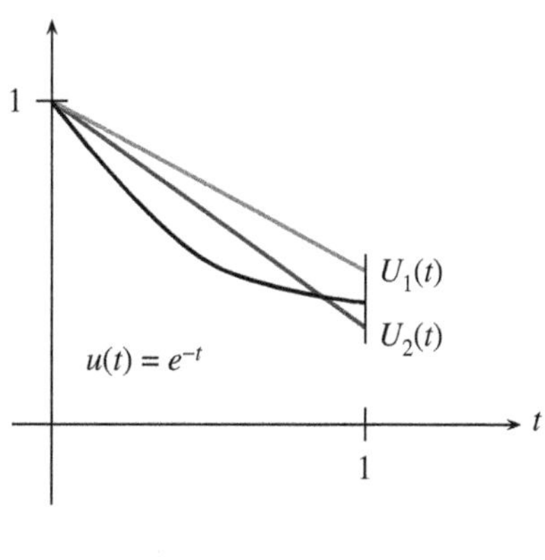

Figure 6.2 Two continuous linear Galerkin approximations of e^{-t}.

Inserting into (6.3.5), we get

$$U(T) - U(0) + \int_0^T a\left(U(0)\frac{T-t}{T} + U(T)\frac{t}{T}\right)\,dt = \int_0^T f\,dt, \tag{6.3.7}$$

which gives an equation for the unknown quantity $U(T)$. Thus, using (6.3.6) with a given $U(0)$, we get the linear approximation $U(t)$ for all $t \in [0, T]$. Below we generalize this example to piecewise linear approximation and demonstrate the iteration procedure for the cG(1) scheme.

6.3.1.1 The cG(1) Algorithm

To reformulate problem (6.3.2) for a partition $\mathcal{T}_k$ of the interval $[0, T]$ into subintervals $I_k = (t_{k-1}, t_k],\ k = 1, 2, \ldots, N$, we perform the following steps:

(1) Given $U(0) = U_0 = u_0$ and a source term f, apply (6.3.7) to the first subinterval $(0, t_1]$ and compute $U(t_1)$. Then, using (6.3.6) for $[0, t_1]$ one gets $U(t), \forall t \in [0, t_1]$.

(2) Assume that we have already computed $U_{k-1} := U(t_{k-1})$. Hence, U_{k-1} and f are considered as known data. Now, we consider the problem on the subintervals $I_k = (t_{k-1}, t_k]$, and compute the unknown $U_k := U(t_k)$ from the local version of (6.3.7):

$$U_k - U_{k-1} + \int_{t_{k-1}}^{t_k} a\left(\frac{t_k - t}{t_k - t_{k-1}}U_{k-1} + \frac{t - t_{k-1}}{t_k - t_{k-1}}U_k\right)\,dt = \int_{t_{k-1}}^{t_k} f\,dt.$$

(3) Having both U_{k-1} and U_k, by linearity, we get $U(t)$ for $t \in I_k$. To get the continuous piecewise linear approximation in the whole interval $[0, t_N]$, step (2) is performed in successive subintervals $I_k, k = 2, \ldots, N$.

6.3.1.2 The cG(q) Method

The continuous Galerkin finite element method of degree q formulated on a partition $\mathcal{T}_k$, $0 = t_0 < t_1 < \cdots < t_N = T$ of the interval $(0, T)$, is referred to as the cG(q) method and reads: find $U(t) \in V_k^{(q)}$, such that $U(0) = u_0$, and

$$\int_0^{t_N} (\dot{U} + aU)w\,dt = \int_0^{t_N} fw\,dt, \qquad \forall w \in W_k^{(q-1)}, \tag{6.3.8}$$

where

$$V_k^{(q)} = \{v\colon v \text{ continuous, } \textit{piecewise polynomials} \text{ of degree } \le q \text{ on } \mathcal{T}_k\},$$

$$W_k^{(q-1)} = \{w\colon w \text{ discontinuous, } \textit{piecewise polynomials} \text{ of degree} \le q-1 \text{ on } \mathcal{T}_k\}.$$

Hence, the difference between the global continuous Galerkin method and cG(q) is that now we have *piecewise polynomials* on a partition of $[0, T]$ rather than global polynomials in the whole interval $[0, T]$.

6.3.2 The Discontinuous Galerkin Method

We start presenting the *global discontinuous Galerkin method of degree q*: find $U(t) \in \mathcal{P}^{(q)}(0, T)$ such that

$$\int_0^T (\dot{U} + aU)v\, dt + \beta(U(0) - u(0))v(0) = \int_0^T fv\, dt, \quad \forall v \in \mathcal{P}^{(q)}(0, T), \quad (6.3.9)$$

where β is a coefficient that weights the relative importance of the residual error for the initial value $U(0) - u(0)$, against the conventional residual $R(U) := \dot{U} + aU - f$ of the approximate solution. This approach gives up the requirement that $U(t)$ satisfies the initial condition. Instead, the initial condition is imposed in a variational sense by the term $(U(0) - u(0))v(0)$. As in the cG(q) case, to derive the *discontinuous Galerkin method of degree q*: the dG(q) scheme, the above strategy can be formulated for the subintervals in a partition $\mathcal{T}_k$. To this end, we recall the notation for the right (+) and left (−) limits: $v_n^{\pm} = \lim_{s \to 0^+} v(t_n \pm s)$ and the corresponding *jump* term $[v_n] = v_n^+ - v_n^-$ at the time level $t = t_n$ (Figure 6.3). Then, the dG(q) method (with $\beta \equiv 1$) for (6.0.1) reads as follows: for $n = 1, \dots, N$; find $U(t) \in \mathcal{P}^{(q)}(t_{n-1}, t_n)$ such that

$$\int_{t_{n-1}}^{t_n} (\dot{U} + aU)v\, dt + U_{n-1}^+ v_{n-1}^+ = \int_{t_{n-1}}^{t_n} fv\, dt + U_{n-1}^- v_{n-1}^+, \quad \forall v \in \mathcal{P}^{(q)}(t_{n-1}, t_n). \quad (6.3.10)$$

Example 6.4 *dG(0).* Let $q = 0$, then v is constant generated by the single basis function: $v \equiv 1$. Further, we have $U(t) = U_n = U_{n-1}^+ = U_n^-$ on $I_n = (t_{n-1}, t_n]$, and $\dot{U} \equiv 0$. Thus, for $q = 0$ (6.3.10) yields the following dG(0) formulation: for $n = 1, \dots, N$; find piecewise constants U_n such that

$$\int_{t_{n-1}}^{t_n} aU_n\, dt + U_n = \int_{t_{n-1}}^{t_n} f\, dt + U_{n-1}. \quad (6.3.11)$$

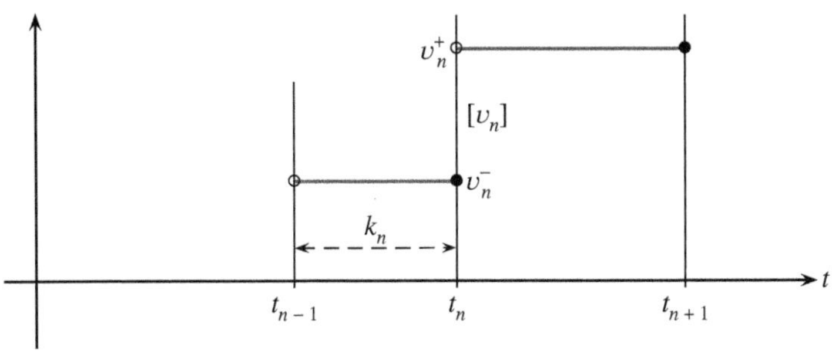

Figure 6.3 The jump $[v_n]$ and the right and left limits $v_n^{\pm}$.

Summing over n in (6.3.10), we get the following global dG(q) formulation: find $U(t) \in W_k^{(q)}$, with $U_0^- = u_0$, such that

$$\sum_{n=1}^{N} \int_{t_{n-1}}^{t_n} (\dot{U} + aU)w\, dt + \sum_{n=1}^{N} [U_{n-1}]w_{n-1}^{+} = \int_0^{t_N} fw\, dt, \quad \forall w \in W_k^{(q)}. \tag{6.3.12}$$

Remark 6.2 one can show that (see Johnson et al. [57]) cG(1) converges faster than dG(0), whereas dG(0) has better stability properties than cG(1): More specifically, in the parabolic case when $a > 0$ is constant and ($f \equiv 0$), we can easily verify that (see Exercise 6.10 at the end of this chapter) the dG(0) solution $\dot{U}_n$ corresponds to the *Backward Euler* scheme

$$U_n = \left(\frac{1}{1+ak}\right)^n u_0,$$

and the cG(1) solution $\tilde{U}_n$ is given by the Crank–Nicolson scheme:

$$\tilde{U}_n = \left(\frac{1 - \frac{1}{2}ak}{1 + \frac{1}{2}ak}\right)^n u_0,$$

where k is the constant time step.

6.4 A Posteriori Error Estimates

In this section, we derive a posteriori error estimates for the most considered Galerkin approaches for time discretization, namely the cG(1) and dG(0) methods. These estimates can be extended to the general cG(q) and dG(q) methods, for $q > 1$. The details are, however, rather involved. Some lower degree polynomial cases (e.g. cG(2)) are left as exercise.

6.4.1 A Posteriori Error Estimate for cG(1)

Recall the IVP (6.0.1) of finding the function u such that

$$\dot{u}(t) + a(t)u(t) = f(t), \quad \forall t \in (0, T], \qquad u(0) = u_0. \tag{6.4.1}$$

A general *variational form* reads as follows: find u such that

$$\int_0^T (\dot{u} + au)v\, dt = \int_0^T fv\, dt, \qquad \text{for all test functions } v.$$

Integrating by parts, we get the equivalent equation

$$u(T)v(T) - u(0)v(0) + \int_0^T u(t)(-\dot{v}(t) + av(t))\, dt = \int_0^T fv\, dt. \tag{6.4.2}$$

If we now choose the test function $v(t)$ to be the solution of *the dual problem*:

$$-\dot{v}(t) + av(t) = 0, \quad \text{for} \quad t \in (0, T), \tag{6.4.3}$$

then (6.4.2) is simplified to

$$u(T)v(T) = u(0)v(0) + \int_0^T fv\, dt, \quad \forall v(t). \tag{6.4.4}$$

In other words, choosing v to be the solution of the dual problem (6.4.3), we may get the final value $u(T)$ of the solution directly coupled to the initial value $u(0)$ and the data f. This type of representation will be crucial in the a posteriori error analysis, e.g. in the proof of Theorem 6.3. Let us state *the dual problem* explicitly:

6.4.1.1 The Dual Problem

The dual problem for (6.4.1) is formulated as follows: find $\varphi(t)$ such that

$$\begin{cases} -\dot{\varphi}(t) + a(t)\varphi(t) = 0, & t_N > t \geq 0, \\ \varphi(t_N) = e_N, & \text{where } e_N = u_N - U_N = u(t_N) - U(t_N). \end{cases} \tag{6.4.5}$$

Note that (6.4.5) runs *backward in time* starting at time $t_N = T$ (Figure 6.4).

Theorem 6.3 A posteriori error estimate for cG(1). *For $N = 1, 2, \ldots,$ the cG(1) solution $U(t)$ satisfies*

$$|e_N| \leq S(t_N) \cdot \max_{t \in [0,t_N]} |k\, r(U)|, \tag{6.4.6}$$

where $k(t)$ is the time step *function: $k(t) = k_n = |I_n|$ for $t \in I_n = (t_{n-1}, t_n]$, and $r(U) = \dot{U} + aU - f$ is the residual error for U. Further, $S(t_N)$, specified below, is the* stability factor *satisfying the bound*

$$S(t_N) := \frac{\int_0^{t_N} |\dot{\varphi}|\, dt}{|e_N|} \leq \begin{cases} e^{\lambda\, t_N}, & \text{if} \quad |a(t)| \leq \lambda, \quad \text{for} \quad 0 \leq t \leq t_N, \\ 1, & \text{if} \quad a(t) \geq 0, \quad \text{for} \quad 0 \leq t \leq t_N. \end{cases} \tag{6.4.7}$$

Proof: Let $e(t) = u(t) - U(t)$. Using the dual problem $-\dot{\varphi}(t) + a(t)\varphi(t) = 0$, we may write

$$e_N^2 = e_N^2 + 0 = e_N^2 + \int_0^{t_N} e(-\dot{\varphi} + a\varphi)\, dt. \tag{6.4.8}$$

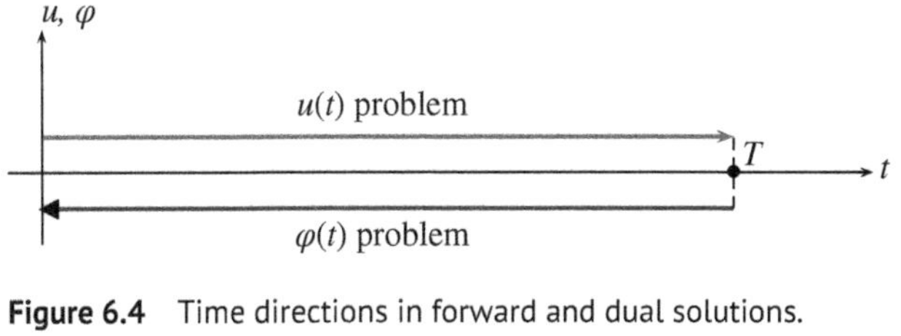

Figure 6.4 Time directions in forward and dual solutions.

Integration by parts yields

$$\int_0^{t_N} e(-\dot{\varphi} + a\varphi)\, dt = [-e(t)\varphi(t)]_{t=0}^{t_N} + \int_0^{t_N} \dot{e}\varphi\, dt + \int_0^{t_N} ea\varphi\, dt$$
$$= -e(t_N)\varphi(t_N) + \int_0^{t_N} (\dot{e} + ae)\varphi\, dt = -e_N^2 + \int_0^{t_N} (\dot{e} + ae)\varphi\, dt,$$

where we used that $e(0) = 0$ and $\varphi(t_N) = e_N$. Note that

$$\dot{e}(t) + a(t)e(t) = \dot{u}(t) - \dot{U}(t) + a(t)u(t) - a(t)U(t)$$
$$= f(t) - \dot{U}(t) - a(t)U(t) := -r(U), \tag{6.4.9}$$

where we used that $f(t) = \dot{u}(t) + a(t)u(t)$ and the residual $r(U) := \dot{U} + aU - f$. Consequently, we get the following *error representation formula*:

$$e_N^2 = -\int_0^{t_N} r(U(t))\varphi(t)\, dt. \tag{6.4.10}$$

To proceed, we use the L_2-projection $\pi_k\varphi = \frac{1}{k_n}\int_{I_n} \varphi(s)\, ds$ of φ onto the space of piecewise constant polynomials $W_k^{(0)}$ and write

$$e_N^2 = -\int_0^{t_N} r(U)(\varphi(t) - \pi_k\varphi(t))\, dt - \int_0^{t_N} r(U)\pi_k\varphi(t)\, dt. \tag{6.4.11}$$

Now, by the cG(1) method (6.3.8),

$$\int_0^{t_N} (\dot{U} + aU)v\, dt = \int_0^{t_N} fv\, dt, \qquad \forall v \in W_k^{(0)}, \tag{6.4.12}$$

where $W_k^{(0)}$ is the space of discontinuous, piecewise constant test functions. This gives the *orthogonality* relation, $r(U) \perp v, \quad \forall v \in W_k^{(0)}$, and implies that

$$\int_0^{t_N} r(U)\pi_k\varphi(t)\, dt = 0, \tag{6.4.13}$$

since $\pi_k\varphi \in W_k^{(0)}$. Thus, the final error representation formula reads as

$$e_N^2 = -\int_0^{t_N} r(U)(\varphi(t) - \pi_k\varphi(t))\, dt. \tag{6.4.14}$$

Next, we shall need the L_2-projection error estimate (proved as the interpolation error estimate) for the function φ in the interval I_n, with $|I_n| = k_n$:

$$\int_0^{t_N} |\varphi - \pi_k\varphi| dt \leq \sum_{n=1}^{N} k_n \int_{I_n} |\dot{\varphi}|\, dt. \tag{6.4.15}$$

To show (6.4.15), let $\overline{\varphi} = \frac{1}{k_n}\int_{I_n}\varphi(s)\,ds$ be the mean value of φ over I_n, then

$$\begin{aligned}\int_{I_n}|\varphi-\pi_k\varphi|\,dt &= \int_{I_n}|\varphi-\overline{\varphi}|\,dt\\ &= \int_{I_n}|\varphi(t)-\varphi(\xi)|\,dt = \int_{I_n}\left|\int_\xi^t \dot{\varphi}(s)\,ds\right|\,dt\\ &\le \int_{I_n}\left|\int_{I_n}\dot{\varphi}(s)\,ds\right|\,dt = k_n\cdot\int_{I_n}|\dot{\varphi}|\,dt,\end{aligned}$$

where we have used the mean value theorem for integrals, with $\xi \in I_n$.

Summing over n yields the global estimate

$$\int_0^{t_N}|\varphi-\pi_k\varphi|\,dt = \sum_{n=1}^{N}\int_{I_n}|\varphi-\pi_k\varphi|\,dt \le \sum_{n=1}^{N}k_n\int_{I_n}|\dot{\varphi}|\,dt. \tag{6.4.16}$$

Now let $|v|_J := \max_{t\in J}|v(t)|$, then using (6.4.16) and the final form of the error representation formula (6.4.14), we have that

$$|e_N|^2 \le \sum_{n=1}^{N}\left(|r(U)|_{I_n}k_n\int_{I_n}|\dot{\varphi}|\,dt\right) \le \max_{1\le n\le N}(k_n|r(U)|_{I_n})\int_0^{t_N}|\dot{\varphi}|\,dt.$$

Since, by the definition of $S(t_N)$, $\int_0^{t_N}|\dot{\varphi}|\,dt = |e_N|\cdot S(t_N)$, we finally get

$$|e_N|^2 \le |e_N|S(t_N)\max_{t\in[0,t_N]}(|kr(U)|). \tag{6.4.17}$$

This completes the proof of the first assertion of the theorem.

To prove the second assertion, we transform the dual problem (6.4.5) to a forward problem, using the change of variables, viz. $s = t_N - t$, $(t = t_N - s)$. Thus, for $\psi(s) := \varphi(t_N - s)$, using the chain rule, we get

$$\frac{d\psi}{ds} = \frac{d\varphi}{dt}\cdot\frac{dt}{ds} = -\dot{\varphi}(t_N - s). \tag{6.4.18}$$

The dual problem (6.4.5) is now reformulated as follows: find $\varphi(t)$ such that

$$-\dot{\varphi}(t_N - s) + a(t_N - s)\varphi(t_N - s) = 0. \tag{6.4.19}$$

The corresponding forward problem for ψ reads as follows: find $\psi(s)$ such that

$$\begin{cases}\dfrac{d\psi(s)}{ds} + a(t_N - s)\psi(s) = 0, & 0 < s \le t_N\\ \psi(0) = \varphi(t_N) = e_N, & e_N = u_N - U_N = u(t_N) - U(t_N).\end{cases}$$

By the solution formula (6.1.1), this problem has the solution $\psi(s)$ given by

$$\begin{aligned}\psi(s) &= \psi(0)\cdot e^{-\int_0^s a(t_N-u)\,du} = [t_N - u = v, -du = dv]\\ &= e_N\cdot e^{\int_{t_N}^{t_N-s}a(v)\,dv} = e_N\cdot e^{A(t_N-s)-A(t_N)}.\end{aligned}$$

Inserting back into the relation $\varphi(t) = \psi(s)$, with $t = t_N - s$, we end up with

$$\varphi(t) = e_N e^{A(t)-A(t_N)}, \quad \text{and} \quad \dot{\varphi}(t) = e_N a(t) e^{A(t)-A(t_N)}. \tag{6.4.20}$$

To estimate $S(t_N)$, we note that for $a(t) \geq 0$, then (6.4.20) yields

$$\begin{aligned}\int_0^{t_N} |\dot{\varphi}(t)|\, dt &= |e_N| \int_0^{t_N} a(t) e^{A(t)-A(t_N)}\, dt = |e_N| [e^{A(t)-A(t_N)}]_{t=0}^{t_N} \\ &= |e_N|(1 - e^{A(0)-A(t_N)}) = |e_N|(1 - e^{-A(t_N)}) \leq |e_N|,\end{aligned}$$

which gives $S(t_N) = \dfrac{\int_0^{t_N} |\dot{\varphi}(t)|\, dt}{|e_N|} \leq 1$.

As for $|a(t)| \leq \lambda$, we use again (6.4.20): $\dot{\varphi}(t) = a(t) e_N e^{A(t)-A(t_N)}$ and write

$$|\dot{\varphi}(t)| \leq \lambda |e_N| e^{A(t)-A(t_N)} = \lambda |e_N| e^{\int_{t_N}^{t} a(s)\, ds} \leq \lambda |e_N| e^{\lambda(t_N - t)}. \tag{6.4.21}$$

Integrating over $(0, t_N)$, we get

$$\int_0^{t_N} |\dot{\varphi}(t)|\, dt \leq |e_N| \int_0^{t_N} \lambda e^{\lambda(t_N-t)}\, dt = |e_N| [-e^{\lambda(t_N-t)}]_0^{t_N} = |e_N|(-1 + e^{\lambda t_N}),$$

which gives that $S(t_N) \leq (-1 + e^{\lambda t_N}) \leq e^{\lambda t_N}$, and completes the proof. □

The convergence rate in (6.4.6) appears to be of order $\mathcal{O}(k)$, but is actually of order $\mathcal{O}(k^2)$ since $|r(U)| \leq Ck$ if f is a smooth function. To prove the convergence of order $\mathcal{O}(k^2)$, we use some further orthogonality observations:

Theorem 6.4 Convergence order $\mathcal{O}(k^2)$. *For $N = 1, 2, \ldots$, and with S_N as in Theorem 6.3, the error for the cG(1) solution $U(t)$ satisfies*

$$|e_N| \leq S(t_N) \max_{t \in [0, t_N]} \left| k^2 \frac{d}{dt}(aU - f) \right|. \tag{6.4.22}$$

Proof: Let $\pi_k g$ be the L_2-projection of g onto the space $W_k^{(0)}$ of piecewise constant polynomials. Then, by the orthogonality relation $(g(t) - \pi_k g(t)) \perp$ (constants) $\forall g(t)$, e.g. as in Figure 6.5. Hence, since $\dot{U}(t)$ is constant on each

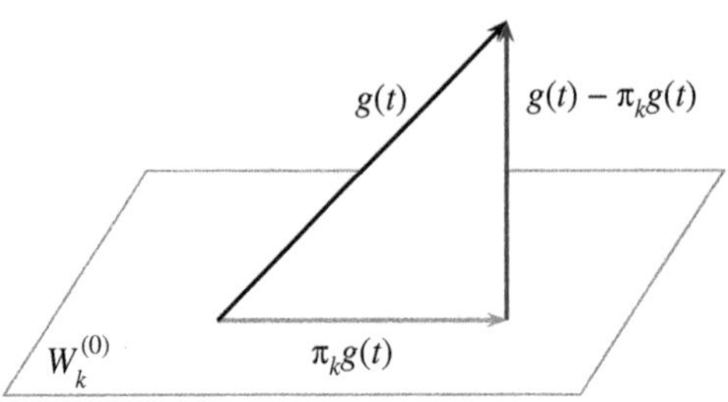

Figure 6.5 Orthogonality: $(g(t) - \pi_k g(t)) \perp$ (constants) $\forall g(t)$.

subinterval I_n, we have that $\int_0^{t_N} \dot{U}(\varphi - \pi_k\varphi)\, dt = 0$. Thus, using the error representation formula (6.4.14), we may write

$$e_N^2 = -\int_0^{t_N} r(U)(\varphi(t) - \pi_k\varphi(t))\, dt = \int_0^{t_N} (f - aU - \dot{U})(\varphi - \pi_k\varphi)\, dt$$
$$= -\int_0^{t_N} (aU - f)(\varphi - \pi_k\varphi)\, dt.$$

Similarly, using the fact that $\pi_k(aU - f)$ is constant on each subinterval I_n,

$$\int_0^{t_N} \pi_k(aU - f)(\varphi - \pi_k\varphi)\, dt = 0. \tag{6.4.23}$$

Consequently, we can write

$$e_N^2 = -\int_0^{t_N} ((aU - f) - \pi_k(aU - f))(\varphi - \pi_k\varphi)\, dt. \tag{6.4.24}$$

(compare (6.4.24) with (6.4.14)). To proceed, recall that for $t \in I_n$:

$$g(t) - \pi_k g(t) = g(t) - \bar{g}(t) = g(t) - g(\xi) = \int_\xi^t \dot{g}(s)\, ds,$$

where $\bar{g}$ is the mean value of g over I_n, $\xi \in I_n$, and we have used the mean value theorem for the integrals. This implies

$$|g(t) - \pi_k g(t)| \leq \int_{I_n} |\dot{g}(t)|\, dt \leq \max_{t\in I_n} |\dot{g}(t)| \cdot k_n. \tag{6.4.25}$$

Now, we apply (6.4.6) with $r(U)$ replaced by $(aU - f) - \pi_k(aU - f)$, (6.4.25) with $g := aU - f$, and the L_2-projection error estimate to get

$$|e_N| \leq S(t_N)|k|(aU - f) - \pi_k(aU - f)||_{[0,t_N]} \leq S(t_N)\left|k^2 \frac{d}{dt}(aU - f)\right|_{[0,t_N]},$$

which is the desired result. □

Remark 6.3 Note that the estimate (6.4.22) is less practical than the estimate (6.4.6) in the sense that it requires the cumbersome procedure of computing the time derivative of the residual.

6.4.2 A Posteriori Error Estimate for dG(0)

We shall now derive an a posteriori error estimate for the dG(0) method. In this case, the residual contains jump discontinuities.

Theorem 6.5 *For* $N = 1, 2, \ldots,$ *the dG(0) solution* $U(t)$ *satisfies*

$$|u(t_N) - U_N| \leq S(t_N)|kR(U)|_{[0,t_N]}, \quad U_N = U(t_N) \tag{6.4.26}$$

where

$$R(U) = |f - aU| + \frac{|U_n - U_{n-1}|}{k_n} \quad \textit{for} \quad t_{n-1} < t \le t_n. \tag{6.4.27}$$

Proof: The proof is similar to that of the cG(1) approach of Theorem 6.3, but we have to take the discontinuities of $U(t)$ at the nodes t_j into account. Here we demonstrate how to rearrange the contribution from these jumps, and in this way derive a corresponding error representation formula. To this end, recalling the dual problem $-\dot{\varphi}(t) + a(t)\varphi(t) = 0$, with $\varphi(t_N) = e_N$, we may write, using integration by parts on each subinterval

$$\begin{aligned}
e_N^2 &= e_N^2 + \sum_{n=1}^{N} \int_{t_{n-1}}^{t_n} e(-\dot{\varphi}(t) + a(t)\varphi(t))\, dt \\
&= e_N^2 + \sum_{n=1}^{N} \left(\int_{t_{n-1}}^{t_n} (\dot{e} + ae)\varphi(t)\, dt - [e\varphi]_{t_{n-1}}^{t_n} \right) \\
&= e_N^2 + \sum_{n=1}^{N} \int_{t_{n-1}}^{t_n} (f - aU)\varphi\, dt - \sum_{n=1}^{N} [e\varphi]_{t_{n-1}}^{t_n},
\end{aligned} \tag{6.4.28}$$

where in the last step, we have used $\dot{e} + ae = \dot{u} - \dot{U} + au - aU = f - aU$ since $\dot{U} = 0$ on each subinterval (where U is constant). For a given function g, we use the notation $g(t_n^-) = g_n^-$ and $g(t_{n-1}^+) = g_{n-1}^+$, and rewrite the last sum as

$$\begin{aligned}
\sum_{n=1}^{N} [e\varphi]_{t_{n-1}}^{t_n} &= \sum_{n=1}^{N} (e(t_n^-)\varphi(t_n^-) - e(t_{n-1}^+)\varphi(t_{n-1}^+)) \\
&= \sum_{n=1}^{N} (e_n^-\varphi_n^- - e_{n-1}^+\varphi_{n-1}^+) = (e_1^-\varphi_1^- - e_0^+\varphi_0^+) + (e_2^-\varphi_2^- - e_1^+\varphi_1^+) \\
&\quad + \cdots + (e_{N-1}^-\varphi_{N-1}^- - e_{N-2}^+\varphi_{N-2}^+) + (e_N^-\varphi_N^- - e_{N-1}^+\varphi_{N-1}^+).
\end{aligned}$$

We use the identity $\varphi_i^- = (\varphi_i^- - \varphi_i^+ + \varphi_i^+)$, $i = 1, \ldots, N-1$, and rewrite the contribution from the jumps as

$$\begin{aligned}
-\sum_{n=1}^{N} [e\varphi]_{t_{n-1}}^{t_n} &= -e_N^-\varphi_N^- + e_0^+\varphi_0^+ - e_1^-(\varphi_1^- - \varphi_1^+ + \varphi_1^+) + e_1^+\varphi_1^+ \\
&\quad - e_2^-(\varphi_2^- - \varphi_2^+ + \varphi_2^+) + e_2^+\varphi_2^+ - \cdots \\
&\quad - e_{N-1}^-(\varphi_{N-1}^- - \varphi_{N-1}^+ + \varphi_{N-1}^+) + e_{N-1}^+\varphi_{N-1}^+.
\end{aligned}$$

Here, the general ith term is written as

$$\begin{aligned}
-e_i^-(\varphi_i^- - \varphi_i^+ + \varphi_i^+) + e_i^+\varphi_i^+ &= -e_i^-\varphi_i^- + e_i^-\varphi_i^+ - e_i^-\varphi_i^+ + e_i^+\varphi_i^+ \\
&= e_i^-(\varphi_i^+ - \varphi_i^-) + \varphi_i^+(e_i^+ - e_i^-) = e_i^-[\varphi_i] + \varphi_i^+[e_i],
\end{aligned}$$

with $[g] = g^+ - g^-$ representing the jump. Hence, we have

$$-\sum_{n=1}^{N}[e\varphi]_{t_{n-1}}^{t_n} = -e_N^2 + e_0^+\varphi_0^+ + \sum_{n=1}^{N-1}[e_n]\varphi_n^+ + \sum_{n=1}^{N-1}e_n^-[\varphi_n]. \tag{6.4.29}$$

Inserting into (6.4.28) and using the fact that φ and u are smooth functions: $[\varphi_n] = [u_n] = 0$ and $[e_n] = [-U_n]$, we get the following error representation:

$$\begin{aligned}
e_N^2 &= e_N^2 + \sum_{n=1}^{N}\int_{t_{n-1}}^{t_n}(f - aU)\varphi\,dt - e_N^2 + e_0^+\varphi_0^+ + \sum_{n=1}^{N-1}[e_n]\varphi_n^+ + \sum_{n=1}^{N-1}[\varphi_n]e_n^- \\
&= e_0^+\varphi_0^+ + \sum_{n=1}^{N}\int_{t_{n-1}}^{t_n}(f - aU)\varphi\,dt - \sum_{n=1}^{N-1}[U_n]\varphi_n^+ \\
&= \sum_{n=1}^{N}\left(\int_{t_{n-1}}^{t_n}(f - aU)\varphi\,dt - [U_{n-1}]\varphi_{n-1}^+\right) \\
&= \sum_{n=1}^{N}\int_{t_{n-1}}^{t_n}(f - aU)(\varphi - \pi_k\varphi)\,dt - [U_{n-1}](\varphi - \pi_k\varphi)_{n-1}^+,
\end{aligned}$$

where we have used $e_0^+ = u_0^+ - U_0^+ = u_0 - U_0^+ = U_0^- - U_0^+ = -[U_0]$, and in the last step, once again, we use definition (6.3.12) of dG(q) with $q = 0$ and the L_2 projection $w = \pi_k\varphi \in W_k^{(0)}$ (space of piecewise constants) of φ. The remaining part of the proof is not substantially different from the proof of Theorem 6.3, and therefore omitted. □

6.4.3 Adaptivity for dG(0)

To guarantee a desired *local* bound for the error of the dG(0) approximation $U(t)$, such as

$$|e_n| = |u(t_n) - U(t_n)| \leq \text{TOL}, \tag{6.4.30}$$

where TOL is a given tolerance, we seek to determine the time step $k_n = t_n - t_{n-1}$ so that, using (6.4.26), locally, we have

$$S(t_N)\max_{t\in I_n}|k_n R(U)| \leq \text{TOL}, \quad n = 1, 2, \ldots, N. \tag{6.4.31}$$

6.4.3.1 An Adaptivity Algorithm

(i) Compute U_n from U_{n-1} using a predicted step k_n, and the relation

$$\int_{t_{n-1}}^{t_n} aU_n\,dt + U_n = \int_{t_{n-1}}^{t_n} f\,dt + U_{n-1}. \tag{6.4.32}$$

(ii) Compute $|kR(U)|_{I_n} := \max_{t\in I_n} |k_n R(U)|$, where $R(U) = |f - aU| + \frac{|U_n - U_{n-1}|}{k_n}$ in I_n, and follow the chart:

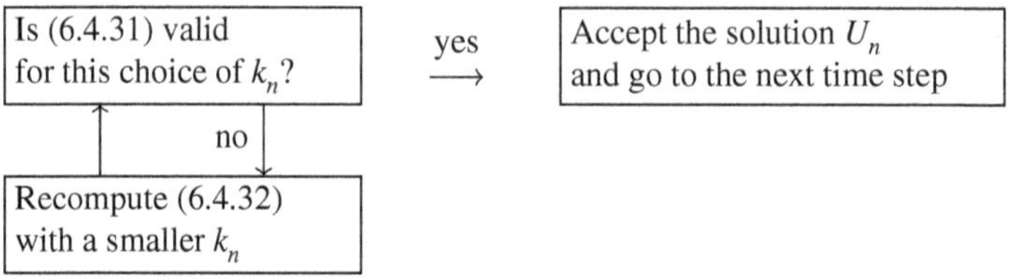

Remark 6.4 This is a rather simple and straightforward adaptive algorithm for the purpose of giving a general view. There are of course more involved adaptivity procedures, some particularly designed for the considered problem and/or the underlying numerical approach.

6.5 A Priori Error Analysis

In this section, we shall derive a priori error estimates for the dG(0) method. For simplicity, first we assume that a is a constant, and then we study the parabolic case $a(t) \geq 0$, where we perform a refined analysis. The corresponding a priori error estimate for the cG(1) method is obtained similarly. The weaker stability properties of cG(1), however, do not allow a refined analysis in the parabolic case.

6.5.1 A Priori Error Estimates for the dG(0) Method

The dG(0) method for $\dot{u} + au = f$, with a constant a, is formulated as follows: find $U = U(t), \quad t \in I_n$, such that

$$\int_{t_{n-1}}^{t_n} \dot{U}\, dt + a \int_{t_{n-1}}^{t_n} U\, dt = \int_{I_n} f\, dt, \quad n = 1, 2, \dots \tag{6.5.1}$$

Note that $U(t) = U_n$ is constant for $t \in I_n$. Let $U_n = U(t_n), \quad U_{n-1} = U(t_{n-1})$, and $k_n = t_n - t_{n-1}$, then

$$\int_{t_{n-1}}^{t_n} \dot{U}\, dt + a \int_{t_{n-1}}^{t_n} U\, dt = U_n - U_{n-1} + ak_n U_n.$$

Hence, with a given initial data $u(0) = u_0$, Eq. (6.5.1) is written as

$$U_n - U_{n-1} + ak_n U_n = \int_{I_n} f\, dt, \quad n = 1, 2, \dots \qquad U_0 = u_0. \tag{6.5.2}$$

For the *exact* solution $u(t)$ of $\dot{u} + au = f$, the same procedure yields

$$u(t_n) - u(t_{n-1}) + ak_n u(t_n) = \int_{I_n} f\, dt + ak_n u(t_n) - a\int_{t_{n-1}}^{t_n} u(t)\, dt, \tag{6.5.3}$$

where we have moved the term $a\int_{t_{n-1}}^{t_n} u(t)\, dt$ to the right-hand side and added $ak_n u(t_n)$ to both sides. Thus, from (6.5.2) and (6.5.3) we have that, denoting $u_n = u(t_n)$,

$$\begin{cases} (1 + k_n a)U_n = U_{n-1} + \displaystyle\int_{I_n} f\, dt, \\ (1 + k_n a)u_n = u_{n-1} + \displaystyle\int_{I_n} f\, dt + ak_n u_n - a\int_{t_{n-1}}^{t_n} u(t)\, dt. \end{cases} \tag{6.5.4}$$

Let now $e_n = u_n - U_n$ and $e_{n-1} = u_{n-1} - U_{n-1}$, then subtracting the first equation in (6.5.4) from the second one yields

$$e_n = (1 + k_n a)^{-1}(e_{n-1} + \rho_n), \qquad \textit{where} \quad \rho_n := ak_n u_n - a\int_{t_{n-1}}^{t_n} u(t)\, dt. \tag{6.5.5}$$

To estimate e_n, we use an iteration procedure and an estimate for ρ_n.

Lemma 6.1 *We have that*

$$|\rho_n| \le \frac{1}{2}|a|k_n^2 \max_{I_n} |\dot{u}(t)| \tag{6.5.6}$$

Proof: Evidently,

$$\begin{aligned} |\rho_n| &= \left| ak_n u_n - a\int_{t_{n-1}}^{t_n} u(t)\, dt \right| = |a|k_n \left| u_n - \frac{1}{k_n}\int_{I_n} u(t)\, dt \right| \\ &= |a| \left| \int_{t_{n-1}}^{t_n} (u_n - u(t))\, dt \right| = |a| \left| \int_{t_{n-1}}^{t_n} \int_t^{t_n} \dot{u}(s)\, ds\, dt \right| \\ &\le |a| \max_{I_n} |\dot{u}(t)| \cdot \int_{t_{n-1}}^{t_n} (t_n - t)\, dt = \frac{1}{2}|a|k_n^2 \cdot \max_{I_n} |\dot{u}(t)|. \end{aligned} \tag{6.5.7}$$

□

To simplify the estimate for e_n, we split the proof of the technical details in

Lemma 6.2 *For $k_n|a| \le 1/2$, and $n \ge 1$, we have that*

(i) $(1 - k_n|a|)^{-1} \le e^{2k_n|a|}$,

(ii) $|e_N| \le \dfrac{1}{2}\displaystyle\sum_{n=1}^{N} (e^{2|a|\tau_n}|a|k_n)(\max_{1\le n\le N} k_n|\dot{u}|_{I_n})$, $\qquad \tau_n = t_N - t_{n-1}$.

(iii) $\displaystyle\sum_{n=1}^{N} e^{2|a|\tau_n}|a|k_n \le e\int_0^{t_N} |a|e^{2|a|\tau}\, d\tau.$

We postpone the proof of this lemma and first show that using (i)–(iii), we can prove the following a priori error estimate.

Theorem 6.6 *If $k_n|a| \le \frac{1}{2}, n \ge 1$, then the error for the dG(0) approximation U satisfies*

$$|u(t_N) - U(t_N)| = |e_N| \le \frac{e}{4}(e^{2|a|t_N} - 1) \max_{1\le n\le N} k_n|\dot{u}(t)|_{I_n}. \tag{6.5.8}$$

Proof: Using estimates (ii) and (iii) of Lemma 6.2, we have that

$$|e_N| \le \frac{1}{2}\sum_{n=1}^{N}(e^{2|a|\tau_n}|a|k_n) \max_{1\le n\le N} k_n|\dot{u}|_{I_n} \le \frac{1}{2}\left(e\int_0^{t_N} |a|e^{2|a|\tau}\,d\tau\right) \max_{1\le n\le N} k_n|\dot{u}|_{I_n}$$
$$= \frac{1}{2}e\left[\frac{e^{2|a|\tau}}{2}\right]_0^{t_N} \max_{1\le n\le N} k_n|\dot{u}(t)|_{I_n} = \frac{e}{4}(e^{2|a|t_N} - 1) \max_{1\le n\le N} k_n|\dot{u}(t)|_{I_n},$$

and the proof is complete. □

Note that the stability factor $\frac{e}{4}(e^{2|a|t_N} - 1)$ grows exponentially depending on $|a|$ and t_N, hence, this result may not be satisfactory at all. We deal with this question below in the parabolic case $a(t) \ge 0$.

Remark 6.5 The a priori error estimate for the cG(1) method is similar to that of Theorem 6.6, with $k_n|\dot{u}|$ replaced by $k_n^2|\ddot{u}|$ (if u is sufficiently smooth).

We now return to the proof of the technical results (i)–(iii):

Proof of Lemma 6.2: (i) For $0 < x := k_n|a| \le 1/2$, we have that $1/2 \le 1 - x < 1$ implies $0 \le 1 - 2x < 1$. We may multiply both sides of $\frac{1}{1-x} < e^{2x}$ (which is the result we want to prove) by $1 - x \ge 1/2 > 0$ to obtain the equivalent relation

$$f(x) := (1 - x)e^{2x} - 1 > 0. \tag{6.5.9}$$

Note that since $f(0) = 0$ and $f'(x) = (1 - 2x)e^{2x} \ge 0$ (equality for $x = \frac{1}{2}$), thus (6.5.9), and hence, the relation (i) is valid.

To prove (ii), we recall that $e_n = (1 + k_n a)^{-1}(e_{n-1} + \rho_n)$. To deal with the factor $(1 + k_n a)^{-1}$, we use $k_n|a| \le \frac{1}{2}$, then by (i) $(1 + k_n a)^{-1} \le (1 - k_n|a|)^{-1} \le e^{2k_n|a|}$, $n \ge 1$. Thus,

$$|e_N| \le \frac{1}{1 - k_N|a|}|e_{N-1}| + \frac{1}{1 - k_N|a|}|\rho_N| \le |e_{N-1}|e^{2k_N|a|} + |\rho_N|e^{2k_N|a|}. \tag{6.5.10}$$

Relabeling, N to $N-1$,

$$|e_{N-1}| \le |e_{N-2}|e^{2k_{N-1}|a|} + |\rho_{N-1}|e^{2k_{N-1}|a|} = e^{2k_{N-1}|a|}(|e_{N-2}| + |\rho_{N-1}|),$$

and inserting into (6.5.10) yields

$$|e_N| \le e^{2k_N|a|}e^{2k_{N-1}|a|}(|e_{N-2}| + |\rho_{N-1}|) + |\rho_N|e^{2k_N|a|}. \tag{6.5.11}$$

Similarly, we have $|e_{N-2}| \le e^{2k_{N-2}|a|}(|e_{N-3}| + |\rho_{N-2}|)$. Now iterating (6.5.11) and using the fact that $e_0 = 0$, we get

$$\begin{aligned}|e_N| \le& e^{2k_N|a|}e^{2k_{N-1}|a|}e^{2k_{N-2}|a|}|e_{N-3}| + e^{2k_N|a|}e^{2k_{N-1}|a|}e^{2k_{N-2}|a|}|\rho_{N-2}| \\ &+ e^{2k_N|a|}e^{2k_{N-1}|a|}|\rho_{N-1}| + |\rho_N|e^{2k_N|a|} \le \cdots \le \\ \le& e^{2|a|\sum_{n=1}^N k_n}|e_0| + \sum_{n=1}^N e^{2|a|\sum_{m=n}^N k_m}|\rho_n| = \sum_{n=1}^N e^{2|a|\sum_{m=n}^N k_m}|\rho_n|.\end{aligned}$$

Recalling (6.5.6): $|\rho_n| \le \frac{1}{2}|a|k_n^2 \max_{I_n}|\dot{u}(t)|$, we have the error estimate

$$|e_N| \le \left(\sum_{n=1}^N e^{2|a|\sum_{m=n}^N k_m}\right)\frac{1}{2}|a|k_n^2 \max_{I_n}|\dot{u}(t)|. \tag{6.5.12}$$

Note that

$$\sum_{m=n}^N k_m = (t_n - t_{n-1}) + (t_{n+1} - t_n) + (t_{n+2} - t_{n+1}) + \cdots + (t_N - t_{N-1}) = t_N - t_{n-1}.$$

Hence, since $\tau_n = t_N - t_{n-1}$, we have shown assertion (ii), i.e.

$$|e_N| \le \sum_{n=1}^N e^{2|a|(t_N - t_{n-1})}\frac{1}{2}|a|k_n^2 \max_{I_n}|\dot{u}(t)| \le \frac{1}{2}\sum_{n=1}^N (e^{2|a|\tau_n}|a|k_n)(\max_{1\le n\le N} k_n|\dot{u}|_{I_n}).$$

(iii) To prove this part, we note that

$$\tau_n = t_N - t_{n-1} = (t_N - t_n) + (t_n - t_{n-1}) = \tau_{n+1} + k_n, \tag{6.5.13}$$

and since $|a|k_n \le 1/2$, we have $2|a|\tau_n = 2|a|\tau_{n+1} + 2|a|k_n \le 2|a|\tau_{n+1} + 1$. Further, for $\tau_{n+1} \le \tau \le \tau_n$, we can write

$$\begin{aligned}e^{2|a|\tau_n}k_n &= \int_{\tau_{n+1}}^{\tau_n} e^{2|a|\tau_n}\,d\tau \le \int_{\tau_{n+1}}^{\tau_n} e^{(2|a|\tau_{n+1}+1)}\,d\tau \\ &= \int_{\tau_{n+1}}^{\tau_n} e^1 \cdot e^{2|a|\tau_{n+1}}\,d\tau \le e\int_{\tau_{n+1}}^{\tau_n} e^{2|a|\tau}\,d\tau.\end{aligned} \tag{6.5.14}$$

Multiplying (6.5.14) by $|a|$ and summing over n, we get

$$\sum_{n=1}^{N} e^{2|a|\tau_n}|a|k_n \leq e\left(\sum_{n=1}^{N}\int_{\tau_{n+1}}^{\tau_n} e^{2|a|\tau}\, d\tau\right)|a|$$

$$= e\int_{\tau_{N+1}}^{\tau_1} e^{2|a|\tau}|a|\, d\tau = e\int_0^{t_N}|a|e^{2|a|\tau}\, d\tau, \qquad (6.5.15)$$

which is the desired result and the proof of (iii) is complete. □

6.6 The Parabolic Case ($a(t) \geq 0$)

The interest in parabolic case is due to the fact that, for a bounded total source ($f \in L_1(0,t)$), it does not accumulate the error. This can be seen from the solution formula (6.1.1) with the presence of the deceasing multiplicative factor $e^{-A(t)}$. Below we state and proof some basic estimate for this case:

Theorem 6.7 *Consider the dG(0) approximation U for* $\dot{u} + au = f$, *with* $a(t) \geq 0$. *Assume that* $k_j|a|_{I_j} \leq \frac{1}{2}$; $\forall j$, *then we have the error estimates*

$$|u(t_N) - U_N| \leq \begin{cases} 3e^{2\lambda t_N} \max\limits_{0\leq t\leq t_N}|k\dot{u}|, & \textit{for} \quad |a(t)| \leq \lambda, \\ 3\max\limits_{0\leq t\leq t_N}|k\dot{u}|, & \textit{for} \quad a(t) \geq 0. \end{cases} \qquad (6.6.1)$$

Main steps in the proof: We shall not give a proof of this theorem, rather derive the crucial steps. Let $e = u - U = (u - \pi_k u) + (\pi_k u - U) := \tilde{e} + \bar{e}$, where $\tilde{e}$ is the projection error with $\pi_k u$ being the L_2-projection of u into $W_k^{(0)}$. Since the L_2-projection error is of the form (6.6.1), hence, it suffices to estimate the *discrete error* $\bar{e}$. To this end, we shall use the following *discrete dual problem* (DDP):

Find $\Phi \in W_k^{(0)}$, such that for $n = N, N-1, \dots, 1$,

$$(DDP) \qquad \begin{cases} \displaystyle\int_{t_{n-1}}^{t_n}(-\dot{\Phi} + a(t)\Phi)v\, dt - [\Phi_n]v_n = 0, \quad \forall v \in W_k^{(0)} \\ \Phi_N^+ = \Phi_{N+1} = (\pi_k u - U)_N :\equiv \bar{e}_N. \end{cases} \qquad (6.6.2)$$

Let now $v = \bar{e}$, then

$$|\bar{e}_N|^2 = \sum_{n=1}^{N}\int_{t_{n-1}}^{t_n}(-\dot{\Phi} + a(t)\Phi)\bar{e}\, dt - \sum_{n=1}^{N-1}[\Phi_n]\bar{e}_n + \Phi_N\bar{e}_N. \qquad (6.6.3)$$

We use $\bar{e} = (\pi_k u - U) = (\pi_k u - u + u - U)$ and rewrite (6.6.3) as

$$|\bar{e}_N|^2 = \sum_{n=1}^{N} \int_{t_{n-1}}^{t_n} (-\dot{\Phi} + a(t)\Phi)(\pi_k u - u + u - U)\, dt$$
$$- \sum_{n=1}^{N-1} [\Phi_n](\pi_k u - u + u - U)_n + \Phi_N(\pi_k u - u + u - U)_N.$$

By Galerkin orthogonality, the total contribution from $u - U$ terms vanishes, and we end up with the error terms involving only $\pi_k u - u$. Hence, due to the fact that $\dot{\Phi} = 0$ on each Subinterval, we have the following error representation formula

$$|\bar{e}_N|^2 = \sum_{n=1}^{N} \int_{t_{n-1}}^{t_n} (-\dot{\Phi} + a(t)\Phi)(\pi_k u - u)\, dt - \sum_{n=1}^{N-1} [\Phi_n](\pi_k u - u)_n + \Phi_N(\pi_k u - u)_N$$
$$= - \int_0^{t_N} (a(t)\Phi)(u - \pi_k u)\, dt + \sum_{n=1}^{N-1} [\Phi_n](u - \pi_k u)_n - \Phi_N(u - \pi_k u)_N. \qquad \square$$

Below are some further important estimates:

Lemma 6.3 *If $|a(t)| \leq \lambda, \forall t \in (0, t_N)$ and $k_j |a|_{I_j} \leq \frac{1}{2}$, for $j = 1, \ldots, N$, then the solution of the DDP satisfies*

(i) $|\Phi_n| \leq e^{2\lambda(t_N - t_{n-1})} |\bar{e}_N|$,

(ii) $\sum_{n=1}^{N-1} |[\Phi_n]| \leq e^{2\lambda t_N} |\bar{e}_N|$,

(iii) $\sum_{n=1}^{N} \int_{t_{n-1}}^{t_n} |a(t)\Phi_n|\, dt \leq e^{2\lambda t_N} |\bar{e}_N|$,

(iv) $Max\left(|\Phi_n|, \sum_{n=1}^{N} |[\Phi_n]|, \sum_{n=1}^{N} \int_{t_{n-1}}^{t_n} a(t)|\Phi_n|\, dt \right) \leq |\bar{e}_N|, \quad a(t) \geq 0.$

Proof: We show the last estimate (iv), (the proofs of (i)–(iii) are similar to that of the stability factor in Theorems 6.4–6.6). Consider the discrete, local dual problem with $v \equiv 1$:

$$\begin{cases} \int_{t_{n-1}}^{t_n} (-\dot{\Phi} + a(t)\Phi)\, dt - [\Phi_n] = 0, \\ \Phi_{N+1} = (\pi_k u - U)_N :\equiv \bar{e}_N. \end{cases} \tag{6.6.4}$$

For dG(0) and for a constant Φ, this becomes

$$\begin{cases} -\Phi_{n+1} + \Phi_n + \Phi_n \int_{t_{n-1}}^{t_n} a(t)\, dt = 0, \quad n = N, N-1, \ldots, 1 \\ \Phi_{N+1} = \bar{e}_N, \qquad \Phi_n = \Phi|_{I_n}. \end{cases} \tag{6.6.5}$$

By iterating, we get

$$\Phi_n = \prod_{j=n}^{N} \left(1 + \int_{I_j} a(t)\, dt\right)^{-1} \Phi_{N+1}. \tag{6.6.6}$$

For $a(t) \geq 0$, we have $\left(1 + \int_{I_j} a(t)\, dt\right)^{-1} \leq 1$, thus (6.6.6) implies that

$$|\Phi_n| \leq \Phi_{N+1} = |\bar{e}_N|. \tag{6.6.7}$$

Further, we have using (6.6.6) that

$$\Phi_{n-1} = \prod_{j=n-1}^{N} \left(1 + \int_{I_j} a(t)\, dt\right)^{-1} \Phi_{N+1} = \left(1 + \int_{I_{n-1}} a(t)\, dt\right)^{-1} \Phi_n \leq \Phi_n$$

which implies that

$$[\Phi_n] = \Phi_n^+ - \Phi_n^- = \Phi_{n+1} - \Phi_n \geq 0. \tag{6.6.8}$$

Now, since by (6.6.6), $\Phi_1 \leq \Phi_{N+1}$, thus

$$\begin{aligned} \sum_{n=1}^{N} |[\Phi_n]| &= \Phi_{N+1} - \Phi_N + \Phi_N - \Phi_{N-1} + \cdots + \Phi_2 - \Phi_1 \\ &= \Phi_{N+1} - \Phi_1 \leq \Phi_{N+1} = |\bar{e}_N|. \end{aligned} \tag{6.6.9}$$

Finally, in the discrete equation:

$$\int_{t_{n-1}}^{t_n} (-\dot{\Phi} + a(t)\Phi) v\, dt - [\Phi_n] v_n = 0, \qquad \forall v \in W_k^{(0)}, \tag{6.6.10}$$

we have $v \equiv 1$ and $\dot{\Phi} \equiv 0$ for the dG(0). Hence, (6.6.10) can be rewritten as

$$\int_{t_{n-1}}^{t_n} a(t)\Phi_n\, dt = [\Phi_n]. \tag{6.6.11}$$

Summing over n, this gives that

$$\sum_{n=1}^{N} \int_{t_{n-1}}^{t_n} a(t)\Phi_n\, dt = \sum_{n=1}^{N} [\Phi_n] = \sum_{n=1}^{N} |[\Phi_n]| \leq |\bar{e}_N|. \tag{6.6.12}$$

Combining (6.6.7), (6.6.9), and (6.6.12) completes the proof of (iv). □

Below we state some useful results (their proofs are standard, however, lengthy and therefore are omitted).

Quadrature rule. Assume that a is constant. Then the error representation formula in the proof of the Theorem 6.5 for dG(0), combined with the quadrature rule for f reads as follows:

$$e_N^2 = \sum_{n=1}^{N} \left(\int_{t_{n-1}}^{t_n} (f - aU)(\varphi - \pi_k\varphi)\, dt - [U_{n-1}](\varphi - \pi_k\varphi)_{n-1}^+ \right.$$
$$\left. + \int_{t_{n-1}}^{t_n} f\pi_k\varphi\, dt - (\overline{f\pi_k\varphi})_n k_n \right), \tag{6.6.13}$$

where the last two terms represent the quadrature error. Here, e.g. for the endpoint-rule $\bar{g}_n = g(t_n)$, whereas for the midpoint-rule $\bar{g}_n := g(t_{(n-1/2)})$. We also recall the *weak stability factor* $\tilde{S}(t_N) := \int_0^{t_N} |\varphi|\, dt/|e_N|$, where φ is the solution of the dual problem

$$-\dot{\varphi} + a\varphi = 0, \quad \text{for} \quad t_N > t \geq 0 \qquad \varphi(t_N) = e_N.$$

Recall that $\pi_k\varphi$ is piecewise constant with the *mass-preserving* property:

$$\int_{I_n} |\pi_k\varphi(t)|\, dt = \int_{I_n} |\varphi(t)|\, dt.$$

It is easy to prove the following relations between the two stability factors:

$$\tilde{S}(t_N) \leq t_N(1 + S(t_N)),$$

and if $a > 0$ is sufficiently small, then we have indeed $\tilde{S}(t_N) \gg S(t_N)$. Not so glad news!

Finally, we state (without proof, see Johnson et al. [57] for further details) the following dG(0) estimate.

Theorem 6.8 The modified a posteriori estimate for dG(0). *For each $N = 1, 2, \ldots,$ the dG(0) approximation $U(t)$ computed using a quadrature rule on the source term f satisfies*

$$|u(t_n) - U_n| \leq S(t_n)|kR(U)|_{(0,t_N)} + \tilde{S}(t_N)C_j|k^j f^{(j)}|_{(0,t_N)}, \tag{6.6.14}$$

where

$$R(U) = \frac{|U_n - U_{n-1}|}{k_n} + |f - aU|, \quad on \quad I_n \tag{6.6.15}$$

and $j = 1$ for the rectangle rule, $j = 2$ for the midpoint rule, $C_1 = 1$, $C_2 = 1/2$, $f^{(1)} := \dot{f} := df/dt$ and $f^{(2)} := \ddot{f} := d^2f/dt^2$.

6.6.1 An Example of Error Estimate

Here we derive a simple versions of the cG(1) error estimate above.

Example 6.5 Let U be the cG(1) approximation of the solution u for the IVP

$$\dot{u} + u = f, \quad t > 0, \quad u(0) = u_0. \tag{6.6.16}$$

Then, with k being the time step, we have that

$$|(u-U)(T)| \leq \max_{[0,T]} |k(f-\dot{U}-U)|. \qquad (6.6.17)$$

Proof: Evidently, the error $e = u - U$ satisfies the Galerkin orthogonality

$$\int_0^T (\dot{e}+e)v\, dt = 0, \qquad \text{for all piecewise constants } v(t). \qquad (6.6.18)$$

Let φ be the solution of the dual equation

$$-\dot{\varphi}+\varphi = 0, \quad t < T, \quad \varphi(T) = e(T), \qquad (6.6.19)$$

then $\varphi(t) = e(T)e^{t-T}$. Further,

$$|e(T)|^2 = e(T)\cdot e(T) + \int_0^T e(-\dot{\varphi}+\varphi)\, dt = e(T)\cdot e(T) - \int_0^T e\dot{\varphi}\, dt + \int_0^T e\varphi\, dt.$$

Integration by parts yields

$$\int_0^T e\dot{\varphi}\, dt = [e\cdot\varphi]_{t=0}^T - \int_0^T \dot{e}\varphi\, dt = e(T)\varphi(T) - e(0)\varphi(0) - \int_0^T \dot{e}\varphi\, dt.$$

Hence, using $\varphi(T) = e(T)$, and $e(0) = 0$, we have

$$|e(T)|^2 = e(T)\cdot e(T) - e(T)\cdot e(T) + \int_0^T \dot{e}\varphi\, dt + \int_0^T e\varphi\, dt = \int_0^T (\dot{e}+e)\varphi\, dt$$

$$= \int_0^T (\dot{e}+e)(\varphi - v)\, dt = \int_0^T (\dot{u}+u-\dot{U}-U)(\varphi - v)\, dt.$$

We have that $r(U) := \dot{U}+U-f$, is the residual and

$$|e(T)|^2 = -\int_0^T r(U)\cdot(\varphi - v)\, dt \leq \max_{[0,T]} |k\cdot r(U)| \int_0^T \frac{1}{k}|\varphi - v|\, dt. \qquad (6.6.20)$$

Recall that

$$\int_I k^{-1}|\varphi - v|\, dx \leq \int_I |\dot{\varphi}|\, dx. \qquad (6.6.21)$$

Further, $-\dot{\varphi}+\varphi = 0$ implies $\dot{\varphi} = \varphi$, and $\varphi(t) = e(T)e^{t-T}$. Thus,

$$|e(T)|^2 \leq \max_{[0,T]} |k\, r(U)| \int_0^T |\dot{\varphi}|\, dt = \max_{[0,T]} |k\, r(U)| \int_0^T |\varphi(t)|\, dt$$

$$\leq \max_{[0,T]} |kr(U)||e(T)| \int_0^T e^{t-T}\, dt. \qquad (6.6.22)$$

Now, since

$$\int_0^T e^{t-T}\,dt = [e^{t-T}]_0^T = e^0 - e^{-T} = 1 - e^{-T} \leq 1, \quad T > 0,$$

finally, we end up with the desired result:

$$|e(T)| \leq \max_{[0,T]} |k\, r(U)|. \qquad \square$$

6.7 Exercises

Problem 6.1 Generalize the Example 6.5 to the problem $\dot{u} + au = f$, with $a =$ positive constant. Is the statement of this example valid for $\dot{u} - u = f$?

Problem 6.2 Study the dG(0)-case for $\dot{u} + au = f, \quad a > 0$

Problem 6.3

(a) Derive the stiffness matrix (coefficient matrix) and load vector in piecewise polynomial (of degree q) approximation for the following ODE in population dynamics,

$$\begin{cases} \dot{u}(t) = \lambda u(t), & \text{for } 0 < t \leq 1, \\ u(0) = u_0. \end{cases}$$

(b) Let $\lambda = 1$ and $u_0 = 1$ and determine the approximate solution $U(t)$, for $q = 1$ and $q = 2$.

Problem 6.4 Consider the initial value problem

$$\dot{u}(t) + a(t)u(t) = f(t), \quad 0 < t \leq T, \qquad u(0) = u_0.$$

Show that if $a(t) = 1, f(t) = 2\sin(t)$, then we have
$u(t) = \sin(t) - \cos(t) = \sqrt{2}\sin(t - \pi/2) + (u_0 + 1)e^{-t}$.

Problem 6.5 Compute the solution for

$$\dot{u}(t) + a(t)u(t) = t^2, \quad 0 < t \leq T, \qquad u(0) = 1, \qquad \text{corresponding to}$$

(a) $a(t) = 4$, (b) $a(t) = -t$.

Problem 6.6 Compute the cG(1) approximation for the differential equations in the above problem. In each case, determine the condition on the step size that guarantees that U exists.

Problem 6.7 Without using fundamental theorem 6.1, prove that if $a(t) \geq 0$ then, a continuously differentiable solution of (6.0.1) is unique.

Problem 6.8 Assume that

$$\int_{I_j} f(s)\, ds = 0, \qquad \text{for } j = 1, 2, \dots,$$

where $I_j = (t_{j-1}, t_j)$, $t_j = jk$ with k being a positive constant. Prove that if $a(t) \geq 0$, then the solution for (6.0.1) satisfies

$$|u(t)| \leq e^{-A(t)}|u_0| + \max_{0 \leq s \leq t} |kf(s)|.$$

Problem 6.9 Let U be the cG(1) approximation of u satisfying the initial value problem

$$\dot{u} + au = f, \quad t > 0, \qquad u(0) = u_0.$$

Let k be the time step and show that for $a = 1$,

$$|(u - U)(T)| \leq \min(||k(f - \dot{U} - U)||_{L^\infty[0,T]}, \; T||k^2\ddot{u}||_{L^\infty[0,T]}).$$

Problem 6.10 Consider the scalar boundary value problem

$$\dot{u}(t) + a(t)u(t) = f(t), \quad t > 0, \qquad u(0) = u_0.$$

(a) Show that for $a(t) \geq a_0 > 0$, we have the stability estimate

$$|u(t)| \leq e^{-a_0 t}\left(|u_0| + \int_0^t e^{a_0 s}|f(s)|\, ds\right)$$

(b) Formulate the cG(1) method for this problem, and show that the condition $\frac{1}{2}a_0 k > -1$, where k is the time step, guarantees that the method is operational, i.e. no zero division occurs.

(c) Assume that $a(t) \geq 0$, $f(t) \equiv 0$, and estimate the quantity $\int_0^T |\dot{u}|\, dt/|u_0|$.

Problem 6.11 Consider the IVP ($u = u(x, t)$)

$$\dot{u} + Au = f, \quad t > 0; \qquad u(t = 0) = u_0.$$

Show that if there is a constant $\alpha > 0$ such that

$$(Av, v) \geq \alpha||v||^2, \qquad \forall v,$$

then the solution u of the IVP satisfies the stability estimate

$$||u(t)||^2 + \alpha \int_0^t ||u(s)||^2\, ds \leq ||u_0||^2 + \frac{1}{\alpha}\int_0^t ||f(s)||^2\, ds.$$

Problem 6.12 Formulate a continuous Galerkin method using piecewise polynomials based on the original global Galerkin method.

Problem 6.13 Use the a priori error bound to show that the residual of the dG(0) approximation satisfies $\mathcal{R}(U) = \mathcal{O}(1)$.

Problem 6.14 Prove the following stability estimate for the dG(0) method described by (6.3.12),

$$|U_N|^2 + \sum_{n=0}^{N-1} |[U_n]|^2 \leq |u_0|^2.$$

7

Initial Boundary Value Problems in 1*d*

A large class of phenomena in nature, science, and technology, such as heat distribution, wave propagation, convection–diffusion–advection processes, etc., are varying both in space and time. To describe these phenomena in a physical domain requires the knowledge of their initial status, as well as information on the boundary of the domain (relevant status of the surrounding environment), or asymptotic behavior in the case of unbounded domains. Problems that model such properties are called initial boundary value problems. In this chapter, we study the two most common equations of this type: namely, the heat equation and the wave equation in one-space dimension. We also address (briefly) the one-space dimensional time-dependent convection–diffusion problem.

7.1 The Heat Equation in 1*d*

In this section, we focus on some basic L_2-stability and finite element error estimates for the one-space dimensional heat equation. A general discussion on the heat equation can be found, e.g. in *Lecture Notes in Fourier Analysis* [12]. Here, to illustrate, we consider an example of an initial boundary value problem (IBVP) for the one-dimensional heat equation, viz.

$$\begin{cases} \dot{u}(x,t) - u''(x,t) = f(x,t), & 0 < x < 1, \quad t > 0, \\ u(x,0) = u_0(x), & 0 < x < 1, \\ u(0,t) = u_x(1,t) = 0, & \qquad\qquad t > 0. \end{cases} \tag{7.1.1}$$

We recall the following notation:

$$\dot{u} := u_t = \frac{\partial u}{\partial t}, \quad u' := u_x = \frac{\partial u}{\partial x}, \quad u'' := u_{xx} = \frac{\partial^2 u}{\partial x^2}.$$

Note that the partial differential equation in (7.1.1) contains three derivatives, two derivatives in space and one derivative in time. This yields 3 degrees of freedom

An Introduction to the Finite Element Method for Differential Equations, First Edition.
Mohammad Asadzadeh.

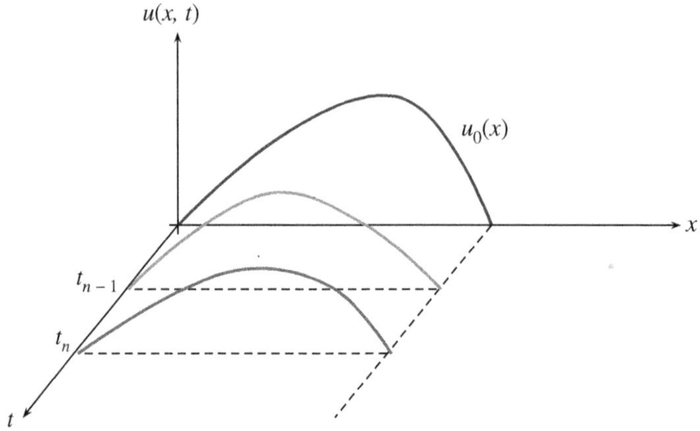

Figure 7.1 A decreasing temperature profile with data $u(0, t) = u(1, t) = 0$.

(in the sense that, to recover the solution u from Eq. (7.1.1), it is necessary to integrate twice in space and once in time, where each integration introduces an arbitrary "constant)." Detailed discussion on this can be found in Chapter 1. Therefore, to determine a unique solution to the heat equation, it is necessary to supply three data: here two boundary conditions corresponding to the two spatial derivatives in u'' and an initial condition corresponding to the time derivative in $\dot{u}$. Figure 7.1 illustrates an example of heat conduction with homogeneous Dirichlet boundary conditions.

Following is an example, where these concepts are described in physical terms.

Example 7.1 Describe the physical meaning of the functions and parameters in Problem (7.1.1), when $f = 21 - u$.

Solution: The problem is an example of heat conduction where

$u(x, t)$,	means the temperature at the point x and time t.
$u(x, 0) = u_0(x)$,	is the initial temperature distribution at time $t = 0$.
$u(0, t) = 0$,	means fixed temperature 0 at the boundary point $x = 0$.
$u'(1, t) = 0$,	means isolated boundary at the boundary point $x = 1$ (where no heat flux occurs).
$f = 21 - u$,	is the heat source, in this case a control system to force $u \to 21$.

Remark 7.1 Observe that it is possible to generalize (7.1.1) to a u dependent source term f, e.g. as in the above example, where $f = 21 - u$.

The analytic approach for this problem is either through the separation of variables technique or the Green's functions approach, viz. Chapter 2.

7.1.1 Stability Estimates

We shall derive a general stability estimate for the mixed (Dirichlet at one end point and Neumann at the other) initial boundary value problem mentioned above, where we rely on a one-dimensional version of the *Poincare inequality.*

Theorem 7.1 *The solution for the IBVP (7.1.1) satisfies the stability estimates*

$$||u(\cdot,t)|| \leq ||u_0|| + \int_0^t ||f(\cdot,s)||\, ds, \tag{7.1.2}$$

$$||u_x(\cdot,t)||^2 \leq ||u_0'||^2 + \int_0^t ||f(\cdot,s)||^2\, ds, \tag{7.1.3}$$

where u_0 and u_0' are assumed to be $L_2(I)$ functions with $I = (0,1)$, further, we have assumed that $f \in L_1([0,t], L_2(I)) \cap L_2([0,t], L_2(I))$. Note further that, here $|| \bullet (\cdot,t)||$ is the time-dependent, spatial, *L_2 norm:*

$$||w(\cdot,s)|| := ||w(\cdot,s)||_{L_2(0,1)} = \left(\int_\Omega ||w(x,s)||^2\, dx\right)^{1/2}.$$

Proof: Multiply the equation in (7.1.1) by u and integrate over $(0,1)$ to get

$$\int_0^1 \dot{u}u\, dx - \int_0^1 u''u\, dx = \int_0^1 fu\, dx. \tag{7.1.4}$$

Note that $\dot{u}u = \frac{1}{2}\frac{d}{dt}(u^2)$. Integrating by parts in the second integral yields

$$\frac{1}{2}\frac{d}{dt}\int_0^1 u^2\, dx + \int_0^1 (u')^2\, dx - u'(1,t)u(1,t) + u'(0,t)u(0,t) = \int_0^1 fu\, dx.$$

Thus, using the boundary conditions and Cauchy–Schwarz' inequality, we end up with

$$||u||\frac{d}{dt}||u|| + ||u'||^2 = \int_0^1 fu\, dx \leq ||f||\; ||u||. \tag{7.1.5}$$

Now, since $||u'||^2 \geq 0$, consequently $||u||\frac{d}{dt}||u|| \leq ||f||\; ||u||$, and thus

$$\frac{d}{dt}||u|| \leq ||f||. \tag{7.1.6}$$

Relabeling the variable from t to s, and integrating over time, we get

$$||u(\cdot,t)|| - ||u(\cdot,0)|| \leq \int_0^t ||f(\cdot,s)||\, ds, \tag{7.1.7}$$

which yields the first assertion (7.1.2) of the theorem. To prove (7.1.3), we multiply the differential equation by $\dot{u}$, integrate over $(0, 1)$, and use integration by parts so that we have on the left-hand side

$$\int_0^1 (\dot{u})^2\, dx - \int_0^1 u''\dot{u}\, dx = \|\dot{u}\|^2 + \int_0^1 u'\dot{u}'\, dx - u'(1,t)\dot{u}(1,t) + u'(0,t)\dot{u}(0,t).$$

Then, since $u_x(1, t) = 0$ and $u(0, t) = 0 \Rightarrow \dot{u}(0, t) = 0$, we have that

$$\|\dot{u}\|^2 + \frac{1}{2}\frac{d}{dt}\|u'\|^2 = \int_0^1 f\dot{u}\, dx \le \|f\|\, \|\dot{u}\| \le \frac{1}{2}(\|f\|^2 + \|\dot{u}\|^2), \tag{7.1.8}$$

where in the last step, we used Cauchy–Schwarz' inequality. Hence,

$$\frac{1}{2}\|\dot{u}\|^2 + \frac{1}{2}\frac{d}{dt}\|u'\|^2 \le \frac{1}{2}\|f\|^2. \tag{7.1.9}$$

and therefore,

$$\frac{d}{dt}\|u'\|^2 \le \|f\|^2. \tag{7.1.10}$$

Finally, integrating over $(0, t)$, we get the second assertion of the theorem:

$$\|u'(\cdot,t)\|^2 - \|u'(\cdot,0)\|^2 \le \int_0^t \|f(\cdot,s)\|^2\, ds, \tag{7.1.11}$$

and the proof is complete. □

Theorem 7.2 (Stability of the homogeneous heat equation) *The initial boundary value problem for the heat equation*

$$\begin{cases} \dot{u} - u'' = 0, & 0 < x < 1, \quad t > 0 \\ u(0,t) = u_x(1,t) = 0, & \qquad\qquad\quad t > 0 \\ u(x,0) = u_0(x), & 0 < x < 1, \end{cases} \tag{7.1.12}$$

satisfies the following stability estimates

(a) $\frac{d}{dt}\|u\|^2 + 2\|u'\|^2 = 0,$

(b) $\|u(\cdot,t)\| \le e^{-t}\|u_0\|.$

Remark 7.2 Note in particular that the stability estimate (b) in above means that in the homogeneous case (i.e. in the absence of a source term in (7.1.1)), the temperature will decrease exponentially in time (although it might increase pointwise). Note further that we do not impose homogeneous Neumann boundary condition at both end points (can you see why?).

Proof: (a) Multiply the equation by u and integrate over $x \in (0, 1)$, to get

$$0 = \int_0^1 (\dot{u} - u'')u\, dx = \int_0^1 \dot{u}u\, dx + \int_0^1 (u')^2\, dx - u'(1,t)u(1,t) + u'(0,t)u(0,t),$$

where we used integration by parts. Using the boundary data, we then have

$$\frac{1}{2}\frac{d}{dt}\int_0^1 u^2\,dx + \int_0^1 (u')^2\,dx = \frac{1}{2}\frac{d}{dt}\|u\|^2 + \|u'\|^2 = 0.$$

This gives the proof of (a). As for the proof of (b), using (a) and the Poincare inequality, with $L = 1$, i.e. $\|u\| \le \|u'\|$, we get

$$\frac{d}{dt}\|u\|^2 + 2\|u\|^2 \le 0. \tag{7.1.13}$$

Multiplying both sides of (7.1.13) by the integrating factor e^{2t} yields

$$\frac{d}{dt}(\|u\|^2 e^{2t}) = \left(\frac{d}{dt}\|u\|^2 + 2\|u\|^2\right) e^{2t} \le 0. \tag{7.1.14}$$

We replace t by s and integrate in s over $(0, t)$, to obtain

$$\int_0^t \frac{d}{ds}(\|u\|^2 e^{2s})\,ds = \|u(\cdot, t)\|^2 e^{2t} - \|u(\cdot, 0)\|^2 \le 0. \tag{7.1.15}$$

This yields

$$\|u(\cdot, t)\|^2 \le e^{-2t}\|u_0\|^2 \Rightarrow \|u(\cdot, t)\| \le e^{-t}\|u_0\|, \tag{7.1.16}$$

and completes the proof. □

Remark 7.3 The stability estimate (7.1.2) may be used to show the uniqueness of solution for the heat equation (7.1.1) as well as the continuous dependence of the solution on the data. If we assume that u_1 and u_2 are two solutions corresponding to the data $u_{1,0}$, f_1, and $u_{2,0}$, f_2, respectively, then $u_1 - u_2$ would satisfy the heat equation (7.1.1) with the data $u_{1,0} - u_{2,0}$ and $f_1 - f_2$ and hence (7.1.2) would read as

$$\|u_1(\cdot, t) - u_2(\cdot, t)\| \le \|u_{1,0} - u_{2,0}\| + \int_0^t \|f_1(s) - f_2(s)\|\,ds. \tag{7.1.17}$$

Hence, we get the uniqueness, letting $u_{1,0} = u_{2,0}$ and $f_1 = f_2$.

Below, for the sake of generality, we shall denote the spatial domain by Ω and the boundary of Ω by $\partial\Omega$. Thus, the corresponding formulation for the homogeneous heat equation reads as follows: given $u_0(x)$, find $u(x, t)$ satisfying

$$\begin{cases} \dot{u} - u'' = 0, & \text{for } x \in \Omega, \quad t > 0, \\ u(x, t) = 0, & \text{for } x \in \partial\Omega_1, \quad t > 0, \\ u_x(x, t) = 0, & \text{for } x \in \partial\Omega_2, \quad t > 0, \\ u(x, 0) = u_0(x), & \text{for } x \in \Omega, \end{cases} \tag{7.1.18}$$

where $\partial\Omega = \partial\Omega_1 \cup \partial\Omega_2$ (e.g. for $\Omega = [a, b]$, $u(a, t) = u'(b, t) = 0$).

Theorem 7.3 (An energy estimate) *For any small positive constant ε, the solution of the homogeneous heat equation (7.1.18) satisfies the estimate*

$$\int_\varepsilon^t \|\dot{u}\|(s)\,ds \le \frac{1}{2}\sqrt{\ln\frac{t}{\varepsilon}}\|u_0\|. \tag{7.1.19}$$

Proof: We multiply the differential equation: $\dot{u} - u'' = 0$, by $-tu''$, and integrate over Ω to obtain

$$-t\int_\Omega \dot{u}u''\,dx + t\int_\Omega (-u'')^2\,dx = 0. \tag{7.1.20}$$

Integrating by parts and using the boundary data (note that $u = 0$ on $\partial\Omega_1$ implies that $\dot{u} = 0$ on $\partial\Omega_1$), we get

$$\int_\Omega \dot{u}u''\,dx = -\int_\Omega \dot{u}'\,u'\,dx = -\frac{1}{2}\frac{d}{dt}\|u'\|^2, \tag{7.1.21}$$

so that (7.1.20) may be rewritten as

$$t\frac{1}{2}\frac{d}{dt}\|u'\|^2 + t\|u''\|^2 = 0. \tag{7.1.22}$$

Using the identity $t\frac{d}{dt}\|u'\|^2 = \frac{d}{dt}(t\|u'\|^2) - \|u'\|^2$, we can write (7.1.22) as

$$\frac{d}{dt}(t\|u'\|^2) + 2t\|u''\|^2 = \|u'\|^2. \tag{7.1.23}$$

Now relabeling t as s and integrating in s over $(0, t)$, we end up with

$$\int_0^t \frac{d}{ds}(s\|u'\|^2(s))\,ds + 2\int_0^t s\|u''\|^2(s)\,ds = \int_0^t \|u'\|^2(s)\,ds \le \frac{1}{2}\|u_0\|^2,$$

where in the last inequality, we just integrate the stability estimate (a) in Theorem 7.2 over the interval $(0, t)$. Consequently,

$$t\|u'\|^2(t) + 2\int_0^t s\|u''\|^2(s)\,ds \le \frac{1}{2}\|u_0\|^2. \tag{7.1.24}$$

In particular, we have

$$\text{(I)}\quad \|u'\|(t) \le \frac{1}{\sqrt{2t}}\|u_0\|, \qquad \text{(II)}\quad \left(\int_0^t s\|u''\|^2(s)\,ds\right)^{1/2} \le \frac{1}{2}\|u_0\|. \tag{7.1.25}$$

Finally, in the differential equation $\dot{u} = u''$, integrating, the L_2-norm of both sides, over (ε, t) for $\varepsilon > 0$, we obtain

$$\begin{aligned}\int_\varepsilon^t \|\dot{u}\|(s)\,ds &= \int_\varepsilon^t \|u''\|(s)\,ds = \int_\varepsilon^t 1\cdot\|u''\|(s)\,ds = \int_\varepsilon^t \frac{1}{\sqrt{s}}\cdot\sqrt{s}\|u''\|(s)\,ds\\ &\le \left(\int_\varepsilon^t s^{-1}\,ds\right)^{1/2}\cdot\left(\int_\varepsilon^t s\|u''\|^2(s)\,ds\right)^{1/2} \le \frac{1}{2}\sqrt{\ln\left(\frac{t}{\varepsilon}\right)}\|u_0\|,\end{aligned}$$

where in the last inequality, we have used the estimate (II) in (7.1.25). □

Remark 7.4 To continue with second-order derivative, one can show that

$$\|u''\|(t) \leq \frac{1}{\sqrt{2}t}\|u_0\|. \tag{7.1.26}$$

Thus, gathering (7.1.2), the estimate (I) in (7.1.25) and (7.1.26), we deduce

$$\|u^{(k)}\|(t) \leq \frac{1}{t^{k/2}}\|u_0\|, \quad \text{for} \quad k = 0, 1, 2. \tag{7.1.27}$$

Hint: To prove (7.1.26), multiply $\dot{u} - u''$ by $t^2(u'')^2$, and note that $u'' = \dot{u} = 0$ on $\partial\Omega$, or alternatively: differentiate $\dot{u} - u'' = 0$ with respect to t and multiply the resulting equation by $t^2\dot{u}$.

7.1.2 FEM for the Heat Equation

We consider the one-dimensional heat equation with Dirichlet data:

$$\begin{cases} \dot{u} - u'' = f, & 0 < x < 1, \quad t > 0, \\ u(0,t) = u(1,t) = 0, & \qquad\qquad\quad t > 0, \\ u(x,0) = u_0(x), & 0 < x < 1. \end{cases} \tag{7.1.28}$$

The *variational formulation* for this problem reads as follows: for each time interval $I_n = (t_{n-1}, t_n]$, find $u(x,t)$, $x \in (0,1)$, $t \in I_n$, such that

$$\int_{I_n}\int_0^1 (\dot{u}v + u'v')\, dx\, dt = \int_{I_n}\int_0^1 fv\, dx\, dt, \quad \forall v : \ v(0,t) = v(1,t) = 0. \tag{VF}$$

A continuous piecewise linear Galerkin finite element method (FEM): cG(1)–cG(1). For each time interval $I_n := (t_{n-1}, t_n]$, with $t_n - t_{n-1} = k_n$, let

$$U(x,t) = U_{n-1}(x)\Psi_{n-1}(t) + U_n(x)\Psi_n(t), \tag{7.1.29}$$

where

$$\Psi_n(t) = \frac{t - t_{n-1}}{k_n}, \qquad \Psi_{n-1}(t) = \frac{t_n - t}{k_n}, \tag{7.1.30}$$

and

$$U_{\tilde{n}}(x) = U_{\tilde{n},1}\varphi_1(x) + U_{\tilde{n},2}\varphi_2(x) + \cdots + U_{\tilde{n},m}\varphi_m(x), \quad \tilde{n} = n-1,\ n \tag{7.1.31}$$

with φ_j being the usual piecewise linear, continuous finite element basis functions (hat-functions) corresponding to a partition of $\Omega = (0,1)$, with $0 = x_0 < \cdots < x_\ell < x_{\ell+1} < \cdots < x_{m+1} = 1$, and $\varphi_j(x_i) := \delta_{ij}$. Now the Galerkin method (FEM) is to determine the unknown coefficients $U_{n,\ell}$ in the above representation for U (U is a continuous, piecewise linear function, both in space and time variables) that satisfies the following discrete variational formulation: find $U(x,t)$ given by (7.1.29) such that

$$\int_{I_n}\int_0^1 (\dot{U}\varphi_i + U'\varphi_i')\, dx\, dt = \int_{I_n}\int_0^1 f\varphi_i\, dx\, dt, \quad i = 1, 2, \ldots, m. \tag{7.1.32}$$

Note that on $I_n = (t_{n-1}, t_n]$ and with $U_n(x) := U(x, t_n)$ and $U_{n-1}(x) := U(x, t_{n-1})$,

$$\dot{U}(x,t) = U_{n-1}(x)\dot{\Psi}_{n-1}(t) + U_n(x)\dot{\Psi}_n(t) = \frac{U_n - U_{n-1}}{k_n}. \tag{7.1.33}$$

Further differentiating (7.1.29) with respect to x, we have

$$U'(x,t) = U'_{n-1}(x)\Psi_{n-1}(t) + U'_n(x)\Psi_n(t). \tag{7.1.34}$$

Inserting (7.1.33) and (7.1.34) into (7.1.32), we get using the identities, $\int_{I_n} dt = k_n$ and $\int_{I_n} \Psi_n \, dt = \int_{I_n} \Psi_{n-1} \, dt = k_n/2$ that,

$$\begin{aligned} &\underbrace{\int_0^1 U_n\varphi_i \, dx}_{\mathbf{M}\cdot U_n} - \underbrace{\int_0^1 U_{n-1}\varphi_i \, dx}_{\mathbf{M}\cdot U_{n-1}} + \underbrace{\int_{I_n} \Psi_{n-1} \, dt}_{k_n/2} \underbrace{\int_0^1 U'_{n-1}\varphi'_i \, dx}_{\mathbf{S}\cdot U_{n-1}} \\ &\quad + \underbrace{\int_{I_n} \Psi_n \, dt}_{k_n/2} \underbrace{\int_0^1 U'_n\varphi'_i \, dx}_{\mathbf{S}\cdot U_n} = \underbrace{\int_{I_n}\int_0^1 f\varphi_i \, dxdt}_{F_n}. \end{aligned} \tag{7.1.35}$$

This can be written in a compact form as the *Crank–Nicolson system*

$$\left(\mathbf{M} + \frac{k_n}{2}\mathbf{S}\right) U_n = \left(\mathbf{M} - \frac{k_n}{2}\mathbf{S}\right) U_{n-1} + F_n,$$

with the solution U_n given by the data U_{n-1} and F, viz.

$$U_n = \underbrace{\left(\mathbf{M} + \frac{k_n}{2}\mathbf{S}\right)^{-1}}_{B^{-1}} \underbrace{\left(\mathbf{M} - \frac{k_n}{2}\mathbf{S}\right)}_{A} U_{n-1} + \underbrace{\left(\mathbf{M} + \frac{k_n}{2}\mathbf{S}\right)^{-1}}_{B^{-1}} F_n, \tag{7.1.36}$$

where **M** and **S**, computed in Chapter 3 (and Chapter 5) are known as the *mass-matrix* and *stiffness-matrix*, respectively, and

$$U_n = \begin{bmatrix} U_{n,1} \\ U_{n,2} \\ \vdots \\ U_{n,m} \end{bmatrix}, \qquad F_n = \begin{bmatrix} F_{n,1} \\ F_{n,2} \\ \vdots \\ F_{n,m} \end{bmatrix}, \qquad F_{n,i} = \int_{I_n}\int_0^1 f\varphi_i \, dx \, dt. \tag{7.1.37}$$

Thus, given the source term f, we can determine the vector F_n and then, for each $n = 1, \dots N$, given the vector U_{n-1} (the initial value is given by $U_{0,j} := u_0(x_j)$), we may use the Crank–Nicolson System (CNS) above to compute $U_{n,\ell}$, $\ell = 1, 2, \dots, m$ (m nodal values of U at the x_j:s, and at the time level t_n).

We now return to the computation of the matrix entries for **M** and **S**, for a uniform partition (all subintervals are of the same length) of the interval $I = (0, 1)$.

Note that differentiating (7.1.31) with respect to x, yields

$$U_n'(x) = U_{n,1}\varphi_1'(x) + U_{n,2}\varphi_2'(x) + \cdots + U_{n,m}\varphi_m'(x). \tag{7.1.38}$$

Hence, for $i = 1, \ldots, m$, the contribution of the stiffness matrix to the left-hand side is

$$\int_0^1 U_n'\varphi_i' dx = \left(\int_0^1 \varphi_i'\varphi_1' dx\right) U_{n,1} + \left(\int_0^1 \varphi_i'\varphi_2' dx\right) U_{n,2} + \cdots + \left(\int_0^1 \varphi_i'\varphi_m' dx\right) U_{n,m}$$

which can be written in matrix form as

$$\mathbf{S}U_n = \begin{bmatrix} \int_0^1 \varphi_1'\varphi_1' & \int_0^1 \varphi_1'\varphi_2' & \cdots & \int_0^1 \varphi_1'\varphi_m' \\ \int_0^1 \varphi_2'\varphi_1' & \int_0^1 \varphi_2'\varphi_2' & \cdots & \int_0^1 \varphi_2'\varphi_m' \\ \vdots & \vdots & \cdots & \vdots \\ \int_0^1 \varphi_m'\varphi_1' & \int_0^1 \varphi_m'\varphi_2' & \cdots & \int_0^1 \varphi_m'\varphi_m' \end{bmatrix} \begin{bmatrix} U_{n,1} \\ U_{n,2} \\ \vdots \\ U_{n,m} \end{bmatrix}. \tag{7.1.39}$$

Thus, $\mathbf{S}$ is just the stiffness matrix $\mathbf{A}_{\text{unif}}$ computed in Chapter 3:

$$\mathbf{S} = \frac{1}{h}\begin{bmatrix} 2 & -1 & 0 & 0 & \cdots & 0 \\ -1 & 2 & -1 & 0 & \cdots & 0 \\ 0 & -1 & 2 & -1 & \cdots & 0 \\ \cdots & \cdots & \cdots & \cdots & \cdots & \cdots \\ 0 & \cdots & \cdots & -1 & 2 & -1 \\ 0 & \cdots & \cdots & \cdots & -1 & 2 \end{bmatrix}. \tag{7.1.40}$$

A nonuniform partition yields a matrix of the form $\mathbf{A}$ in Chapter 3.

Similarly, recalling the notation for the mass matrix $\mathbf{M}$ in (7.1.35), we have

$$[\mathbf{M}U_n]_i = \int_0^1 U_n\varphi_i dx, \quad i = 1, \ldots, m. \tag{7.1.41}$$

Hence, to compute the mass matrix $\mathbf{M}$ one drops all derivatives from the general form of the matrix for $\mathbf{S}$ given by (7.1.39). In other words, unlike in $\mathbf{S}U_n$, the $[\mathbf{M}U_n]_i = \int_0^1 U_n\varphi_i dx$, does not involve any derivatives. Consequently,

$$\mathbf{M} = \begin{bmatrix} \int_0^1 \varphi_1\varphi_1 & \int_0^1 \varphi_1\varphi_2 & \cdots & \int_0^1 \varphi_1\varphi_m \\ \int_0^1 \varphi_2\varphi_1 & \int_0^1 \varphi_2\varphi_2 & \cdots & \int_0^1 \varphi_2\varphi_m \\ \vdots & \vdots & \cdots & \vdots \\ \int_0^1 \varphi_m\varphi_1 & \int_0^1 \varphi_m\varphi_2 & \cdots & \int_0^1 \varphi_m\varphi_m \end{bmatrix}. \tag{7.1.42}$$

For a uniform partition, we have computed this mass matrix in Chapter 5:

$$\mathbf{M} = h\begin{bmatrix} \frac{2}{3} & \frac{1}{6} & 0 & 0 & \cdots & 0 \\ \frac{1}{6} & \frac{2}{3} & \frac{1}{6} & 0 & \cdots & 0 \\ 0 & \frac{1}{6} & \frac{2}{3} & \frac{1}{6} & \cdots & 0 \\ \cdots & \cdots & \cdots & \cdots & \cdots & \cdots \\ 0 & \cdots & \cdots & \frac{1}{6} & \frac{2}{3} & \frac{1}{6} \\ 0 & \cdots & \cdots & \cdots & \frac{1}{6} & \frac{2}{3} \end{bmatrix} = \frac{h}{6}\begin{bmatrix} 4 & 1 & 0 & 0 \cdots & 0 \\ 1 & 4 & 1 & 0 \cdots & 0 \\ 0 & 1 & 4 & 1 \cdots & 0 \\ \cdots & \cdots & \cdots & \cdots & \cdots \\ 0 \cdots & \cdots & 1 & 4 & 1 \\ 0 \cdots & \cdots & \cdots & 1 & 4 \end{bmatrix}.$$

7.1.3 Error Analysis

In this section, we outline a general framework for the error estimation procedure (this procedure is applicable for higher-space dimensions as well). We denote the spatial domain by Ω with the boundary $\partial\Omega$. In the one-dimensional case, treated here, $\Omega := (a, b)$ with $\partial\Omega = \{a, b\}$. For each $n = 1, 2, \ldots$, we consider the slab $\mathbf{S}_n := \Omega \times I_n$, with $I_n = (t_{n-1}, t_n]$, and rewrite the variational formulation for Eq. (7.1.28) as follows: find the exact solution $u(x, t) \in \mathcal{H}_0^1 := C((0, T]; H_0^1((a, b)))$ such that for each $n = 1, 2, \ldots$,

$$\int_{I_n}\int_{\Omega} (\dot{u}v + u'v')\, dx\, dt = \int_{I_n}\int_{\Omega} fv\, dx\, dt, \quad \forall v \in C(I_n; H_0^1((a, b))). \tag{7.1.43}$$

We can formulate a corresponding FEM for Eq. (7.1.28) as the continuous, piecewise linear space–time approximation in (7.1.29)–(7.1.32). However, we should have an upper limit for n, say $n = 1, 2, \ldots, N$, in order to get a finite dimensional (computable) problem.

We may alternatively construct the cG(1)dG(q) method, e.g. a FEM based on approximation using continuous piecewise linear functions in space and discontinuous piecewise polynomials of degree q in time, viz. We let $\mathbf{S}_n = \Omega \times I_n$, and define the trial space as

$$W_k^{(q)} = \{v(x, t) \colon v|_{\mathbf{S}_n} \in W_{k,n}^{(q)}, \quad n = 1, \ldots, N\}, \tag{7.1.44}$$

where

$$W_{k,n}^{(q)} = \left\{ v(x, t) \colon v(x, t) = \sum_{j=0}^{q} t^j \varphi_j(x) : \ \varphi_j \in V_n^0, \ (x, t) \in \mathbf{S}_n \right\}.$$

Here, $V_n^0 = V_{h_n}^0$ is the space of continuous piecewise linear functions on a subdivision $\mathcal{T}_n$ of Ω with mesh function h_n, vanishing at the boundary. The functions in $W_k^{(q)}$ may be discontinuous in time at the discrete time levels t_n. To account for this, we use the same notation as in Chapter 6:

$$w_n^{\pm} = \lim_{s\to 0^{\pm}} w(t_n + s), \qquad [w_n] := w_n^+ - w_n^-. \tag{7.1.45}$$

Now, we may formulate the cG(1)dG(q) method as follows. Let $U_0^- := u_0$, and find $U \in W_k^{(q)}$ such that, for $n = 1, \ldots, N$,

$$\int_{I_n}\int_{\Omega}(\dot{U}v + U'v')\,dx\,dt + \int_{\Omega}[U_{n-1}]v_{n-1}^+\,dx = \int_{I_n}\int_{\Omega} fv\,dx\,dt \quad \text{for all } v \in W_{k,n}^{(q)}. \tag{7.1.46}$$

Subtracting (7.1.46) from (7.1.43), we obtain the *time-discontinuous Galerkin orthogonality* relation for the error $e := u - U$ as

$$\int_{I_n}\int_{\Omega}(\dot{e}v + e'v')\,dx\,dt + \int_{\Omega}[e_{n-1}]v_{n-1}^+\,dx = 0, \quad \text{for all } v \in W_{kh}^{(q)}. \tag{7.1.47}$$

Obviously, in the time-continuous case the jump term in (7.1.46) disappears, and we end up with cG(1)cG(q) method:

$$\int_{I_n}\int_{\Omega}(\dot{U}v + U'v')\,dx\,dt = \int_{I_n}\int_{\Omega} fv\,dx\,dt \quad \text{for all } v \in W_{k,n}^{(q)}. \tag{7.1.48}$$

Subtracting (7.1.48) from (7.1.43), we obtain the *Galerkin orthogonality* relation for the error $e = u - U$, for the cG(1)cG(q) method, as

$$\int_{I_n}\int_{\Omega}(\dot{e}v + e'v')\,dxdt = 0, \quad \text{for all } v \in W_{kn}^{(q)}. \tag{7.1.49}$$

A priori and a posteriori error estimates for the heat equation are obtained combining the results in Chapters 5 and 6. To become familiar with some standard techniques, below we shall demonstrate the a posteriori approach.

Theorem 7.4 (cG(1)cG(1) a posteriori error estimate). *Assume*

- $h_n = h, \quad \forall n$
- $u_0 \in V_{h_1}^0 = V_h^0$ *(so that* $e(0) = 0$*).*

Let $\varepsilon = k_N$ *be small, and let* $0 < t < T$. *Let* u *and* U *be the exact and approximate solutions for the heat equation (7.1.28), respectively, i.e.* U *is the finite element approximation satisfying in either one of the equations (7.1.32): cG(1)cG(1) or (7.1.48): cG(1)cG(q). Then, the error* $e = u - U$ *is bounded as*

$$\|e(T)\| \le \left(2\sqrt{\ln\frac{T}{\varepsilon}}\right)\max_{[0,T]}\|(k + h^2)r(U)\|, \tag{7.1.50}$$

where $r(U) := f + U'' - \dot{U}$ *is the residual and* k *and* h *are temporal and spatial mesh functions, respectively.*

Proof: Following the general framework for a posteriori error estimates, we let $\varphi(x,t)$ be the solution of the *dual problem*

$$\begin{cases} -\dot{\varphi}(x,t) - \varphi''(x,t) = 0, & (x,t) \in \Omega \times (0,T), \\ \varphi(x,t) = 0, & (x,t) \in \partial\Omega \times (0,T), \\ \varphi(x,T) = e(T), & x \in \Omega, \end{cases} \tag{7.1.51}$$

where $e := e(T) = e(\cdot,T) = u(\cdot,T) - U(\cdot,T), \quad T = t_N$. By a change of variables, letting $w(x,s) = \varphi(x,T-s), \quad (s>0)$; the *backward dual problem* (7.1.51) is rewritten in the *forward form* as

$$\begin{cases} \dot{w}(x,s) - w''(x,s) = 0, & (x,s) \in \Omega \times (0,T), \\ w(x,s) = 0, & (x,s) \in \partial\Omega \times (0,T), \\ w(x,0) = e, & x \in \Omega. \end{cases} \tag{7.1.52}$$

For this problem, we have shown in the energy estimate Theorem 7.3 that

$$\int_\varepsilon^T \|\dot{w}\|\,ds \le \frac{1}{2}\sqrt{\ln\frac{T}{\varepsilon}}\;\|e\|. \tag{7.1.53}$$

With $s = T - t$, $ds = -dt$, and we have $\dot{w}(x,s) = -\dot{\varphi}(x,T-s)$. Hence, since $-\varphi'' = \dot{\varphi}$, we have for φ, that

$$\max\left(\int_0^{T-\varepsilon} \|\dot{\varphi}\|\,dt\,,\,\int_0^{T-\varepsilon} \|\varphi''\|\,dt\right) \le \frac{1}{2}\sqrt{\ln\frac{T}{\varepsilon}}\;\|e\|. \tag{7.1.54}$$

Now let $u_0 \in V^0_{h_1} = V^0_h$, then, since $-\dot{\varphi} - \varphi'' = 0$, integration by parts in t and x, and using the boundary data, yields

$$\begin{aligned} \|e(T)\|^2 &= \int_\Omega e(T)\cdot e(T)\,dx + \int_0^T\int_\Omega e(-\dot{\varphi} - \varphi'')\,dx\,dt \\ &= \int_\Omega e(T)\cdot e(T)\,dx - \int_\Omega e(T)\cdot\varphi(T)\,dx + \underbrace{\int_\Omega e(0)\cdot\varphi(0)\,dx}_{=0;\ (u_0\in V^0_{h_1})} \\ &\quad + \int_0^T\int_\Omega (\dot{e}\varphi + e'\varphi')\,dx\,dt. \end{aligned}$$

Thus, by Galerkin orthogonality (7.1.49) (for $v \in W_k^{(0)}$, i.e. piecewise constant in time and continuous, piecewise linear in space), we get

$$\|e(T)\|^2 = \int_0^T\int_\Omega \dot{e}(\varphi - v) + e'(\varphi - v)'\,dx\,dt. \tag{7.1.55}$$

On the other hand, using local (elementwise) integration by parts in the spatial variable, we have that

$$\int_\Omega \dot{e}(\varphi - v) + e'(\varphi - v)'\,dx$$

$$
\begin{aligned}
&= \int_\Omega (\dot{u} - \dot{U})(\varphi - v)\, dx \\
&\quad + \sum_{i=1}^{M+1} \int_{x_{i-1}}^{x_i} (u' - U')(\varphi - v)'\, dx \\
&= \int_\Omega (\dot{u} - \dot{U})(\varphi - v)\, dx + \sum_{i=1}^{M+1} [(u' - U')(\varphi - v)]_{x_{i-1}}^{x_i} \\
&\quad - \sum_{i=1}^{M+1} \int_{x_{i-1}}^{x_i} (u'' - U'')(\varphi - v)\, dx \\
&= \int_\Omega (f - \dot{U})(\varphi - v)\, dx + \sum_{i=1}^{M} [U_i'](\varphi - v)(x_i).
\end{aligned}
$$

Here, the contribution from u' yields a telescoped finite series with sum $u'(x_{M+1}) - u'(x_0) = 0 - 0 = 0$. Thus,

$$
\begin{aligned}
\|e(T)\|^2 &= \int_0^T \int_\Omega (f - \dot{U})(\varphi - v)\, dx + \int_0^T \sum_{i=1}^{M} [U_i'](\varphi - v)(x_i) dx \\
&= \int_0^T \int_\Omega r_1(U)(\varphi - v)\, dx\, dt + \int_0^T \sum_{i=1}^{M} [U_i'](\varphi - v)(x_i) dx, \qquad (7.1.56)
\end{aligned}
$$

where we have used $\dot{e} - e'' = \dot{u} - u'' - \dot{U} + U'' = f - \dot{U} + U'' = f - \dot{U} := r_1(U)$ (residual). Next, with mesh functions $h = h(x)$ and $k = k(t)$ in x and t, respectively, we may derive (using the fact that v is piecewise constant in t and continuous, piecewise linear in x) an interpolation estimate of the form:

$$
\left\|(k + h^2)^{-1}(\varphi - v)\right\|_{L_2} \le C_i(\|\dot{\varphi}\|_{L_2} + \|\varphi''\|_{L_2}). \qquad (7.1.57)
$$

To estimate the first term in (7.1.56) (see Eriksson et al. [57] Chapter 16), we have using the estimates (7.1.54) that

$$
\begin{aligned}
\int_0^T \int_\Omega r_1(U)(\varphi - v)\, dx\, dt &\le C_i \int_0^{T-k_N} \|(k + h^2) r_1(U)\|(\|\dot{\varphi}\| + \|\varphi''\|)\, dt \\
&\quad + \int_{T-k_N}^T \|r_1(U)\| \|\varphi - v\|\, dt \\
&\le C_i \max_{[0,T-k_N]} \|(k + h^2) r_1(U)\| \cdot \sqrt{\ln \frac{T}{k_N}} \|e(T)\| \\
&\quad + \max_{[T-k_N,T]} (\|r_1(U)\| \cdot \|\varphi - v\|) \cdot k_N,
\end{aligned}
$$

where the last term can be estimated using the stability estimate (7.1.2) for the dual problem with $f = 0$, e.g. for v being a linear interpolant of φ, see Figure 7.2, as

$$
(\|r_1(U)\| \cdot \|\varphi - v\|) \cdot k_N \le C\|k_N r_1(U)\| \|\varphi\| \le C\|k_N r_1(U)\| \|e(T)\|. \qquad (7.1.58)
$$

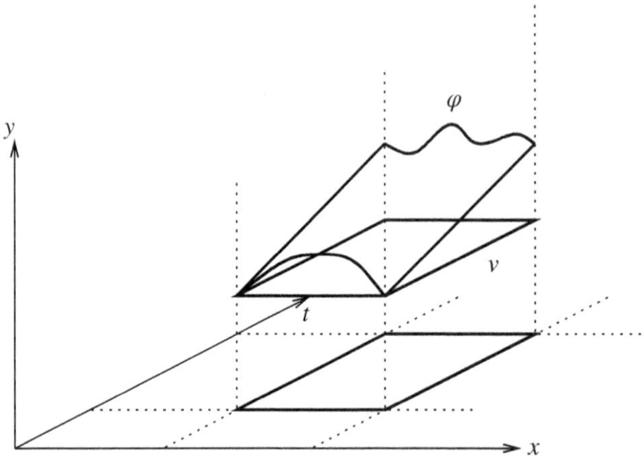

Figure 7.2 φ and its linear interpolant (both in x and t) v.

Summing up, we get

$$\int_0^T \int_\Omega r_1(U)(\varphi - v)\, dx\, dt \leq C_i \max_{[0,T]} \|(k + h^2) r_1(U)\| \cdot \sqrt{\ln \frac{T}{k_N}} \cdot \|e(T)\|. \tag{7.1.59}$$

The second term in (7.1.56) can be estimated similarly

$$\int_0^T \sum_{i=1}^{M} [U_i'](\varphi - v)(x_i) \leq C_i \max_{[0,T]} \|(k + h^2) r_2(U)\| \cdot \sqrt{\ln \frac{T}{k_N}} \cdot \|e(T)\| dx, \tag{7.1.60}$$

where $r_2(U) := h_i^{-1} |[U_i']|$, for $x \in I_i$. This gives our final estimate, and the proof is complete:

$$\|e(T)\| \leq C \left(\sqrt{\ln \frac{T}{k_N}} \right) \max_{[0,T]} \|(k + h^2) r(U)\|, \tag{7.1.61}$$

where $r(U) = r_1(U) + r_2(U)$. □

Adaptivity algorithm. For the a posteriori error estimates, this can, roughly, be outlined as follows: consider, as an example, the equation

$$-u'' = f \quad \text{in } \Omega, \qquad u = 0, \quad \text{on } \partial\Omega \tag{7.1.62}$$

where for the error $u - U$, we have derived the a posteriori estimate

$$\|e'\| \leq C \|h\, r(U)\|, \tag{7.1.63}$$

with $r(U) = |f| + \max_{I_k} |[u']|$ and $[\cdot]$ denoting the jump (over the endpoints of a partition interval I_k). Then, an adaptivity can be formulated as follows:

(1) Choose an arbitrary mesh-size $h = h_k(x)$ for $x \in I_k$ and a tolerance TOL > 0.
(2) Given h, compute the corresponding U (also denoted as u_h).
(3) If $C\|h\, r(U)\| \leq \text{TOL}$, then accept U. Otherwise, choose a new (refined) $h = h(x)$ and return to the step (2) above. □

Examples of higher-order elements cG(2). In the one-dimensional case, approximate solutions with piecewise polynomials of degree 2 are determined by the values of the *approximate solution* at the, e.g. mid-point and end-points of each subinterval (note that this is not! interpolant which uses the values of *exact solution*). The construction is through using the Lagrange interpolation basis $\lambda_0(x), \lambda_1(x), \dots, \lambda_4(x)$, see Figure 7.3.

Example 7.2 *(Error estimates in a simple case for cG(2))* For the Poisson equation $-u'' = f$, $0 < x < 1$, associated with a Dirichlet (or Neumann) boundary condition, we have the following cG(2) error bounds

$$\|(u - U)'\| \leq C\|h^2 d^3 u\|. \tag{7.1.64}$$

$$\|u - U\| \leq C \max(h\|h^2 d^3 u\|). \tag{7.1.65}$$

$$\|u - U\| \leq C\|h^2 r(U)\|, \quad \text{where} \quad |r(U)| \leq Ch. \tag{7.1.66}$$

The proof of the estimates (7.1.64)–(7.1.66) are rather involved (see Eriksson et al. [57] for the details) and therefore are omitted. These estimates can be extended to the space–time discretizations for the heat equation.

Example 7.3 *(The equation of an elastic beam)*

$$\begin{cases} (au'')'' = f, & \Omega = (0,1), \\ u(0) = 0, & u'(0) = 0, \quad \text{(Dirichlet)} \\ u''(1) = 0, & (au'')'(1) = 0, \quad \text{(Neumann)} \end{cases} \tag{7.1.67}$$

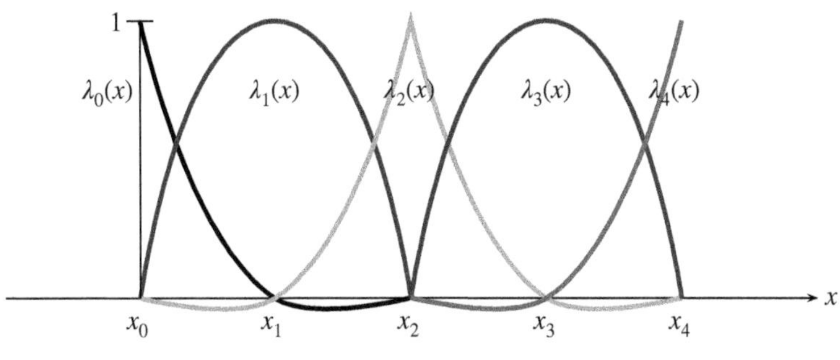

Figure 7.3 Quadratic Lagrange basis functions $\lambda_0(x), \dots, \lambda_4(x)$.

where a is the bending stiffness, au'' is the moment, f is the load function, and $u = u(x)$ is the vertical deflection.

A variational formulation for this equation can be written as

$$\int_0^1 au''v''\,dx = \int_0^1 fv\,dx, \quad \forall v, \quad \text{such that} \quad v(0) = v'(0) = 0. \tag{7.1.68}$$

Here, considering piecewise linear finite element functions (with vanishing second derivatives) is inadequate.

7.1.4 Exercises

Problem 7.1 Derive a system of equations, as (7.1.36) for the cG(1)–dG(0) method: with the continuous piecewise linear spatial discretization cG(1) and discontinuous piecewise constants dG(0) in time.

Problem 7.2 Work out the details with piecewise cubic polynomials having continuous first derivatives, i.e. 2 degrees of freedom on each node.
Hint: A cubic polynomial in (a, b) is uniquely determined by $\varphi(a)$, $\varphi'(a)$, $\varphi(b)$, and $\varphi'(b)$.

Problem 7.3 Let $\|\cdot\|$ denote the $L_2(0,1)$-norm. Consider the problem

$$\begin{cases} -u'' = f, & 0 < x < 1, \\ u'(0) = v_0, & u(1) = 0. \end{cases}$$

(a) Show that $|u(0)| \le \|u'\|$ and $\|u\| \le \|u'\|$.
(b) Use (a) to show that $\|u'\| \le \|f\| + |v_0|$.

Problem 7.4 Assume that $u = u(x)$ satisfies

$$\int_0^1 u'v'\,dx = \int_0^1 fv\,dx, \quad \text{for all } v(x) \quad \text{such that } v(0) = 0. \tag{7.1.69}$$

Show that $-u'' = f$ for $0 < x < 1$ and $u'(1) = 0$.

Problem 7.5 (Generalized Poincare) Show that for a continuously differentiable function v defined on $(0, 1)$, we have that

$$\|v\|^2 \le v(0)^2 + v(1)^2 + \|v'\|^2.$$

Hint: Use partial integration for $\int_0^{1/2} v(x)^2\,dx$ and $\int_{1/2}^1 v(x)^2\,dx$ and note that $(x - 1/2)$ has the derivative 1.

Problem 7.6 Let $\|\cdot\| := \|\cdot\|_{L_2(0,1)}$. Consider the following heat equation

$$\begin{cases} \dot{u} - u'' = 0, & 0 < x < 1, \quad t > 0, \\ u(0,t) = u_x(1,t) = 0, & \qquad\qquad t > 0, \\ u(x,0) = u_0(x), & 0 < x < 1. \end{cases}$$

(a) Show that the norms: $\|u(\cdot,t)\|$ and $\|u_x(\cdot,t)\|$ are non-increasing in time.
(b) Show that $\|u_x(\cdot,t)\| \to 0$, as $t \to \infty$.
(c) Give a physical interpretation for (a) and (b).

Problem 7.7 Consider the inhomogeneous problem 7.6:

$$\begin{cases} \dot{u} - \varepsilon u'' = f, & 0 < x < 1, \quad t > 0, \\ u(0,t) = u_x(1,t) = 0, & \qquad\qquad t > 0, \\ u(x,0) = u_0(x), & 0 < x < 1. \end{cases}$$

where $f = f(x,t)$.
Show that for the corresponding stationary ($\dot{u} \equiv 0$) problem, we have

$$\|u'\| \le \frac{1}{\varepsilon}\|f\|.$$

Problem 7.8 Give an a priori error estimate for the cG(2) scheme for the following problem:

$$(au_{xx})_{xx} = f, \quad 0 < x < 1, \qquad u(0) = u'(0) = u(1) = u'(1) = 0,$$

where $a(x) > 0$ on the interval $I = (0,1)$.

7.2 The Wave Equation in 1*d*

The wave equation, being of hyperbolic type, is fundamentally different from the heat equation which is a parabolic equation. Nevertheless, in our study here, the structure of the finite element procedure for the wave equation is, mainly, the same as that of the heat equation outlined in Section 7.1. This is due to the fact that both equations have the Laplace operator in the spatial variable. To proceed, we start with an example of the homogeneous wave equation, by considering the initial-boundary value problem

$$\begin{cases} \ddot{u} - u'' = 0, & 0 < x < 1 & t > 0 & \text{(DE)} \\ u(0,t) = 0, & u(1,t) = 0 & t > 0 & \text{(BC)} \\ u(x,0) = u_0(x), & \dot{u}(x,0) = v_0(x), & 0 < x < 1. & \text{(IC)} \end{cases} \tag{7.2.1}$$

An analytic approach to the solution for this problem is the representation given by *d'Alembert formula*.

The most important property of the wave equation is the *conservation of energy*.

Theorem 7.5 *For the wave equation (7.2.1), we have that*

$$\frac{1}{2}\|\dot{u}(\cdot,t)\|^2 + \frac{1}{2}\|u'(\cdot,t)\|^2 = \frac{1}{2}\|v_0\|^2 + \frac{1}{2}\|u_0'\|^2 = \textit{Constant}. \tag{7.2.2}$$

Proof: We multiply the equation by $\dot{u}$ and integrate over $I = (0,1)$ to get

$$\int_0^1 \ddot{u}\,\dot{u}\,dx - \int_0^1 u''\,\dot{u}\,dx = 0. \tag{7.2.3}$$

Using integration by parts and the boundary data, we have that

$$\begin{aligned}
&\int_0^1 \frac{1}{2}\frac{\partial}{\partial t}(\dot{u})^2\,dx + \int_0^1 u'\,(\dot{u})'\,dx - [u'(x,t)\dot{u}(x,t)]_0^1 \\
&\quad = \int_0^1 \frac{1}{2}\frac{\partial}{\partial t}(\dot{u})^2\,dx + \int_0^1 \frac{1}{2}\frac{\partial}{\partial t}(u')^2\,dx \\
&\quad = \frac{1}{2}\frac{d}{dt}(\|\dot{u}\|^2 + \|u'\|^2) = 0.
\end{aligned} \tag{7.2.4}$$

Thus, integrating (7.2.4) over t, we get that the quantity

$$\frac{1}{2}\|\dot{u}\|^2 + \frac{1}{2}\|u'\|^2 = \frac{1}{2}\|v_0\|^2 + \frac{1}{2}\|u_0'\|^2 = \text{Constant, independent of } t. \tag{7.2.5}$$

Therefore, the total energy is conserved. We recall that $\frac{1}{2}\|\dot{u}\|^2$ is the kinetic energy, and $\frac{1}{2}\|u'\|^2$ is the potential (elastic) energy. □

7.2.1 Wave Equation as a System of Hyperbolic PDEs

We rewrite the wave equation as a hyperbolic system of differential equations. To this approach, we consider solving

$$\begin{cases} \ddot{u} - u'' = 0, & 0 < x < 1, \quad t > 0, \\ u(0,t) = 0, & u'(1,t) = g(t), \quad t > 0, \\ u(x,0) = u_0(x), & \dot{u}(x,0) = v_0(x), \quad 0 < x < 1, \end{cases} \tag{7.2.6}$$

where we let $\dot{u} = v$, and reformulate the problem as follows:

$$\begin{cases} \dot{u} - v = 0, & \text{(Convection)} \\ \dot{v} - u'' = 0, & \text{(Diffusion)}. \end{cases} \tag{7.2.7}$$

We may now set $w = (u, v)^t$ and rewrite the system (7.2.7) as $\dot{w} + Aw = 0$:

$$\dot{w} + Aw = \begin{pmatrix} \dot{u} \\ \dot{v} \end{pmatrix} + \begin{pmatrix} 0 & -1 \\ -\frac{\partial^2}{\partial x^2} & 0 \end{pmatrix} \begin{pmatrix} u \\ v \end{pmatrix} = \begin{pmatrix} 0 \\ 0 \end{pmatrix}. \tag{7.2.8}$$

where the matrix differential operator is given by

$$A = \begin{pmatrix} 0 & -1 \\ -\frac{\partial^2}{\partial x^2} & 0 \end{pmatrix}.$$

Note that, in scalar form, this equation was studied in full detail in Chapter 6. Therefore, in this setting, the study of the wave equation is strongly connected with that of *systems of initial value problems.*

7.2.2 The Finite Element Discretization Procedure

We follow the same procedure as in the case of the heat equation, and consider the slabs $\mathbf{S}_n = \Omega \times I_n$, $n = 1, 2, \ldots, N$, with $I_n = (t_{n-1}, t_n]$. Then, for each n, we define, on $\mathbf{S}_n$, the piecewise linear approximations

$$\begin{cases} U(x,t) = U_{n-1}(x)\Psi_{n-1}(t) + U_n(x)\Psi_n(t), \\ V(x,t) = V_{n-1}(x)\Psi_{n-1}(t) + V_n(x)\Psi_n(t), \end{cases} \quad 0 < x < 1, \quad t \in I_n. \tag{7.2.9}$$

We may choose the spatial basis functions as $\varphi_j(x)$, $j = 1, \ldots, m$, and write

$$\begin{cases} U_{\tilde{n}}(x) = U_{\tilde{n},1}\varphi_1(x) + \cdots + U_{\tilde{n},m}\varphi_m(x), & \tilde{n} = n-1, \ n \\ V_{\tilde{n}}(x) = V_{\tilde{n},1}\varphi_1(x) + \cdots + V_{\tilde{n},m}\varphi_m(x), & \tilde{n} = n-1, \ n. \end{cases} \tag{7.2.10}$$

Figure 7.4 shows examples of typical linear time and spatial basis functions.

For $\dot{u} - v = 0$ and $t \in I_n$, we write the general variational formulation

$$\int_{I_n}\int_0^1 \dot{u}\varphi \, dx \, dt - \int_{I_n}\int_0^1 v\varphi \, dx \, dt = 0, \quad \text{for all } \varphi(x,t). \tag{7.2.11}$$

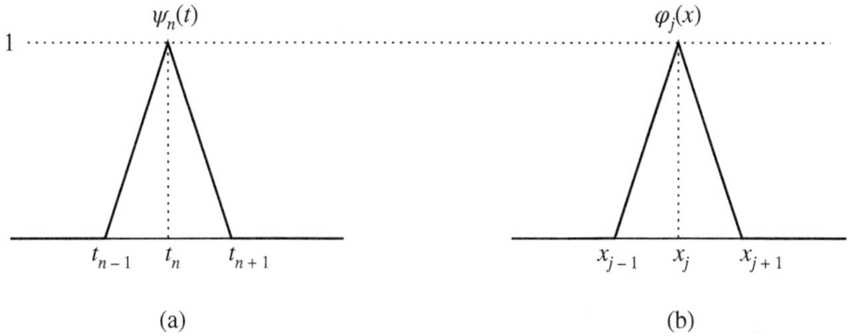

Figure 7.4 The linear time (a) and space (b) basis functions.

Likewise, $\dot{v} - u'' = 0$ yields a variational formulation, viz.

$$\int_{I_n}\int_0^1 \dot{v}\varphi \, dx \, dt - \int_{I_n}\int_0^1 u''\varphi \, dx \, dt = 0. \tag{7.2.12}$$

Integrating by parts in x, in the second term, and using the boundary condition $u'(1,t) = g(t)$, we get

$$\int_0^1 u''\varphi dx = [u'\varphi]_0^1 - \int_0^1 u'\varphi' \, dx = g(t)\varphi(1,t) - u'(0,t)\varphi(0,t) - \int_0^1 u'\varphi' \, dx.$$

Inserting the right-hand side in (7.2.12), we get for all φ with $\varphi(0,t) = 0$:

$$\int_{I_n}\int_0^1 \dot{v}\varphi dx \, dt + \int_{I_n}\int_0^1 u'\varphi' \, dxdt = \int_{I_n} g(t)\varphi(1,t) \, dt. \tag{7.2.13}$$

The corresponding cG(1)cG(1) FEM reads as follows: for each n, $n = 1, 2, \dots, N$, find continuous piecewise linear functions $U(x,t)$ and $V(x,t)$, in a partition, $0 = x_0 < x_1 < \cdots < x_m = 1$ of $\Omega = (0,1)$, such that

$$\begin{aligned} &\int_{I_n}\int_0^1 \frac{U_n(x) - U_{n-1}(x)}{k_n}\varphi_j(x) \, dx \, dt \\ &\quad - \int_{I_n}\int_0^1 (V_{n-1}(x)\Psi_{n-1}(t) + V_n(x)\Psi_n(t))\varphi_j(x) \, dx \, dt = 0, \\ &\qquad\qquad \text{for } j = 1, 2, \dots, m, \end{aligned} \tag{7.2.14}$$

and

$$\begin{aligned} &\int_{I_n}\int_0^1 \frac{V_n(x) - V_{n-1}(x)}{k_n}\varphi_j(x) \, dx \, dt \\ &\quad + \int_{I_n}\int_0^1 (U'_{n-1}(x)\Psi_{n-1}(t) + U'_n(x)\Psi_n(t))\varphi'_j(x) \, dx \, dt \\ &\quad = \int_{I_n} g(t)\varphi_j(1) \, dt, \qquad \text{for } j = 1, 2, \dots, m, \end{aligned} \tag{7.2.15}$$

where $\dot{U}$, U', $\dot{V}$, and V' are computed using (7.2.9) with

$$\Psi_{n-1}(t) = \frac{t_n - t}{k_n}, \qquad \Psi_n(t) = \frac{t - t_{n-1}}{k_n}, \quad k_n = t_n - t_{n-1}.$$

Thus, Eqs. (7.2.14) and (7.2.15) are reduced to the *iterative forms*:

$$\underbrace{\int_0^1 U_n(x)\varphi_j(x) \, dx}_{\mathbf{M}U_n} - \frac{k_n}{2}\underbrace{\int_0^1 V_n(x)\varphi_j(x) \, dx}_{\mathbf{M}V_n}$$
$$= \underbrace{\int_0^1 U_{n-1}(x)\varphi_j(x) \, dx}_{\mathbf{M}U_{n-1}} + \frac{k_n}{2}\underbrace{\int_0^1 V_{n-1}(x)\varphi_j(x) \, dx}_{\mathbf{M}V_{n-1}}, \quad j = 1, 2, \dots, m,$$

and

$$\underbrace{\int_0^1 V_n(x)\varphi_j(x)\,dx}_{\mathbf{M}V_n} + \underbrace{\tfrac{k_n}{2}\int_0^1 U_n'(x)\varphi_j'(x)\,dx}_{\mathbf{S}U_n}$$
$$= \underbrace{\int_0^1 V_{n-1}(x)\varphi_j(x)\,dx}_{\mathbf{M}V_{n-1}} - \underbrace{\tfrac{k_n}{2}\int_0^1 U_{n-1}'(x)\varphi_j'(x)\,dx}_{\mathbf{S}U_{n-1}} + g_n, \quad j = 1, 2, \dots, m,$$

respectively, where we used (7.2.10) and as we computed earlier

$$\mathbf{S} = \frac{1}{h}\begin{bmatrix} 2 & -1 & \cdots & 0 \\ -1 & 2 & -1 & \cdots \\ \cdots & \cdots & \cdots & \cdots \\ 0 & -1 & 2 & -1 \\ 0 & 0 & -1 & 1 \end{bmatrix}, \quad \mathbf{M} = \frac{h}{6}\begin{bmatrix} 4 & 1 & \cdots & 0 \\ 1 & 4 & 1 & \cdots \\ \cdots & \cdots & \cdots & \cdots \\ \cdots & 1 & 4 & 1 \\ 0 & \cdots & 1 & 2 \end{bmatrix},$$

and

$$g_n = (0, \dots, 0, g_{n,m})^T, \quad \text{with } g_{n,m} = \int_{I_n} g(t)\,dt.$$

In the compact form, the vectors U_n and V_n are determined by solving the linear system of equations:

$$\begin{cases} \mathbf{M}U_n - \frac{k_n}{2}\mathbf{M}V_n = \mathbf{M}U_{n-1} + \frac{k_n}{2}\mathbf{M}V_{n-1} \\ \frac{k_n}{2}\mathbf{S}U_n + \mathbf{M}V_n = -\frac{k_n}{2}\mathbf{S}U_{n-1} + \mathbf{M}V_{n-1} + g_n, \end{cases} \tag{7.2.16}$$

which is a system of $2m$ equations with $2m$ unknowns:

$$\underbrace{\begin{bmatrix} \mathbf{M} & -\frac{k_n}{2}\mathbf{M} \\ \frac{k_n}{2}\mathbf{S} & \mathbf{M} \end{bmatrix}}_{A} \underbrace{\begin{bmatrix} U_n \\ V_n \end{bmatrix}}_{W_n} = \underbrace{\begin{bmatrix} \mathbf{M} & \frac{k_n}{2}\mathbf{M} \\ -\frac{k_n}{2}\mathbf{S} & \mathbf{M} \end{bmatrix} \begin{bmatrix} U_{n-1} \\ V_{n-1} \end{bmatrix} + \begin{bmatrix} 0 \\ g_n \end{bmatrix}}_{b_n},$$

with $W = A^{-1}b_n$, $U_n = W(1 : m)$, and $V_n = W(m+1 : 2m)$.

Note that the final solution vectors are $\hat{U}_n = (0, U_n)$ and $\hat{V}_n = (0, V_n)$.

7.2.3 Exercises

Problem 7.9 Show that $\|(\dot{u})'\|^2 + \|u''\|^2 =$ constant, independent of t.
Hint: Differentiate the equation with respect to x and multiply by $\dot{u}$.
Alternatively, multiply (DE): $\ddot{u} - u'' = 0$, by $-(\dot{u})''$ and integrate over I.

Problem 7.10 Derive a total conservation of energy relation using the Robin type boundary condition: $\dfrac{\partial u}{\partial n} + u = 0$.

Problem 7.11 Derive the corresponding linear system of equations in the case of time discretization with dG(0).

Problem 7.12 (*Discrete conservation of energy*) Show that cG(1)–cG(1) for the wave equation in system form, with $g(t) = 0$, conserves energy, i.e.

$$\|U_n'\|^2 + \|V_n\|^2 = \|U_{n-1}'\|^2 + \|V_{n-1}\|^2. \tag{7.2.17}$$

Hint: Multiply the first equation by $(U_{n-1} + U_n)^t\mathbf{S}\mathbf{M}^{-1}$ and the second equation by $(V_{n-1} + V_n)^t$ and add up. Use then, e.g. the fact that $U_n^t\mathbf{S}U_n = \|U_n'\|^2$, where

$$U_n = \begin{pmatrix} U_{n,1} \\ U_{n,2} \\ \vdots \\ U_{n,m} \end{pmatrix}, \quad \text{and} \quad U_n = U_n(x) = U_{n,1}(x)\varphi_1(x) + \cdots + U_{n,m}(x)\varphi_m(x).$$

Problem 7.13 Consider the wave equation

$$\begin{cases} \ddot{u} - u'' = 0, & x \in \mathbb{R}, \quad t > 0, \\ u(x,0) = u_0(x), & x \in \mathbb{R}, \\ \dot{u}(x,0) = v_0(x), & x \in \mathbb{R}. \end{cases} \tag{7.2.18}$$

Plot the graph of $u(x, 2)$ in the following cases:
(a) $v_0 = 0$ and

$$u_0(x) = \begin{cases} 1, & x < 0, \\ 0, & x > 0. \end{cases}$$

(b) $u_0 = 0$, and

$$v_0(x) = \begin{cases} -1, & -1 < x < 0, \\ 1, & 0 < x < 1, \\ 0, & |x| > 0. \end{cases}$$

Problem 7.14 Compute the solution for the wave equation

$$\begin{cases} \ddot{u} - 4u'' = 0, & x > 0, & t > 0, \\ u(0,t) = 0, & & t > 0, \\ u(x,0) = u_0(x), & \dot{u}(x,0) = 0, & x > 0. \end{cases} \tag{7.2.19}$$

Plot the solutions for the three cases $t = 0.5$, $t = 1$, $t = 2$, and with

$$u_0(x) = \begin{cases} 1, & x \in [2,3] \\ 0, & \text{else} \end{cases} \tag{7.2.20}$$

Problem 7.15 Apply cG(1) time discretization directly to the wave equation by letting

$$U(x,t) = U_{n-1}\Psi_{n-1}(t) + U_n(x)\Psi_n(t), \quad t \in I_n. \tag{7.2.21}$$

Note that $\dot{U}$ is piecewise constant in time and comment on:

$$\underbrace{\int_{I_n}\int_0^1 \ddot{U}\varphi_j \, dx\, dt}_{=0} + \underbrace{\int_{I_n}\int_0^1 u'\varphi_j' \, dx\, dt}_{\frac{k}{2}\mathbf{S}(U_{n-1}+U_n)} = \underbrace{\int_{I_n} g(t)\varphi_j(1)\, dt}_{g_n}, \quad j = 1, 2, \dots, m.$$

Problem 7.16 Construct a FEM for Problem

$$\begin{cases} \ddot{u} + \dot{u} - u'' = f, & 0 < x < 1, \quad t > 0, \\ u(0,t) = 0, & u'(1,t) = 0, \quad t > 0, \\ u(x,0) = 0, & \dot{u}(x,0) = 0, \quad 0 < x < 1. \end{cases} \tag{7.2.22}$$

Problem 7.17 Determine the solution for the wave equation

$$\begin{cases} \ddot{u} - c^2u'' = f, & x > 0, \qquad t > 0, \\ u(x,0) = u_0(x), & u_t(x,0) = v_0(x), \quad x > 0, \\ u_x(1,t) = 0, & \qquad t > 0, \end{cases}$$

in the following cases:

(a) $f = 0$.

(b) $f = 1, \quad u_0 = 0, \quad v_0 = 0$.

7.3 Convection–Diffusion Problems

Several multiphysical phenomena are described by the convection, diffusion, and absorption procedures. Problems of fluid- and gas dynamics, chemical reaction-diffusion, electromagnetic fields, charged particles collision, etc., are often modeled as convection–diffusion–absorption type problems. In Chapter 5, we illustrated the finite element procedure for the one-dimensional *stationary* absorption–diffusion and convection–diffusion problems (see Examples 5.1 and 5.2). Here, we shall give the derivation for a *time-dependent* convection

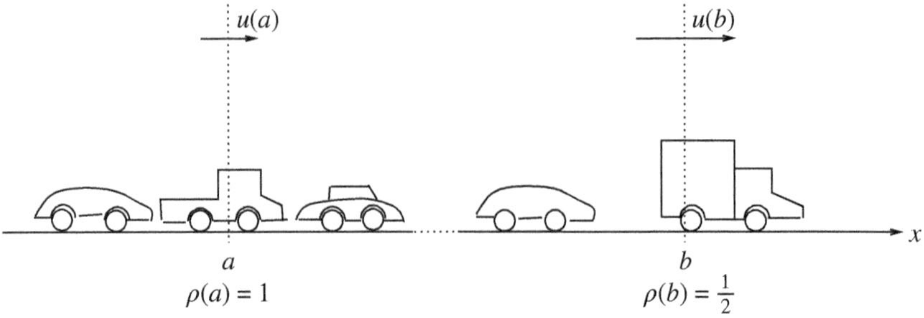

Figure 7.5 Flow of cars with on highway.

diffusion problem in one-dimensional case. We also introduce the *Streamline Diffusion Method* (SD) that is of vital interest for convection-dominated convection–diffusion problems: the SD method corresponds to generating extra diffusion in the equation which helps, e.g. to enhance the stability. SD method is also adequate for discontinuous Galerkin in time.

A convection–diffusion model problem. We illustrate the convection–diffusion phenomenon by an example:

Example 7.4 Consider the traffic flow in a highway, e.g. in Figure 7.5. Let $\rho = \rho(x,t)$ be the density of cars ($0 \le \rho \le 1$) and $u = u(x,t)$ the velocity (speed vector) of the cars at the position $x \in (a,b)$ and time t. For a highway path (a,b), the difference between the traffic inflow $u(a)\rho(a)$ at the point $x = a$ and outflow $u(b)\rho(b)$ at $x = b$ gives the density variation at time t on (a,b):

$$\frac{d}{dt}\int_a^b \rho(x,t)\,dx = \int_a^b \dot{\rho}(x,t)\,dx = \rho(a)u(a) - \rho(b)u(b) = -\int_a^b (u\rho)'\,dx$$

or equivalently

$$\int_a^b (\dot{\rho} + (u\rho)')\,dx = 0. \tag{7.3.1}$$

Since a and b can be chosen arbitrary, we have

$$\dot{\rho} + (u\rho)' = 0. \tag{7.3.2}$$

Let now $u = 1 - \rho$ (motivate this choice), then (7.3.2) is rewritten as

$$\dot{\rho} + ((1-\rho)\rho)' = \dot{\rho} + (\rho - \rho^2)' = 0. \tag{7.3.3}$$

Hence,

$$\dot{\rho} + (1 - 2\rho)\rho' = 0 \quad \text{(A nonlinear convection equation).} \tag{7.3.4}$$

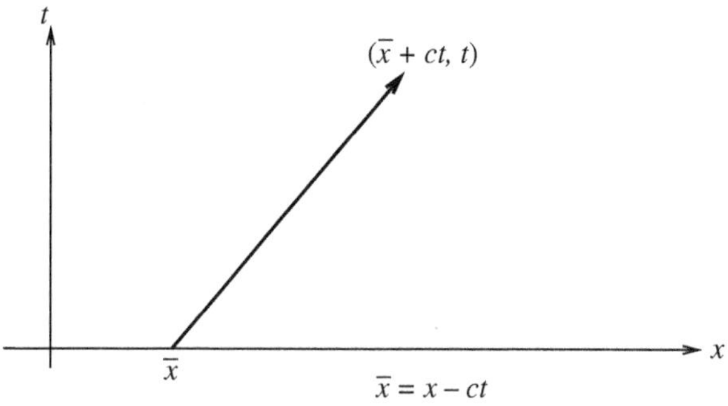

Figure 7.6 Characteristic line.

Alternatively, (to obtain *a convection–diffusion model*), we may assume that $u = c - \varepsilon \cdot (\rho'/\rho),\;\; c > 0,\;\; \varepsilon > 0$, (motivate). Then, we get from (7.3.2) that

$$\dot{\rho} + \left(\left(c - \varepsilon\frac{\rho'}{\rho}\right)\rho\right)' = 0, \tag{7.3.5}$$

i.e.

$$\dot{\rho} + c\rho' - \varepsilon\rho'' = 0 \quad \text{(A convection–diffusion equation).} \tag{7.3.6}$$

Equation (7.3.6) is convection-dominated if $c > \varepsilon$.

For $\varepsilon = 0$, the solution is given by the exact transport $\rho(x, t) = \rho_0(x - ct)$, because then ρ is constant along the characteristics; $(c, 1)$-direction (Figure 7.6).

Note that differentiating $\rho(x, t) = \rho(\bar{x} + ct, t)$ with respect to t, we get

$$\frac{\partial\rho}{\partial x} \cdot \frac{\partial x}{\partial t} + \frac{\partial\rho}{\partial t} = 0, \quad \Longleftrightarrow \quad c\rho' + \dot{\rho} = 0. \tag{7.3.7}$$

Finally, let us rewrite the convection–diffusion equation (7.3.6), for ρ, by changing the notation from ρ to u, and replacing c by β, i.e.

$$\dot{u} + \beta\, u' - \varepsilon\, u'' = 0. \tag{7.3.8}$$

Remark 7.5 Compare Eq. (7.3.8) with the Navier–Stokes equations for the incompressible flows:

$$\dot{u} + (\beta \cdot \nabla)u - \varepsilon\Delta u + \nabla P = 0, \quad \wedge \quad \text{div}\, u = 0, \tag{7.3.9}$$

where $\beta = u$, $u = (u_1, u_2, u_3)$ is the velocity vector, with u_1 representing the mass, u_2 momentum, and u_3 the energy. Further, P stands for the pressure and $\varepsilon = 1/Re$ with Re denoting the Reynold's number. For the Navier–Stokes equations, a typical range for the Reynold's number is between 10^5 and 10^7. Therefore, for $\varepsilon > 0$ and *small*, because of difficulties related to boundary layer and turbulence, the Navier–Stokes equations are not easily solvable.

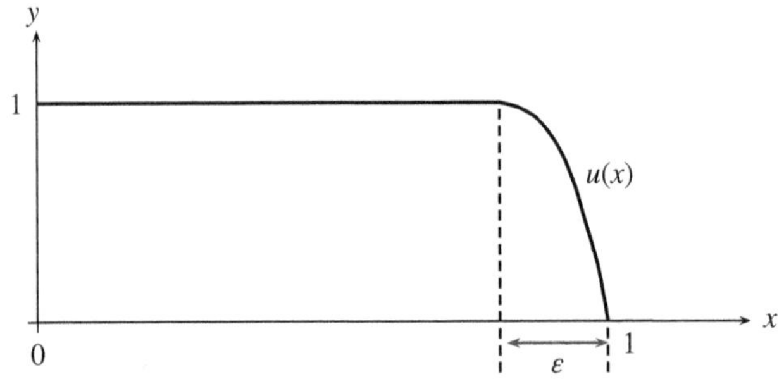

Figure 7.7 A ε boundary layer.

Example 7.5 *(The boundary layer)* Consider the boundary value problem

$$\text{(BVP)} \quad \begin{cases} u' - \varepsilon u'' = 0, & 0 < x < 1 \\ u(0) = 1, & u(1) = 0. \end{cases} \tag{7.3.10}$$

The exact solution to this problem is given by

$$u(x) = C(e^{1/\varepsilon} - e^{x/\varepsilon}), \quad \text{with} \quad C = \frac{1}{e^{1/\varepsilon} - 1}, \tag{7.3.11}$$

which has an outflow boundary layer of width $\sim \varepsilon$, as seen in Figure 7.7.

7.3.1 Finite Element Method

We shall now study the finite element approximation of problem (7.3.10). To this end, first we represent, as usual, the finite element solution by

$$U(x) = \varphi_0(x) + U_1\varphi_1(x) + \cdots + U_n\varphi_n(x), \tag{7.3.12}$$

where the φ_j:s are the basis functions, here continuous piecewise linears (hat-functions). The corresponding variational formulation yields the FEM:

$$\int_0^1 (U'\varphi_j + \varepsilon U'\varphi_j')\, dx = 0, \quad j = 1, 2, \ldots, n, \tag{7.3.13}$$

which, for piecewise linear Galerkin approximation (see the entries in convection and stiffness matrices), can be represented by the equations

$$\frac{1}{2}(U_{j+1} - U_{j-1}) + \frac{\varepsilon}{h}(2U_j - U_{j-1} - U_{j+1}) = 0, \quad j = 1, 2, \ldots, n, \tag{7.3.14}$$

where $U_0 = 1$ and $U_{n+1} = 0$.

Note that, using *central-differencing*, we may also discretize (7.3.10), viz.

$$\underbrace{\frac{U_{j+1} - U_{j-1}}{2h}}_{\text{corresp. to } u'(x_j)} - \varepsilon \underbrace{\frac{U_{j+1} - 2U_j + U_{j-1}}{h^2}}_{\text{corresp. to } u''(x_j)} = 0 \quad \left(\Longleftrightarrow \frac{1}{h} \times \text{Eq. (7.3.14)}\right).$$

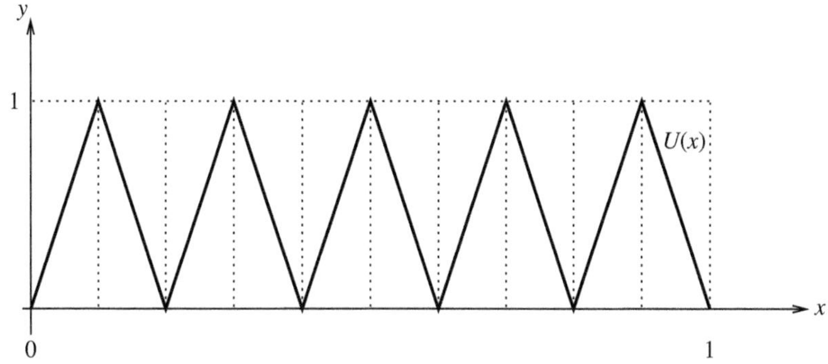

Figure 7.8 Upstreams transported oscillatory approximate solution.

Now for ε being very small, this gives that $U_{j+1} \approx U_{j-1}$, which results, for even n values, alternating 0 and 1 as the solution values at the nodes:

i.e. oscillations in U are transported "upstreams" making U a "globally bad approximation" of u, viz. Figure 7.8.

A better approach would be to approximate $u'(x_j)$ by an *upwind derivative*:

$$u'(x_j) \approx \frac{U_j - U_{j-1}}{h}, \tag{7.3.15}$$

which, formally, gives a better stability, however, with low accuracy.

Remark 7.6 The Example 7.5 above demonstrates that a high accuracy without stability is indeed useless.

A more systematic method of making the finite element solution for the fluid problems stable is through using the streamline diffusion method which we, formally, introduce below.

7.3.2 The Streamline-Diffusion Method (SDM)

The idea is to choose, in the variational formulation, the test functions of the form $(v + \frac{1}{2}\beta h v')$, instead of just v (this would finally correspond to adding an extra diffusion to the original equation in the direction of the streamlines). Then, e.g. for our model problem, we obtain the equation ($\beta \equiv 1$)

$$\int_0^1 \left[u'\left(v + \frac{1}{2}hv'\right) - \varepsilon \cdot u''\left(v + \frac{1}{2}hv'\right)\right] dx = \int_0^1 f\left(v + \frac{1}{2}hv'\right) dx. \tag{7.3.16}$$

In the case of approximation with piecewise linears, in the discrete version of the variational formulation, we should interpret the term $\int_0^1 U''v' \, dx$ as

$$\int_0^1 U''v' \, dx := \sum_j \int_{I_j} U''v' \, dx = 0. \tag{7.3.17}$$

Then, with piecewise linear test functions, choosing $v = \varphi_j$ we get the discrete term corresponding to the second term in the first integral in (7.3.16) as

$$\int_0^1 U' \frac{1}{2} h\varphi_j' \, dx = U_j - \frac{1}{2} U_{j+1} - \frac{1}{2} U_{j-1}, \tag{7.3.18}$$

which adding to the obvious approximation of first term in the first integral:

$$\int_0^1 U'\varphi_j \, dx = \frac{U_{j+1} - U_{j-1}}{2}, \tag{7.3.19}$$

we end up with $(U_j - U_{j-1})$, (using, e.g. convection and stiffness matrices) as an approximation of the first integral in (7.3.16), corresponding to the *upwind scheme*.

Remark 7.7 The streamline-diffusion method (SDM) can also be viewed as a *least-square method*:

Let $A = \frac{d}{dx}$, then $A^t = -\frac{d}{dx}$. Now u minimizes the expression $\|w' - f\|$ if $u' = Au = f$. This can be written as

$$A^tAu = A^tf \iff -u'' = -f', \quad \text{(the continuous form)}. \tag{7.3.20}$$

While multiplying $u' = Au = f$ by v and integrating over $(0, 1)$, we have

$$\int_0^1 U'v \, dx = \int_0^1 fv \, dx \quad \text{(the weak form)}, \tag{7.3.21}$$

where we replaced u' by U'. Thus, the weak form of the discretized equation may be written as

$$(AU, v) = (f, v), \quad \forall v \in V_h, \tag{7.3.22}$$

and

$$(AU, Av) = (f, Av), \quad \forall v \in V_h. \tag{7.3.23}$$

Thus, adding (7.3.22) by δ times (7.3.23), we have the SDM form:

$$(AU, v + \delta Av) = (f, v + \delta Av), \quad \forall v \in V_h. \tag{7.3.24}$$

For the time-dependent convection equation, the oriented time–space elements are used. Consider the time-dependent problem

$$\dot{u} + \beta u' - \varepsilon u'' = f. \tag{7.3.25}$$

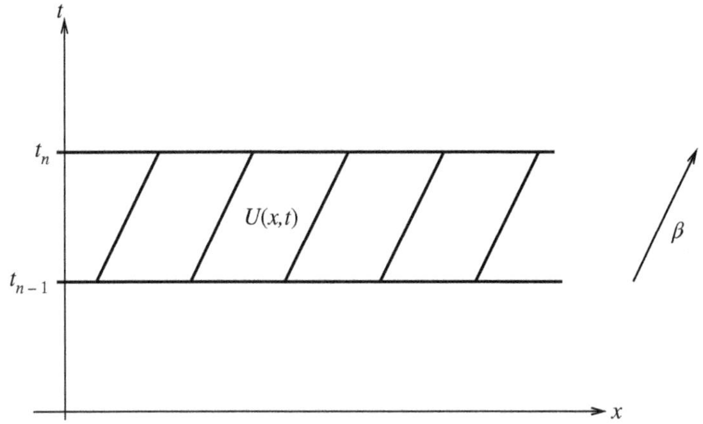

Figure 7.9 A β-telted mesh in a slab $\mathbf{S}_n$.

Consider a slab $\mathbf{S}_n = I_n \times \Omega$ as in Figure 7.9. Set $U(x,t)$ such that U is piecewise linear in x and piecewise constant in the $(\beta, 1)$-direction. Combine with SDM and add up some artificial viscosity, $\hat{\varepsilon}$, depending on the residual term and get for each time interval I_n, the scheme:

$$\int_{I_n}\int_{\Omega}\left[(\dot{U}+\beta U)\left(v+\frac{\beta}{2}h\dot{v}\right)+\hat{\varepsilon}\;U'v'\right]\,dx\,dt=\int_{I_n}\int_{\Omega}f\left(v+\frac{\beta}{2}hv'\right)\,dxdt.$$

□

7.3.3 Exercises

Problem 7.18 Prove that the solution u of the convection–diffusion problem

$$-u_{xx}+u_x+u=f,\quad \text{in } I=(0,1),\quad u(0)=u(1)=0,$$

satisfies the following estimate

$$\left(\int_I u^2\phi\,dx\right)^{1/2}\le\left(\int_I f^2\phi\,dx\right)^{1/2}.$$

where $\phi(x)$ is a positive weight function defined on $(0,1)$ satisfying $\phi_x(x)\le 0$ and $-\phi_x(x)\le\phi(x)$ for $0\le x\le 1$.

Problem 7.19 Let ϕ be a solution of the problem

$$-\varepsilon\phi''-3\phi'+2\phi=e,\qquad \phi'(0)=\phi(1)=0.$$

Let $\|\cdot\|$ denote the L_2-norm on I. Show that there is a constant C such that

$$|\phi'(0)|\le C\|e\|,\qquad \|\varepsilon\phi''\|\le C\|e\|.$$

Problem 7.20 Use relevant interpolation theory estimates and prove an a priori and an a posteriori error estimate for the cG(1) finite element method for the problem

$$-u'' + u' = f, \quad \text{in } I = (0,1), \quad u(0) = u(1) = 0.$$

Problem 7.21 Prove an a priori and an a posteriori error estimate for the cG(1) finite element method of the problem

$$-u'' + u' + u = f, \quad \text{in } I = (0,1), \quad u(0) = u(1) = 0.$$

Problem 7.22 Consider the problem

$$-\varepsilon u'' + xu' + u = f, \quad \text{in } I = (0,1), \quad u(0) = u'(1) = 0,$$

where ε is a positive constant, and $f \in L_2(I)$. Prove that

$$\|\varepsilon u''\| \leq \|f\|.$$

Problem 7.23 We modify Problem 7.22 above according to

$$-\varepsilon u'' + c(x)u' + u = f(x) \quad 0 < x < 1, \qquad u(0) = u'(1) = 0,$$

where ε is a positive constant, the function c satisfies $c(x) \geq 0$, $c'(x) \leq 0$, and $f \in L_2(I)$. Prove that there are positive constants C_1, C_2, and C_3 such that

$$\sqrt{\varepsilon}\|u'\| \leq C_1\|f\|, \quad \|cu'\| \leq C_2\|f\|, \quad \text{and } \varepsilon\|u''\| \leq C_3\|f\|,$$

where $\|\cdot\|$ is the $L_2(I)$-norm.

Problem 7.24 Consider the convection–diffusion–absorption problem

$$-\varepsilon u_{xx} + u_x + u = f, \quad \text{in } I = (0,1), \quad u(0) = 0, \qquad \sqrt{\varepsilon}u_x + u(1) = 0,$$

where ε is a positive constant, and $f \in L_2(I)$. Prove the following stability estimates for the solution u

$$\|\sqrt{\varepsilon}u_x\| + \|u\| + |u(1)| \leq C\|f\|,$$

$$\|u_x\| + \|\varepsilon u_{xx}\| \leq C\|f\|,$$

where $\|\cdot\|$ denotes the $L_2(I)$-norm, $I = (0,1)$, and C is an appropriate constant.

8

Approximation in Several Dimensions

8.1 Introduction

This chapter is devoted to, mainly piecewise linear, finite element approximations in higher dimensions. Starting with the two-dimensions, we shall prove a simple version of the Green's formula that is, basically, the two-dimensional integration by parts. We shall demonstrate the finite element structure through a canonical example, introduce some common finite element spaces, and extend the interpolation concept to higher dimensions. In this setting, most higher dimensional approaches are direct extensions of the study of one-dimensional problems in Chapters 6 and 7. Then, the *mathematical calculus*, being in several variables, becomes somewhat involved (we assume the reader is familiar with calculus of several variables). On the other hand, the two- and three-dimensional cases are the most relevant models both from physical point of views as well as applications. A general problem to study is the convection–diffusion–absorption equation:

$$\begin{cases} -\Delta u(\mathbf{x}) + \beta(\mathbf{x}) \cdot \nabla u(\mathbf{x}) + a(\mathbf{x})u(\mathbf{x}) = f(\mathbf{x}), & \mathbf{x} \in \Omega \subset \mathbb{R}^n, \\ \qquad +BCs, e.g. \quad u(\mathbf{x}) = 0, & \mathbf{x} \in \partial\Omega, \end{cases} \tag{8.1.1}$$

where $n \geq 2$, $\beta(\mathbf{x})$, and $a(\mathbf{x})$ are convection and absorption coefficients, respectively, and $f(\mathbf{x})$ is a source term. The discretization procedure, e.g. approximating with piecewise polynomials and deriving the approximation error in certain norm, would require extension of the interpolation estimates from the intervals in $\mathbb{R}$, the one-dimensional ($1d$) case, to higher-dimensional domains in $\mathbb{R}^n$, $n \geq 2$. Here, for simplicity, most estimates are carried out in two space dimensions. Then, to compute double integrals that appear, e.g. in the weak formulations, the frequently used partial integration in the $1d$ case is now replaced by the *Green's formula* below, which we state and prove for a rectangular domain in two-dimensions ($2d$). The classical proof for a general domain Ω, in arbitrary dimension, can be found in any textbook in the calculus of several variables.

An Introduction to the Finite Element Method for Differential Equations, First Edition.
Mohammad Asadzadeh.

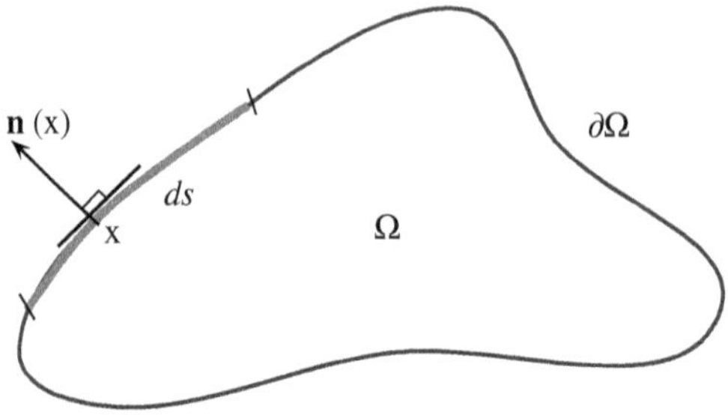

Figure 8.1 A smooth domain Ω with an outward unit normal **n**.

Lemma 8.1 (Green's formula) *Let Ω be an open-bounded subset of $\mathbb{R}^2$, with a piecewise smooth boundary and set $u \in C^2(\Omega)$ and $v \in C^1(\Omega)$, then*

$$\int_{\Omega} (\Delta u) v \, dx \, dy = \int_{\partial\Omega} (\nabla u \cdot \boldsymbol{n}) v \, ds - \int_{\Omega} \nabla u \cdot \nabla v \, dx \, dy, \tag{8.1.2}$$

where $\boldsymbol{n} := \mathbf{n}(x, y)$, as in Figure 8.1, is the outward unit normal at the boundary point $\boldsymbol{x} = (x, y) \in \partial\Omega$ and ds is a curve element on the boundary $\partial\Omega$.

Proof: Since we shall work with polygonal domains, we give a simple proof when Ω is a rectangle, see Figure 8.2. Then, using integration by parts, we have

$$\begin{aligned}\iint_{\Omega} \frac{\partial^2 u}{\partial x^2} v \, dx \, dy &= \int_0^b \int_0^a \frac{\partial^2 u}{\partial x^2}(x, y) \, v(x, y) \, dx \, dy \\ &= \int_0^b \left(\left[\frac{\partial u}{\partial x}(x, y) \, v(x, y) \right]_{x=0}^{a} - \int_0^a \frac{\partial u}{\partial x}(x, y) \, \frac{\partial v}{\partial x}(x, y) \, dx \right) dy\end{aligned}$$

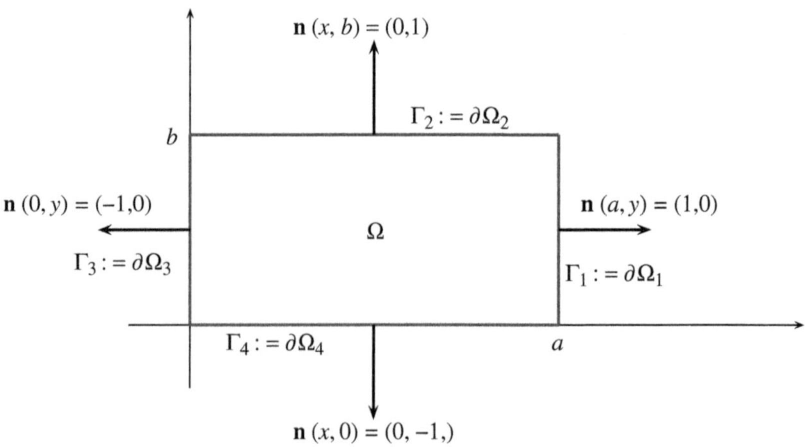

Figure 8.2 A rectangular domain with outward unit normals to its sides.

$$= \int_0^b \left(\frac{\partial u}{\partial x}(a,y)\ v(a,y) - \frac{\partial u}{\partial x}(0,y)\ v(0,y) \right)\ dy$$
$$- \iint_\Omega \frac{\partial u}{\partial x}(x,y)\ \frac{\partial v}{\partial x}(x,y)\ dx\ dy.$$

Now identifying $\mathbf{n}$ on $\partial\Omega$, viz. on Γ_1: $\mathbf{n}(a,y) = (1,0)$,
on Γ_2: $\mathbf{n}(x,b) = (0,1)$,
on Γ_3: $\mathbf{n}(0,y) = (-1,0)$,
on Γ_4: $\mathbf{n}(x,0) = (0,-1)$,

the first integral on the right-hand side above can be written as

$$\left(\int_{\Gamma_1} + \int_{\Gamma_3} \right) \left(\frac{\partial u}{\partial x}, \frac{\partial u}{\partial y} \right) (x,y) \cdot \mathbf{n}(x,y)\ v(x,y)\ ds.$$

Hence,

$$\iint_\Omega \frac{\partial^2 u}{\partial x^2} v\ dx\ dy = \int_{\Gamma_1 \cup \Gamma_3} \left(\frac{\partial u}{\partial x}, \frac{\partial u}{\partial y} \right) \cdot \mathbf{n}(x,y)\ v(x,y)\ ds - \iint_\Omega \frac{\partial u}{\partial x} \frac{\partial v}{\partial x}\ dx\ dy.$$

Similarly, in the y-direction, we get

$$\iint_\Omega \frac{\partial^2 u}{\partial y^2} v\ dx\ dy = \int_{\Gamma_2 \cup \Gamma_4} \left(\frac{\partial u}{\partial x}, \frac{\partial u}{\partial y} \right) \cdot \mathbf{n}(x,y)\ v(x,y)\ ds - \iint_\Omega \frac{\partial u}{\partial y} \frac{\partial v}{\partial y}\ dx\ dy.$$

Adding up these two relations, the proof is complete.

8.2 Piecewise Linear Approximation in 2*d*

In this section, we outline the procedure of the piecewise linear polynomial approximations of the solutions for differential equations in two-dimensional polygonal domains. Hence, we consider partitions (*meshes*) without any concern about curved boundaries. At the end of this chapter, in connection with a formal description of the finite element procedure, we shall briefly discuss some common extensions of the linear approximations to higher-order polynomials both in triangular and quadrilateral (as well as tetrahedrons in the three-dimensional case) elements. See also Tables 8.1–8.4.

8.2.1 Basis Functions for the Piecewise Linears in 2*d*

Recall that in the one-dimensional case a function which is linear on an interval is uniquely determined by its values at two points (e.g. the endpoints of the interval), since there is only one straight line connecting two points, e.g. as in Figure 8.3. Similarly a plane in $\mathbb{R}^3$ is uniquely determined by three points which are not lying

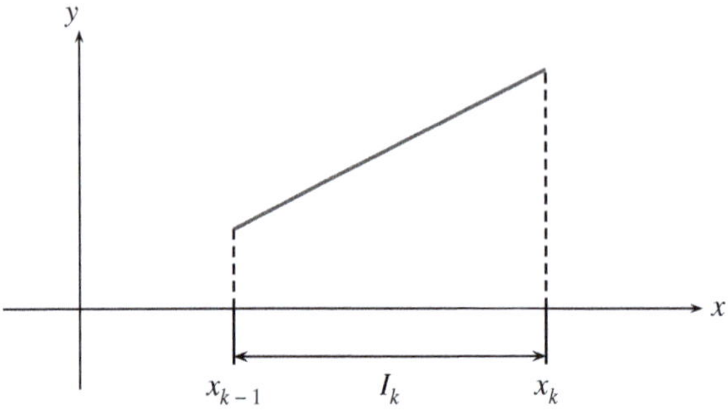

Figure 8.3 A linear function on a subinterval $I_k = (x_{k-1}, x_k)$.

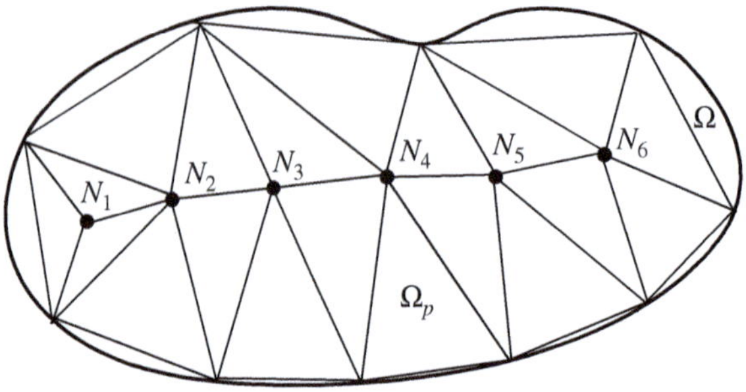

Figure 8.4 Example of triangulation of a 2d domain.

on the same straight line. Therefore, for piecewise linear approximation of a function u defined on a two-dimensional polygonal domain $\Omega_p \subset \mathbb{R}^2$, i.e. $u : \Omega_p \to \mathbb{R}$, it is natural to make partitions of Ω_p into triangular elements and let the sides of the triangles correspond to the endpoints of the intervals in the one-dimensional case.

Figure 8.4 illustrates a "partitioning" (*triangulation*) of a polygonal domain Ω_p generated by domain Ω with curved boundary. Here, we have six internal nodes N_i, $1 \leq i \leq 6$ and Ω_p is the large *polygonal* domain inside Ω, which is triangulated. Figure 8.5 shows a piecewise linear function on a single triangle (element) which is determined by its values, z_i, $i = 1, 2, 3$, at the vertices of the triangle ABC. Below we define the concepts of *subdivision* in any spatial dimension and *triangulation* in the 2d case.

Definition 8.1 A subdivision of a computational domain Ω is a finite collection of open sets $\{K_i\}$ such that

(a) $K_i \bigcap K_j = \emptyset$
(b) $\bigcup_i \overline{K}_i = \overline{\Omega}$.

Definition 8.2 A *triangulation* of a polygonal domain $\Omega \subset \mathbb{R}^2$ is a subdivision of Ω consisting of triangles with the property that no vertex of any triangle lies in the interior of an edge of another triangle.

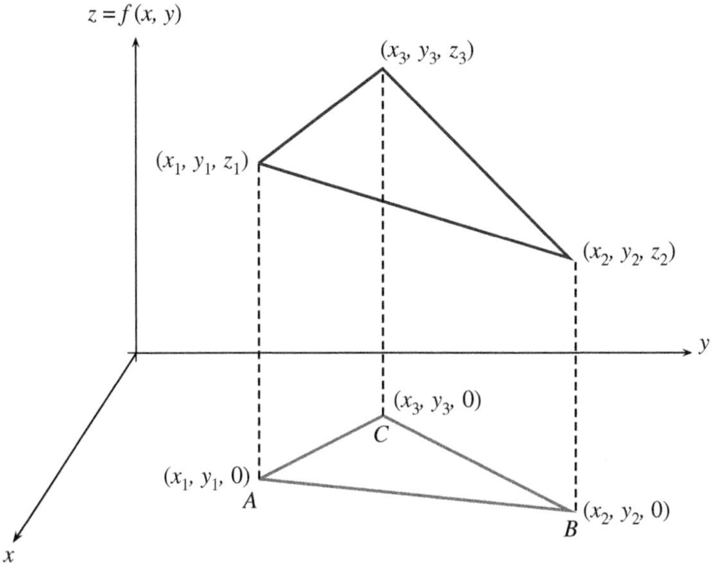

Figure 8.5 A triangle in 3*d* as a piecewise linear function and its projection in 2*d*.

Returning to Figure 8.4, for every linear function U on Ω_p,

$$U(\mathbf{x}) = U_1\varphi_1(\mathbf{x}) + U_1\varphi_2(\mathbf{x}) + \cdots + U_6\varphi_6(\mathbf{x}), \tag{8.2.1}$$

where $U_i = U(N_i)$, $i = 1, 2, \ldots, 6$, are numerical values (nodal values) of U at the nodes N_i (these are not, necessarily, the nodal values of u). φ_i are the basis functions with $\varphi_i(N_i) = 1$, $\varphi_i(N_j) = 0$ for $j \neq i$ and $\varphi_i(\mathbf{x})$ is linear in $\mathbf{x}$ in every triangle/element (as in the Figure 8.6).

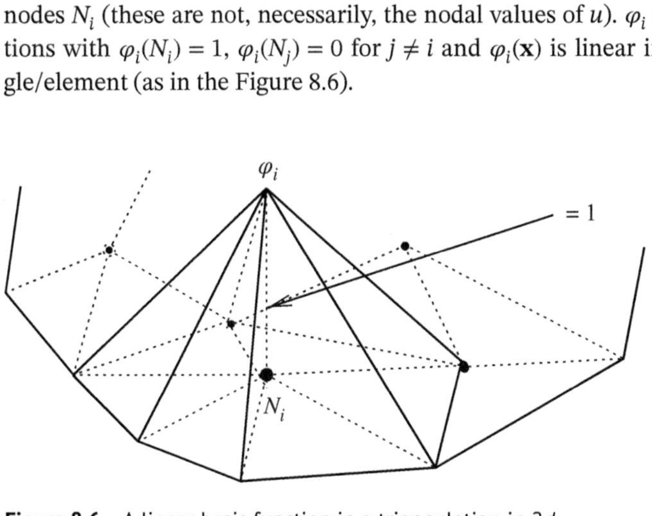

Figure 8.6 A linear basis function in a triangulation in 2*d*.

Note that with the Dirichlet boundary condition, as in the one-dimensional case, the test functions use $\varphi_i(\mathbf{x}) = 0$ for $\mathbf{x} \in \partial\Omega_p$. Hence, for a Dirichlet boundary value problem, to determine the approximate solution U, in Ω_p, is reduced to find the *nodal values* $U_1, U_2, \dots, U_6$, obtained from the corresponding discrete variational formulation.

Example 8.1 Let $\Omega = \{(x, y) : \; 0 < x < 4, \;\; 0 < y < 3\}$ and make a piecewise linear finite element method (FEM) discretization of the following boundary value problem:

$$\begin{cases} -\Delta u = f & \text{in } \Omega \\ u = 0 & \text{on } \partial\Omega. \end{cases} \tag{8.2.2}$$

Solution: Using Green's formula, the variational formulation for problem (8.2.2) reads as follows: find the function $u \in H_0^1(\Omega)$ such that

$$\iint_\Omega (\nabla u \cdot \nabla v)\, dx\, dy = \iint_\Omega f v\, dx\, dy, \quad \forall v \in H_0^1(\Omega). \tag{8.2.3}$$

Recall that $H_0^1(\Omega)$ is the usual Hilbert space

$$H_0^1(\Omega) := \{v : \; v \in L_2(\Omega), \;\; |\nabla v| \in L_2(\Omega), \;\; \text{and } v = 0, \;\; \text{on } \Gamma := \partial\Omega\}.$$

Now we make a test function space of piecewise linears. To this end, we triangulate Ω as in the Figure 8.7 and let $V_h^0 \subset V^0 := H_0^1$ be the space of continuous piecewise linears on the triangulation, $\mathcal{T}_h$ of Ω in Figure 8.7:

$$V_h^0 = \{v \in C(\Omega) : \;\; v \text{ is continuous, linear on each triangle and } v = 0 \text{ on } \partial\Omega\}.$$

A discrete solution u_h in this scheme is uniquely determined by its values (not, necessarily, the nodal values for u!) at the vertices of the triangles and 0 at the boundary of the domain. In our example, we consider a uniform division with only

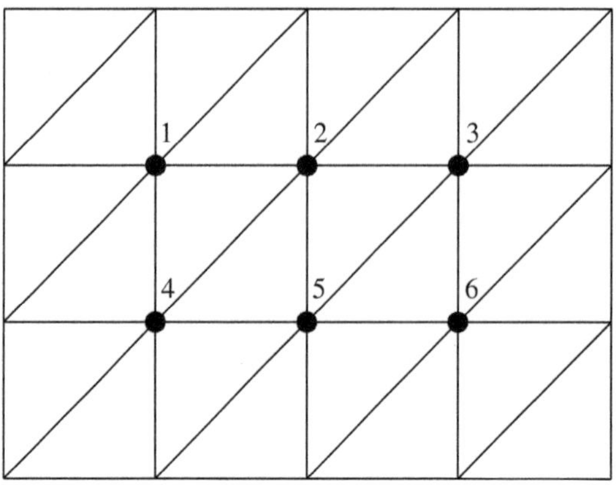

Figure 8.7 Uniform triangulation of Ω with $h = 1$.

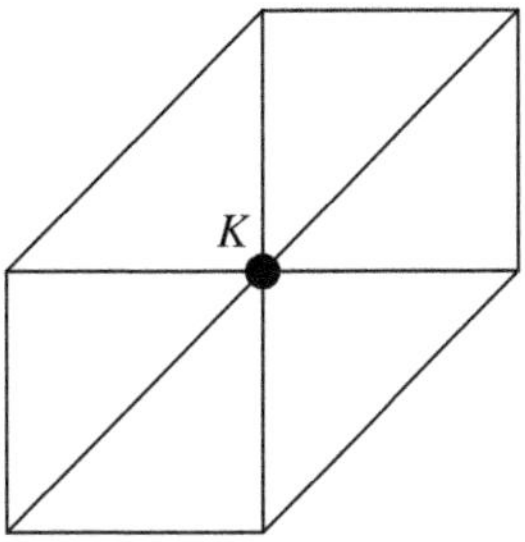

Figure 8.8 Support of a single basis function in Example 8.1.

six inner vertices of interest. Hence, as in the one-dimensional case, we construct continuous piecewise linear basis functions (six of them in this particular case), with values 1 at one of the (interior) nodes and zero at all other neighboring nodes (both interior and boundary nodes). Then, we get the two-dimensional correspondence to the hat-functions φ_j, called *tent functions* as shown in the Figure 8.6. Then, the support of each one of the basis functions in triangulation in Figure 8.7 is of the shape of Figure 8.8.

Then, the FEM for (8.2.2) reads as follows: find $u_h \in V_h^0$ such that

$$\iint_\Omega (\nabla u_h \cdot \nabla v)\, dx\, dy = \iint_\Omega fv\, dx\, dy, \quad \forall v \in V_h^0(\Omega). \tag{8.2.4}$$

The approximation procedure follows the path generalizing the $1d$-case. Below we illustrate the procedure for an example of a mixed boundary value problem in two dimensions and derive the corresponding coefficient matrices.

Example 8.2 Formulate the cG(1) piecewise continuous Galerkin method for the boundary value problem

$$\begin{cases} -\Delta u + u = f, & x \in \Omega; \\ u = 0, & x \in \partial\Omega \setminus (\Gamma_1 \cup \Gamma_2), \\ \nabla u \cdot \mathbf{n} = 0, & x \in \Gamma_1 \cup \Gamma_2, \end{cases} \tag{8.2.5}$$

on the two-dimensional domain Ω below, with outward unit normal $\mathbf{n}$ at the boundary (as in the Figure 8.9). Compute the matrices for the resulting system of equations using the following mesh with nodes at N_1, N_2, and N_3.

The idea is to use a standard triangle-element T with axi-parallel sides of length h.

Solution: Let V be the linear function space defined as

$$V := \{v : v \in H^1(\Omega), \quad v = 0, \quad \text{on } \partial\Omega \setminus (\Gamma_1 \cup \Gamma_2)\}.$$

We denote the scalar product by $(\cdot, \cdot)$, viz.

$$(f, g) = \int_\Omega f(\mathbf{x}) g(\mathbf{x})\, d\mathbf{x}.$$

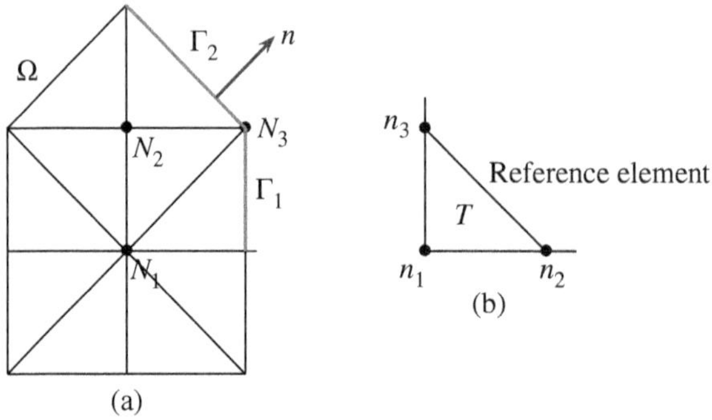

Figure 8.9 (a) Uniform triangulation of Ω and (b) the reference element T.

Multiplying the differential equation by $v \in V$ and integrating over Ω we get

$$-(\Delta u, v) + (u, v) = (f, v), \quad \forall v \in V.$$

Now using Green's formula we have that

$$\begin{aligned}-(\Delta u, v) &= (\nabla u, \nabla v) - \int_{\partial\Omega} (\mathbf{n} \cdot \nabla u) v \, ds \\ &= (\nabla u, \nabla v) - \int_{\partial\Omega \setminus (\Gamma_1 \cup \Gamma_2)} (\mathbf{n} \cdot \nabla u) v \, ds - \int_{\Gamma_1 \cup \Gamma_2} (\mathbf{n} \cdot \nabla u) v \, ds \\ &= (\nabla u, \nabla v), \quad \forall v \in V,\end{aligned}$$

where we used the boundary data and the definition of V. Thus, the variational formulation reads as follows: find $u \in V$ such that

$$(\nabla u, \nabla v) + (u, v) = (f, v), \quad \forall v \in V.$$

Let V_h be the finite element space consisting of continuous piecewise linear functions, on the partition as in the figure in the example, satisfying the boundary condition $v = 0$ on $\partial\Omega \setminus (\Gamma_1 \cup \Gamma_2)$: Then, the cG(1) method for (8.2.5) is formulated as follows: find $u_h \in V_h$ such that

$$(\nabla u_h, \nabla v) + (u_h, v) = (f, v), \quad \forall v \in V_h.$$

Making the "Ansatz" $u_h(x) = \sum_{j=1}^{3} \xi_j \varphi_j(x)$, where φ_j are the standard, two-dimensional, basis functions, we obtain the system of equations

$$\sum_{j=1}^{3} \xi_j \left(\int_\Omega \nabla \varphi_j \cdot \nabla \varphi_i \, dx + \int_\Omega \varphi_j \varphi_i \, dx \right) = \int_\Omega f \varphi_i \, dx, \quad i = 1, 2, 3,$$

or in matrix form,

$$(\mathbf{S} + \mathbf{M})\xi = F,$$

where $\mathbf{S}_{ij} = (\nabla\varphi_i, \nabla\varphi_j)$ is the stiffness matrix, $\mathbf{M}_{ij} = (\varphi_i, \varphi_j)$ is the mass matrix, and $F_i = (f, \varphi_i)$ is the load vector. To proceed, first we compute the mass and stiffness matrices for the standard element T with the side length h (see Figure 8.9). The local basis functions and their gradients are given by

$$\phi_1(x_1,x_2) = 1 - \frac{x_1}{h} - \frac{x_2}{h}, \qquad \nabla\phi_1(x_1,x_2) = -\frac{1}{h}\begin{bmatrix}1\\1\end{bmatrix},$$

$$\phi_2(x_1,x_2) = \frac{x_1}{h}, \qquad \nabla\phi_2(x_1,x_2) = \frac{1}{h}\begin{bmatrix}1\\0\end{bmatrix},$$

$$\phi_3(x_1,x_2) = \frac{x_2}{h}, \qquad \nabla\phi_3(x_1,x_2) = \frac{1}{h}\begin{bmatrix}0\\1\end{bmatrix}.$$

Hence, with $|T| = \int_T dz = h^2/2$,

$$m_{11} = (\phi_1,\phi_1) = \int_T \phi_1^2 \, dx = h^2 \int_0^1 \int_0^{1-x_2} (1-x_1-x_2)^2 \, dx_1 \, dx_2 = \frac{h^2}{12},$$

$$s_{11} = (\nabla\phi_1,\nabla\phi_1) = \int_T |\nabla\phi_1|^2 \, dx = \frac{2}{h^2}|T| = 1.$$

Alternatively, we can use the midpoint rule, which is exact for polynomials of degree 2 (precision 3):

$$m_{11} = (\phi_1,\phi_1) = \int_T \phi_1^2 \, dx = \frac{|T|}{3}\sum_{j=1}^{3} \phi_1(\hat{x}_j)^2 = \frac{h^2}{6}\left(0 + \frac{1}{4} + \frac{1}{4}\right) = \frac{h^2}{12},$$

where $\hat{x}_j$ are the midpoints of the edges. Similarly, we can compute the other elements and thus obtain the local mass and stiffness matrices for the standard element T:

$$m = \frac{h^2}{24}\begin{bmatrix}2 & 1 & 1\\ 1 & 2 & 1\\ 1 & 1 & 2\end{bmatrix}, \qquad s = \frac{1}{2}\begin{bmatrix}2 & -1 & -1\\ -1 & 1 & 0\\ -1 & 0 & 1\end{bmatrix}.$$

We can now assemble the global matrices $\mathbf{M}$ and $\mathbf{S}$ from the local ones m and s, through identifying the contribution from each element sharing the same node to the global matrices. For instance, since the node N_1 is common to eight triangles and represents vertices corresponding to n_2 or n_3 in reference element T so we have

$$\mathbf{M}_{11} = 8m_{22} = \frac{8}{12}h^2, \qquad \mathbf{S}_{11} = 8s_{22} = 4,$$

Similarly

$$\mathbf{M}_{12} = 2m_{12} = \frac{1}{12}h^2, \qquad \mathbf{S}_{12} = 2s_{12} = -1,$$

$$\mathbf{M}_{13} = 2m_{23} = \frac{1}{12}h^2, \qquad \mathbf{S}_{13} = 2s_{23} = 0,$$

$$\mathbf{M}_{22} = 4m_{11} = \frac{4}{12}h^2, \qquad \mathbf{S}_{22} = 4s_{11} = 4,$$

$$\mathbf{M}_{23} = 2m_{12} = \frac{1}{12}h^2, \qquad \mathbf{S}_{23} = 2s_{12} = -1,$$

$$\mathbf{M}_{33} = 3m_{22} = \frac{3}{12}h^2, \qquad \mathbf{S}_{33} = 3s_{22} = 3/2.$$

The remaining matrix elements are obtained by symmetry $\mathbf{M}_{ij} = \mathbf{M}_{ji}$, $\mathbf{S}_{ij} = \mathbf{S}_{ji}$. Hence,

$$\mathbf{M} = \frac{h^2}{12}\begin{bmatrix} 8 & 1 & 1 \\ 1 & 4 & 1 \\ 1 & 1 & 3 \end{bmatrix}, \qquad \mathbf{S} = \begin{bmatrix} 4 & -1 & 0 \\ -1 & 4 & -1 \\ 0 & -1 & 3/2 \end{bmatrix}.$$

8.3 Constructing Finite Element Spaces

The simplest finite element space consists of continuous piecewise linear basis functions on a partition of a bounded domain $\Omega \in \mathbb{R}^d$, $d = 1, 2, 3$. This, however, is not sufficient to capture all important features of a solution, and many problems are better approximated considering higher spectral orders which involves functional values and some first-, second-, and higher-order derivatives on the sides and at the vertices of the elements.

In the one-dimensional case, because of the simplicity of the geometry, we could manage to set up approximation procedures relying on elementary calculus and detailed formalism was unnecessary. However, a general rigorous approach to finite elements sought for a somewhat detailed structure. Below we give a general framework constructing a *finite element method*:

Definition 8.3 By a finite element, we mean a triple (K, P_K, Σ) where

K is a geometric object, i.e. a triangle, a quadrilateral, a tetrahedron
P_K is a finite-dimensional linear (vector) space of functions defined on K
Σ is a set of degrees of freedom

where a function $v \in P_K$ is uniquely determined by the degrees of freedom Σ.

In this setting, $\Sigma = \{\sigma_1, \dots, \sigma_k\}$ is a basis for P'_K (the set of nodal variables), where P'_K stands for the algebraic dual of P_K.

Definition 8.4 Let (K, P_K, Σ) be a finite element, and let $\{\varphi_1, \varphi_2, \dots, \varphi_k\}$ be basis for P_K dual to Σ:

$$\sigma_i(\varphi_j) = \delta_{ij}, \qquad 1 \le i, j \le k.$$

Then, $\{\varphi_j\}_{j=1}^k$ is called a *nodal basis* for P_K.

For the simplest case of the linear approximation in one-dimension, we have that (K, P_K, Σ) is a finite element with

$K = I$ an interval

$P_K = P_1(I)$ is a polynomial of degree 1 on I
Σ is the values at the endpoints of I.

Hence, we can identify our previous, simple, one-dimensional basis as follows:

Example 8.3 For $K = (0, 1)$, $P_K = P_1(K)$, and $\Sigma = \{\sigma_1, \sigma_2\}$ with $\sigma_1(v) = v(0)$ and $\sigma_2(v) = v(1)$ for all $v \in P_1(K)$, $(K, P_1(K), \Sigma)$ is a finite element, with nodal basis $\varphi_1(x) = 1 - x$ and $\varphi_2(x) = x$.

Now consider a subdivision $\mathcal{T}_h = \{K\}$ of a bounded domain $\Omega \subset \mathbb{R}^d$, $d = 2, 3$, into elements K (triangles or quadrilaterals for $d = 2$ and tetrahedrons for $d = 3$). The commonly used finite element spaces V_h consists of piecewise polynomials satisfying

$$
\begin{aligned}
V_h \subset H^1(\Omega) &\Longleftrightarrow V_h \subset C^0(\overline{\Omega}) \\
V_h \subset H^2(\Omega) &\Longleftrightarrow V_h \subset C^1(\overline{\Omega}).
\end{aligned}
\tag{8.3.1}
$$

Remark 8.1 The first equivalence in (8.3.1) is due to the fact that the functions $v \in V_h$ are polynomials on each element $K \in \mathcal{T}_h$ so that for a continuous v across the common boundary $\partial K_i \cap \partial K_j$ of any pair of adjacent elements K_i and K_j we have that $D^\alpha v$, $|\alpha| = 1$, exist and are piecewise continuous, i.e. $v \in H^1(\Omega)$. On the other hand, if v is not continuous across an interelement boundary, then $v \notin C^0(\overline{\Omega})$ and $D^\alpha v, |\alpha| = 1$, is a δ function across the boundary of discontinuity. Hence, $D^\alpha v \notin L_2(\Omega)$, $|\alpha| = 1$, consequently $v \notin H^1(\Omega)$. The second relation in (8.3.1) is justified similarly.

To define a finite element space V_h on a triangulation $\mathcal{T}_h = \{K\}$ of the domain Ω we may need to specify certain quantities for each element, e.g. as

- ● function values
- ∘ values of the first derivatives
- ○ values of the second derivatives
- / values of the normal derivatives
- ↗ value of the mixed derivatives $\partial^2 v/(\partial x_1 \partial x_2)$.

In the following tables, we have collected some of the commonly used finite elements. In the one-dimensional problems, we have considered the first two rows in Table 8.1. As for the two- and three-dimensional cases (Tables 8.2–8.4) we have listed function spaces of fifth spectral order on triangular elements (Table 8.2; 2d case), up to third spectral order on quadrilaterals (Table 8.3; 2d case) and finally up to second spectral order on tetrahedrons (Table 8.4; 3d case).

Table 8.1 Some one-dimensional finite elements.

Degrees of freedom Σ	Function space P_k	Degree of continuity of V_h
 2	$P_1(K)$	C^0
 3	$P_2(K)$	C^0
 4	$P_3(K)$	C^1

Table 8.2 Some two-dimensional finite elements with triangular elements.

Degrees of freedom Σ	Function space P_k	Degree of continuity of V_h
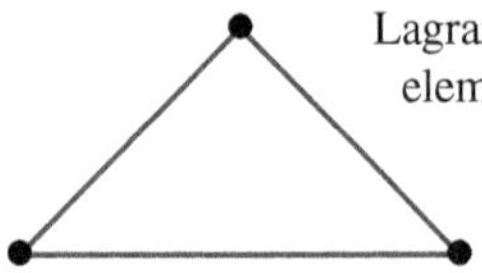 Lagrange element 3	$P_1(K)$	C^0
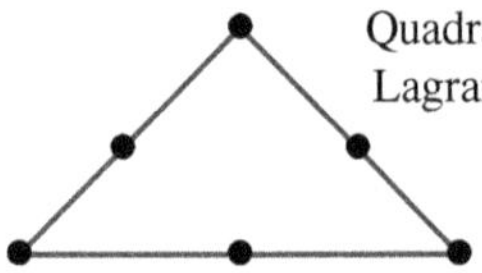 Quadratic Lagrange 6	$P_2(K)$	C^0
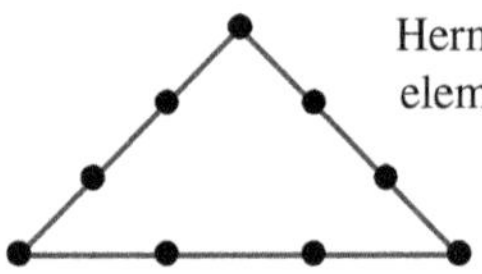 Hermite element 10	$P_3(K)$	C^0
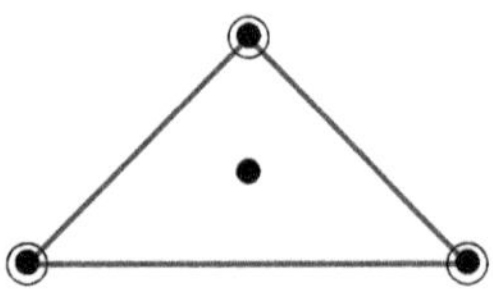 10	$P_3(K)$	C^0
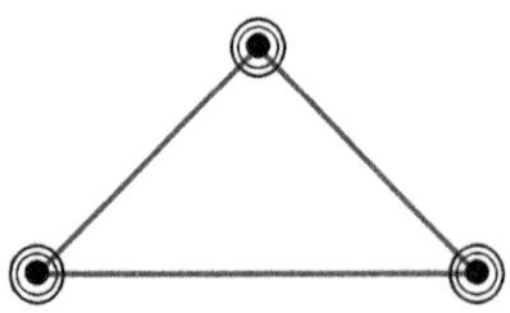 18	$P'_5(K)$	C^1
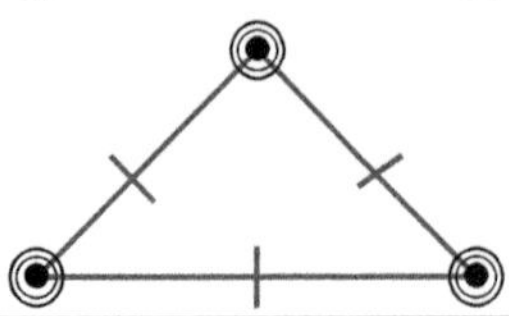 21	$P_5(K)$	C^1

Table 8.3 Some two-dimensional finite elements with quadrilateral elements.

Degrees of freedom Σ	Function space P_k	Degree of continuity of V_h
 4	$Q_1(K)$	C^0
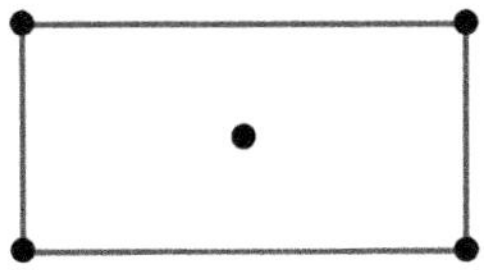 9	$Q_2(K)$	C^0
 16	$Q_3(K)$	C^1

Table 8.4 Some three-dimensional finite elements with tetrahedron elements.

Degrees of freedom Σ	Function space P_k	Degree of continuity of V_h
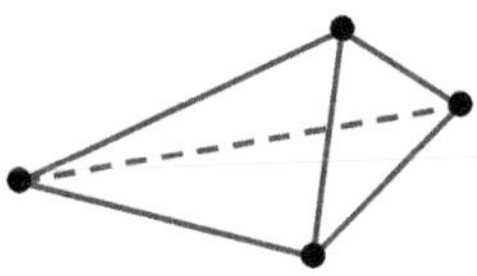 4	$P_1(K)$	C^0
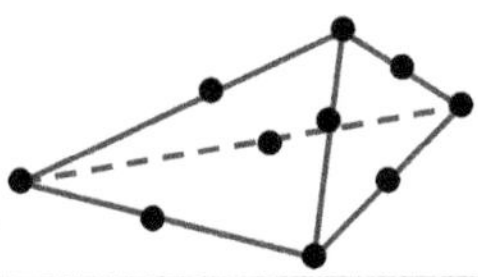 10	$P_2(K)$	C^0

8.4 The Interpolant

An approach to approximation error. Given a differential equation with an unknown exact solution u in a certain infinite dimensional space, e.g. a Sobolev space V. We construct an approximate solution $\tilde{u}$ in an adequate finite dimensional subspace $\tilde{V}$, e.g. in a finite element space V_h. In Chapters 6 and 7, we demonstrated two procedures (a priori and a posteriori) estimating the error $e := u - u_h$ (assuming $\tilde{u} := u_h$) in some energy norm for one-dimensional problems. A general approach reads as follows: given a simply identifiable approximation $\hat{u} \in \tilde{V}$ for which we can derive an error estimate for $\eta := u - \hat{u}$ (theoretical) in an adequate norm $\|\cdot\|$, we may write the approximation error as

$$e := u - \tilde{u} = u - \hat{u} + \hat{u} - \tilde{u} \equiv \eta + \xi,$$

and hence, we have that

$$\|e\| \le \|u - \hat{u}\| + \|\hat{u} - \tilde{u}\| = \|\eta\| + \|\xi\|.$$

The next task is to bound $\|\xi\|$ by $\|\eta\|$ (technical). Note that $\xi \in \tilde{V}$, since both $\tilde{u}$ and $\hat{u}$ are in this finite dimensional space $\tilde{V}$.

The crucial step in this procedure is the estimate of the approximation error η, where $u \in V$ and $\hat{u} \in \tilde{V}$. Generally, the estimation of the error ξ (error between functions belonging the finite dimensional space $\tilde{V}$) is much easier. In constructive approaches, $\hat{u}$ is chosen as the *interpolant* of u (a function in $\tilde{V}$ with the same nodal values as u in some discrete points). Then, one needs to further formalize the concept of interpolation error (in particular give a clue how the interpolation constants may look like). This yields an estimation for $\|\eta\|$, and an estimate of the form $\|\xi\| \le C\|\eta\|$ will finish the job. Here, to derive interpolation error in higher dimensions, we describe the interpolant concept more closely. To this end, we recall some standard definitions.

Definition 8.5 Let (K, P_K, Σ) be a finite element and let $\{\varphi_j\}_{j=1}^k$ be a basis for P_K dual to Σ. For a function v for which all $\sigma_i \in \Sigma$, $i = 1, \dots, k$, are defined, we define the local interpolant $\mathcal{I}_K v$ (on K) of v as

$$\mathcal{I}_K v = \sum_{i=1}^{k} \sigma_i(v)\varphi_i.$$

Example 8.4 Consider the standard triangle of Example 8.2. We want to find the local interpolant $\mathcal{I}_T v$ of the function $v(x, y) = (1 + x^2 + y^2)^{-1}$.

By the definition, we have that

$$\mathcal{I}_T v = \sigma_1(v)\phi_1 + \sigma_2(v)\phi_2 + \sigma_3(v)\phi_3. \tag{8.4.1}$$

As in the Example 8.2, we can easily verify that $\phi_1(x, y) = 1 - x/h - y/h$, $\phi_2(x, y) = x/h$, and $\phi_3(x, y) = y/h$. Further, here we have that $\sigma_1(v) = v(0, 0) = 1$, $\sigma_2(v) = v(h, 0) = (1 + h^2)^{-1}$, and $\sigma_3(v) = v(0, h) = (1 + h^2)^{-1}$. Inserting in (8.4.1), we get

$$\mathcal{I}_T(v) = 1 - \frac{x + y}{h}\left(1 - \frac{1}{1 + h^2}\right).$$

Lemma 8.2 (The properties of the local interpolation).

i) The mapping $v \mapsto \mathcal{I}_K v$ *is linear.*

ii) $\sigma_i(\mathcal{I}_K(v)) = \sigma_i(v), \quad i = 1, \ldots, k.$
iii) $\mathcal{I}_K v = v$ *for* $v \in P_K$, *i.e.* $\mathcal{I}_K$ *is idempotent:* $\mathcal{I}_K^2 = \mathcal{I}_K$.

Proof:

i) Follows from the fact that each $\sigma_i : v \mapsto \sigma_i(v), \ i = 1, \ldots, k$ is linear.
ii) For $i = 1, \ldots, k$, we have that

$$\sigma_i(\mathcal{I}_K(v)) = \sigma_i\left(\sum_{j=1}^{k} \sigma_j(v)\varphi_j\right) = \sum_{j=1}^{k} \sigma_j(v)\sigma_i(\varphi_j) = \sum_{j=1}^{k} \sigma_j(v)\delta_{ij} = \sigma_i(v).$$

iii) That $\mathcal{I}_K v = v$ for $v \in P_K$, follows using (ii), via $\sigma_i(v - \mathcal{I}_K(v)) = 0, \quad i = 1, \ldots, k$. Then, since $\mathcal{I}_K(v) \in P_K$,

$$\mathcal{I}_K^2 v = \mathcal{I}_K(\mathcal{I}_K(v)) = \mathcal{I}_K(v),$$

and the proof is complete.

To extend the local interpolant to a global computational domain, we need an adequate partitioning strategy which we introduce below.

Definition 8.6 Let Ω be a bounded open set in $\mathbb{R}^n$ with subdivision $\mathcal{T}$. Assume that each subdomain K is associated with a finite element (K, P_K, Σ) with the shape function P_K and nodal variables Σ. If m is the order of highest partial derivatives involved in the nodal variables, then, for $v \in C^m(\overline{\Omega})$, we define the *global interpolant* $\mathcal{I}_h v$ on $\overline{\Omega}$ by

$$\mathcal{I}_h v|_{K_i} = \mathcal{I}_{K_i} v, \qquad \forall K_i \in \mathcal{T}.$$

Definition 8.7 An interpolant has *continuity of order r*, if $\mathcal{I}_h v \in C^r(\overline{\Omega})$ for all $v \in C^m(\overline{\Omega})$. The space

$$\{\mathcal{I}_h v : \ v \in C^m(\overline{\Omega})\}$$

is called a C^r finite element space.

Finally, to be able to compare global interpolation operators on different elements, we introduce the following definition of *affine equivalence*.

Definition 8.8 Let $K \subset \Omega \subset \mathbb{R}^n$, suppose that (K, P_K, Σ) is a finite element and $F(x) = \mathbf{A}\mathbf{x} + \mathbf{b}$ where $\mathbf{A}$ is a nonsingular $n \times n$ matrix and $\mathbf{x}$ and $\mathbf{b}$ are n-component column vectors. Then, the finite element $(\hat{K}, \hat{P}_K, \hat{\Sigma})$ is *affine* *equivalence* to (K, P_K, Σ) if:

(a) $F(K) = \hat{K}$;
(b) $F^*\hat{P}_K = P_K$ and
(c) $F_*\Sigma = \hat{\Sigma}$.

Here F^* is the pull-back of F defined by $F^*(\hat{v}) = \hat{v}\circ F$, and F_* is the push-forward of F defined by $(F_*\Sigma)(\hat{v}) = \Sigma(F^*(\hat{v})) = \Sigma(\hat{v}\circ F)$.

Example 8.5 *Lagrange* elements on triangles (see Table 8.2) with appropriate choice of edge and interior nodes are affine equivalent. The same is true for *Hermite elements* on triangles.

8.4.1 Error Estimates for Piecewise Linear Interpolation

Deriving error estimates, both in a priori and a posteriori settings, rely on using, in a certain order, combination of some fundamental concepts as follows: variational formulation, finite element formulation, Galerkin orthogonality, Cauchy–Schwarz' inequality and interpolation estimates. These concepts, except interpolation estimates, obey a universal form regardless of the dimension. In this section, we make a straightforward generalization of the one-dimensional linear interpolation estimates on an interval to a two-dimensional linear interpolation on a triangle.

As in the $1d$ case, our estimates indicate that the *linear* interpolation errors depend on the second order, this time, partial derivatives of the function being interpolated, (i.e. the *curvature* of the function), the mesh size and also the shape of the triangle.

We consider a triangulation $\mathcal{T}_h = \{K\}$ of a two-dimensional polygonal domain Ω. Let v_i, $i = 1, 2, 3$, be the vertices of the triangle K and f a continuous function defined on K. Define the linear interpolant $\mathcal{I}_K f \in \mathcal{P}^{(1)}(K)$ by

$$\mathcal{I}_K f(v_i) = f(v_i), \quad i = 1, 2, 3. \tag{8.4.2}$$

Note that $\mathcal{I}_K$ is used for single element K (as in Figure 8.10), whereas $\mathcal{I}_h$ is employed for a partition $\mathcal{T}_h$.

In the higher-dimensional case, for an elliptic problem, satisfying the conditions in the Riesz representation theorem (weak formulation with a symmetric, coercive, bilinear form and a linear form as in the Definition 2.21), a typical error estimate is a generalization of the a priori error estimate for two-point boundary value problems:

$$\|u - u_h\|_V \le C\|u - v\|_V, \quad \forall v \in V_h. \tag{8.4.3}$$

Choosing v as the interpolant of u: $v = \mathcal{I}_h u \in V_h$ the right-hand side in (8.4.3) is replace by $\|u - \mathcal{I}_h u\|_V$, where $\mathcal{I}_h u \in V_h$ is chosen with the same degrees of

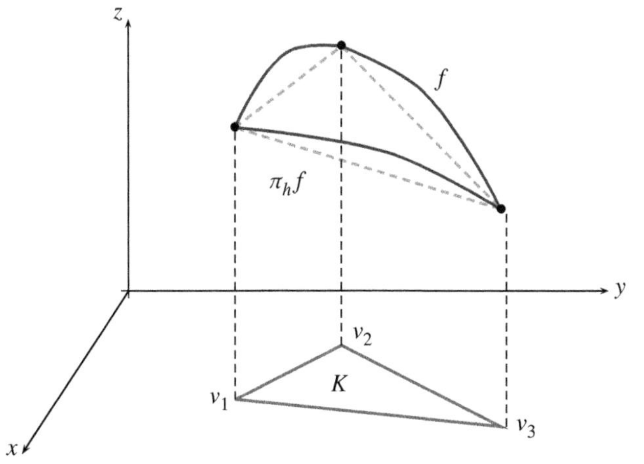

Figure 8.10 The nodal interpolant of f in $2d$ case.

freedom as that of $u_h \in V_h$. This would reduced the global estimate (8.4.3) of the problem to local estimates of the form $\|u - \mathcal{I}_h\|_K, \forall K \in \mathcal{T}_h$.

To proceed, we need to define the concept of *star-shaped* domain:

Definition 8.9 A bounded open set in $\Omega \subset \mathbb{R}^n$ is called *star-shaped* with respect to $B \subset \Omega$, if for all $x \in \Omega$ the closed convex hull of $\{x\} \cup B$ is a subset of Ω.

We now include some basic interpolation results that will be frequently used in the error estimates. We shall only give some sketchy proofs. Detailed proofs can be found in, e.g. [37] or [131].

A key result in finite element error analysis is the *Bramble–Hilbert lemma*. It relies of the following proposition:

Proposition 8.1 (Polynomial approximation). *For every $v \in W_p^m(\Omega)$, there exists a polynomial Q of degree $\leq m - 1$ and a constant Λ_j such that*

$$|v - Q|_{W_p^j(\Omega)} \leq \Lambda_j |v|_{W_p^m(\Omega)}, \quad j = 0, \ldots, m-1.$$

Bramble and Hilbert gave a nonconstructive proof of this proposition based on Hahn–Banach theorem. We skip the proof and refer the reader to a constructive version given by Süli [131].

Lemma 8.3 (Bramble–Hilbert lemma). *Suppose that Ω is a bounded open set in $\mathbb{R}^n$ and assume that Ω is star-shaped with respect to every point in a set $B \subset \Omega$ of positive measure. Let ℓ be a bounded linear functional on the Sobolev space $W_p^m(\Omega)$, $m \geq$*

1, $1 < p < \infty$, *such that* $\ell(Q) = 0$ *for any polynomial* Q *of degree* $\leq m-1$. *Then, there exists a constant* $C > 0$ *such that*

$$|\ell(v)| \leq C|v|_{W_p^m(\Omega)}, \quad \forall v \in W_p^m(\Omega).$$

Proof: By the assumptions, there exists a $C_0 > 0$ such that

$$|\ell(v)| \leq C_0\|v\|_{W_p^m(\Omega)}, \quad \forall v \in W_p^m(\Omega).$$

By the linearity of ℓ and using the fact that $\ell(Q) = 0$ for any polynomial Q of degree $\leq m-1$, we have that

$$\begin{aligned}|\ell(v)| = |\ell(v-Q)| &\leq C_0\|v-Q\|_{W_p^m(\Omega)} \\ &= C_0\left(\sum_{j=0}^{m} |v-Q|^p_{W_p^j(\Omega)}\right)^{1/p} \\ &= C_0\left(\sum_{j=0}^{m-1} |v-Q|^p_{W_p^j(\Omega)} + |v|^p_{W_p^m(\Omega)}\right)^{1/p} \\ &\leq C_0\left(\sum_{j=0}^{m-1} |v-Q|_{W_p^j(\Omega)} + |v|_{W_p^m(\Omega)}\right).\end{aligned}$$

Now, by the above proposition, we have that

$$|\ell(v)| \leq C_0\left(1 + \sum_{j=0}^{m-1}\Lambda_j\right)|v|_{W_p^m(\Omega)},$$

and the proof is complete choosing $C = C_0\left(1 + \sum_{j=0}^{m-1}\Lambda_j\right)$.

A corollary of the Bramble–Hilbert lemma (for a proof see [37]), which is crucial in determining the interpolation constant, reads as follows:

Corollary 8.1 *Let* $\Omega \subset \mathbb{R}^n$ *be a bounded open set of diameter* d *which is star-shaped with respect to every point of an open ball* $B \subset \Omega$ *of diameter* μd, $\mu \in (0,1]$. *Suppose that* $1 < p < \infty$ *and* $0 \leq j < m$, $m \geq 1$. *If* $v \in W_p^m(\Omega)$ *then*

$$\inf_{Q\in\mathcal{P}^{(m-1)}} |v-Q|_{W_p^j(\Omega)} \leq C(m,n,p,j,\mu)d^{m-j}|v|_{W_p^m(\Omega)}, \qquad \text{where}$$

$$C(m,n,p,j,\mu) = \mu^{-n/p}(\#\{\alpha : |\alpha| = j\})\frac{p(m-j)}{p-1}\left(\sum_{|\beta|=m-j}(\beta!)^{-p'}\right)^{1/p}$$

where $1/p + 1/p' = 1$, *and* $\#A$ *denotes the number of elements in* A.

To derive a basic interpolation error estimate, we define the *norm of the local interpolation operator* $\mathcal{I}_K : C^l(\overline{K}) \to W_p^m(\Omega)$ by

$$\sigma(K) = \sup_{v \in C^l(\overline{K})} \frac{\|\mathcal{I}_K v\|_{W_p^m(\Omega)}}{\|v\|_{C^l(\overline{K})}}.$$

Now we are ready to state and prove our interpolation theorem,

Theorem 8.1 *Let (K, P_K, Σ) be a finite element satisfying the conditions:*

i) K is star-shaped with respect to some ball contained in K;
ii) $\mathcal{P}^{(m-1)} \subset P_K \subset W_\infty^m(K)$,
iii) $\Sigma \subset (C^l(\overline{K}))'$.

Suppose that $1 < p < \infty$ and $m - l - (n/p) > 0$. Then, for $0 \le j \le m$,

$$|v - \mathcal{I}_K v|_{W_p^j(K)} \le C(m, n, p, \mu, \sigma(\hat{K}))h_K^{m-j}|v|_{W_p^m(K)}, \quad \forall v \in W_p^m(K),$$

where h_K is the diameter of K, $\hat{K} = \{x/h_K : x \in K\}$ and μ is the largest number in the interval $(0, 1]$ such that all balls of diameter μh_K is contained in K.

Proof: We may assume, without loss of generality that $K = \hat{K}$ and diam $K = 1$, the general case follows by an scaling argument. Further, we note that the local interpolation operator is well defined on $W_p^m(K)$ by the Sobolev embedding theorem and, using Sobolev inequality, there exists a constant $C = C_{m,n,p}$ such that for all $v \in W_p^m(K)$,

$$\|v\|_{C^l(\overline{K})} \le C_{m,n,p}\|v\|_{W_p^m(K)}, \quad \text{for } m - l > n/p, \quad 1 \le p < \infty.$$

Now, given $x \in B$, we define a *Taylor expansion* as

$$P_m(v)(x, y) = \sum_{|\beta| \le m} D^\beta v(x) \frac{(y - x)^\beta}{\beta!}$$

where $(y - x)^\beta := (y_1 - x_1)^{\beta_1} \cdots (y_n - x_n)^{\beta_n}$. Let $Q_m v$ be the averaging operator defined by

$$Q_m(v)(y) = \frac{1}{|B|} \int_B P_m(v)(x, y)\, dx.$$

Then, since $\mathcal{I}_K f = f$ for any $f \in P_K$, and $Q_m v \in \mathcal{P}^{(m-1)} \subset P_K$, we have

$$\mathcal{I}_K Q_m v = Q_m v.$$

Thus, by the Sobolev embedding theorem

$$\begin{aligned} \|v - \mathcal{I}_K v\|_{W_p^m(K)} &\le \|v - Q_m v\|_{W_p^m(K)} + \|Q_m v - \mathcal{I}_K v\|_{W_p^m(K)} \\ &= \|v - Q_m v\|_{W_p^m(K)} + \|\mathcal{I}_K(Q_m v - \mathcal{I}_K v)\|_{W_p^m(K)} \end{aligned}$$

$$\le \|v - Q_m v\|_{W_p^m(K)} + \sigma(K)\|Q_m v - \mathcal{I}_K v\|_{C^l(\overline{K})}$$
$$\le (1 + C_{m,n,p}\sigma(K))\|v - Q_m v\|_{W_p^m(K)}.$$

Finally, by Proposition 8.1 and Corollary 8.1, we have that

$$\|v - \mathcal{I}_K v\|_{W_p^m(K)} \le C(m, n, p, \mu, \sigma(K))|v|_{W_p^m(K)},$$

and we have the desired result.

Under a certain condition on the subdivision $\mathcal{T} = \{K\}$ of the computational domain Ω, one may remove the $\sigma(K)$ dependence from $C(m, n, p, \mu, \sigma(K))$. This, however, is beyond our goal in this text, and for such investigations, we refer the reader to [37] and [131].

Below we introduce a somewhat elaborate alternative to the above theorem which is also useful to derive the convergence of global interpolation estimate on Ω. To this end, we first introduce the concept of *Shape-regularity*:

Definition 8.10 *(Shape-regularity).* A family of meshes $\{\mathcal{T}_h\}_{h>0}$ is said to be shape-regular if there exists an $\alpha_0 > 0$ such that

$$\forall h, \quad \forall K \in \mathcal{T}, \quad \alpha_K = \frac{h_K}{\rho_K} \le \alpha_0,$$

where h_K is the maximal diameter of K and ρ_K is the diameter of the largest ball inscribed in K.

Theorem 8.2 (Local interpolation). *With the assumption of Theorem 8.1 and α_K as defined above, we have that there exists a constant C such that*

$$|v - \mathcal{I}_K v|_{W_p^j(K)} \le C\alpha_K^j h_K^{m-j}|v|_{W_p^m(K)}, \quad \forall v \in W_p^m(K), \quad 1 \le p \le \infty.$$

Proof: For a proof of this theorem, see [58].

Theorem 8.3 (Global interpolation). *Let Ω be a polyhedron and let $\{\mathcal{T}_h\}_{h>0}$ be a shape-regular family of affine meshes on Ω. Then, with the assumption of Theorems 8.1 and 8.2, there exists a constant C such that for the global interpolation operator, we have that for all h and $v \in W_p^m(\Omega)$,*

$$\|v - \mathcal{I}_K v\|_{L_p(\Omega)} + \sum_{j=1}^{m} h^j \left(\sum_{K \in \mathcal{T}_h} |v - \mathcal{I}_K v|^p_{W_p^j(K)} \right)^{1/p} \le Ch^m|v|_{W_p^m(\Omega)}, \quad p < \infty,$$

(8.4.4)

$$\|v - \mathcal{I}_K v\|_{L_\infty(\Omega)} + \sum_{j=1}^{m} h^j \max_{K\in\mathcal{T}_h} |v - \mathcal{I}_K v|_{W^j_\infty(K)} \le Ch^m |v|_{W^m_\infty(\Omega)}. \tag{8.4.5}$$

Further, for $p < \infty$ and $v \in L_p(\Omega)$, we have the density result:

$$\lim_{h\to 0}(\inf_{v_n\in V_h} \|v - v_h\|_{L_p(\Omega)}) = 0, \tag{8.4.6}$$

where V_h is the approximation space associated with $\mathcal{T}_h$.

Proof: The proofs of (8.4.4) and (8.4.5) follow from Theorem 8.2. Let now $v \in L_p(\Omega)$ and $\varepsilon > 0$. Then, since for $p < \infty$, $W^m_p(\Omega)$ is dense in $L_p(\Omega)$, there exists a $v_\varepsilon \in W^m_p(\Omega)$ such that $\|v - v_\varepsilon\|_{L_p(\Omega)} \le \varepsilon$. Moreover, by (8.4.4)

$$\|v_\varepsilon - \mathcal{I}_h v_\varepsilon\|_{L_p(\Omega)} \le Ch^m |v_\varepsilon|_{W^m_p(\Omega)}.$$

This yields

$$\inf_{v_h\in V_h} \|v - v_h\|_{L_p(\Omega)} \le \|v - \mathcal{I}_h v_\varepsilon\|_{L_p(\Omega)} \le \|v - v_\varepsilon\|_{L_p(\Omega)} + \|v_\varepsilon - \mathcal{I}_h v_\varepsilon\|_{L_p(\Omega)}.$$

Consequently,

$$\limsup_{h\to 0}(\inf_{v_n\in V_h} \|v - v_h\|_{L_p(\Omega)}) \le \varepsilon,$$

and (8.4.6) follows because ε is arbitrary.

Theorem 8.4 (Interpolation in $W^r_p(\Omega)$). *Under the assumptions of Theorem 8.3, we have that for W^1_p-conformal (roughly speaking, discrete test and trial spaces are the subsets of the corresponding continuous ones) space V^k_h and $v \in W^s_p(\Omega)$, there exists a constant C such that*

$$|v - \mathcal{I}^k_h v|_{W^1_p(\Omega)} \le Ch^{s-1} |v|_{W^s_p(\Omega)}. \tag{8.4.7}$$

Further, for $v \in H^s(\Omega)$,

$$\|v - \mathcal{I}^k_h v\|_{L_2(\Omega)} + h|v - \mathcal{I}^k_h v|_{H^1(\Omega)} \le Ch^s |v|_{H^s(\Omega)}. \tag{8.4.8}$$

Proof: The proof is a direct consequence of Theorem 8.3.

In the sequel, we also use the following two-dimensional interpolation estimates.

Theorem 8.5 *If f has continuous second-order partial derivatives, then*

$$\|f - \mathcal{I}_K f\|_{L_\infty(K)} \le \frac{1}{2} h_K^2 \|D^2 f\|_{L_\infty(K)}, \tag{8.4.9}$$

$$\|\nabla(f - \mathcal{I}_K f)\|_{L_\infty(K)} \le \frac{3}{\sin(\theta_K)} h_K \|D^2 f\|_{L_\infty(K)}, \tag{8.4.10}$$

where h_K is the largest side of K, θ_K is the smallest angle of K, and

$$D^2 f = \left(\sum_{i,j=1}^{2} \left(\frac{\partial^2 f}{\partial x_i \partial x_j} \right)^2 \right)^{1/2}.$$

Remark 8.2 Note that the gradient estimate (8.4.10) deteriorates for small $\sin(\alpha_K)$; i.e. for a thinner triangle K. This phenomenon is avoided assuming a *quasi-uniform triangulation*, where there is a minimum angle requirement for the triangles, viz.

$$\sin(\theta_K) \geq C, \quad \text{for some constant } C \in (0,1). \tag{8.4.11}$$

For the shape-regular triangle K, we can see that

$$\alpha_K = \frac{h_K}{\rho_K} \leq \frac{2}{\sin\,\theta_K}.$$

8.5 The L_2 (Revisited) and Ritz Projections

In this section, we introduce two other approximation procedures: the L_2 and Ritz projections. Generally, the projection techniques have some advantages and properties compared to the interpolation approach. Here, we outline only the most simple forms of the projections that are used in some of our estimates. For a more elaborate study, we refer the reader to [30].

Definition 8.11 Let V be an arbitrary function space defined over Ω. The L_2-projection $Pu \in V$ of a function $u \in L_2(\Omega)$ is defined by

$$(u - Pu, v) = 0, \quad \forall v \in V. \tag{8.5.1}$$

Usually, as a special case, we shall take $V = V_h$ the space of all continuous piecewise linear functions on a triangulation $\mathcal{T}_h = \{K\}$ of the domain Ω and Eq. (8.5.1) is then written as

$$(u - P_h u, v) = 0, \quad \forall v \in V_h. \tag{8.5.2}$$

Hence, the error $u - P_h u$ is orthogonal to V_h. Equation (8.5.2) yields a linear system of equations for the coefficients of $P_h u$ with respect to the nodal basis of V_h.

Advantages of the L_2-projection to the nodal interpolation:

- The L_2-projection $P_h u$ is well defined for $u \in L_2(\Omega)$, whereas the nodal interpolant $\pi_h u$ in general requires u to be continuous. Therefore, the L_2 projection is an alternative for the nodal interpolation for, e.g. discontinuous L_2-functions.

- Letting $v \equiv 1$ in (8.5.1) we have that

$$\int_\Omega P_h u \, dx = \int_\Omega u \, dx. \tag{8.5.3}$$

Thus, the L_2-projection conserves the *total mass*, whereas in general, the nodal interpolation operator does not preserve the total mass.

- The L_2-projection does not need to satisfy the (Dirichlet) boundary condition of the function, whereas an interpolant has to.
- Finally, we have the following error estimate for the L_2 projection:

Theorem 8.6 *For a u with square integrable second derivative, the L_2 projection P_h satisfies*

$$\|u - P_h u\| \le C_i \|h^2 D^2 u\|. \tag{8.5.4}$$

Proof: For a $v \in V_h$, we have using (8.5.2) and the Cauchy's inequality that

$$\begin{aligned} \|u - P_h u\|^2 &= (u - P_h u, u - P_h u) \\ &= (u - P_h u, u - v) + (u - P_h u, v - P_h u) = (u - P_h u, u - v) \\ &\le \|u - P_h u\| \|u - v\|. \end{aligned} \tag{8.5.5}$$

This yields

$$\|u - P_h u\| \le \|u - v\|, \quad \forall v \in V_h. \tag{8.5.6}$$

Thus, $P_h u$ is at least as close to u (is an at least as good approximation in the L_2-norm) as $\pi_h u$. Now choosing $v = \pi_h u$ and recalling the interpolation Theorem 8.5 (cf. (8.4.9)), we get the desired estimate (8.5.4).

A model problem in two dimensions. Let Ω be a bounded polygonal domain in $\mathbb{R}^2$ and consider the model problem

$$\mathcal{A}u := -\nabla \cdot (a \nabla u) = f \quad \text{in } \Omega \qquad u = 0 \quad \text{on } \partial\Omega. \tag{8.5.7}$$

Assume that the coefficient $a = a(x)$ is smooth with $a(x) \ge \alpha > 0$ in $\overline{\Omega}$ (dissipativity) and $f \in L_2(\Omega)$. Then, the variational formulation for (8.5.7) reads as follows: find $u \in H_0^1(\Omega)$ such that

$$a(u, v) = (f, v), \quad \forall v \in H_0^1(\Omega), \tag{8.5.8}$$

where

$$a(u, v) = \int_\Omega a \nabla u \cdot \nabla v \, dx \quad \text{and} \quad (f, v) = \int_\Omega f v \, dx.$$

We can easily verify that this problem has a unique solution in $H_0^1(\Omega)$ which, for convex domain Ω satisfies the stability estimate

$$||u||_2 \leq C||f||. \tag{8.5.9}$$

As usual, the finite element approximation yields a discrete version of the variational formulation (8.5.8) as

$$a(u_h, v) = (f, v), \quad \forall v \in V_h^0, \tag{8.5.10}$$

and we can prove the following error estimate:

Theorem 8.7 *Let $\Omega \subset \mathbb{R}^2$ be a convex domain and assume that u and u_h are the solutions for (8.5.8) and (8.5.10), respectively. Then*

$$||u - u_h|| \leq Ch^2||u||_2. \tag{8.5.11}$$

Proof: Let $e = u - u_h$. We shall use the dual problem

$$\mathcal{A}\phi = e, \quad \text{in } \Omega \qquad \phi = 0, \quad \text{on } \partial\Omega. \tag{8.5.12}$$

Using the stability estimate (8.5.9), we have that

$$||\phi||_2 \leq C||e||. \tag{8.5.13}$$

Further, by an standard estimate we have that

$$||e|| \leq Ch|e|_1 \leq Ch^2||u||_2. \tag{8.5.14}$$

8.5.1 The Ritz or Elliptic Projection

The Ritz projection is a general form of the L_2-projection, viz.

Definition 8.12 Let $R_h : H_0^1 \to V_h^0$ be the orthogonal projection with respect to the energy inner product, induced by weak formulation for a partial differential equation (PDE), so that

$$a(R_h v - v, \varphi) = 0, \quad \forall \varphi \in V_h^0, \quad v \in H_0^1. \tag{8.5.15}$$

The operator R_h is called the *Ritz projection* or *elliptic projection*. One may verify, using Galerkin orthogonality, that the finite element solution u_h of (8.5.10) is exactly the Ritz projection of the analytic solution u: $u_h = R_h u$.

Using (8.5.11) one can verify that, see, e.g. [92]

$$||u - R_h u|| + h|u - R_h u|_1 \leq Ch^s||u||_s, \quad \forall v \in H^s \cap H_0^1. \tag{8.5.16}$$

8.6 Exercises

Problem 8.1 Show that the function $u : \mathbb{R}^2 \to \mathbb{R}$ given by $u(x) = \log(|x|^{-1})$ for $x \neq 0$ is a solution to the Laplace equation $\Delta u(x) = 0$.

Problem 8.2 Show that the Laplacian of a C^2 function $u : \mathbb{R}^2 \to \mathbb{R}$ in the polar coordinates is written by

$$\Delta u = \frac{1}{r}\frac{\partial}{\partial r}\left(r\frac{\partial u}{\partial r}\right) + \frac{1}{r^2}\frac{\partial^2 u}{\partial \theta^2}. \tag{8.6.1}$$

Problem 8.3 Show using (8.6.1) that the function $u = a\log(r) + b$, where a and b are arbitrary constants is a solution of the Laplace equation $\Delta u(x) = 0$ for $x \neq 0$. Are there any other solutions of the Laplace equation in $\mathbb{R}^2$ which are invariant under rotation (i.e. they depend only on $r = |x|$)?

Problem 8.4 For a given triangle K, determine the relation between the smallest angle τ_K, the triangle diameter h_K and the diameter ρ_K of the largest inscribed circle.

Problem 8.5 Prove that a linear function in $\mathbb{R}^2$ is uniquely determined by its values at three points as long as they do not lie on a straight line.

Problem 8.6 Let K be a triangle with nodes $\{a^i\}$, $i = 1, 2, 3$ and let the midpoints of the edges be denoted by $\{a^{ij}, 1 \leq i < j \leq 3\}$.

(a) Show that a function $v \in \mathcal{P}^{(1)}(K)$ is uniquely determined by the degrees of freedom: $\{v(a^{ij}), 1 \leq i < j \leq 3\}$.

(b) Are functions continuous in the corresponding finite element space of piecewise linears?

Problem 8.7 Prove that if K_1 and K_2 are two neighboring triangles and $w_1 \in \mathcal{P}^{(2)}(K_1)$ and $w_2 \in \mathcal{P}^{(2)}(K_2)$ agree at three nodes on the common boundary (e.g. two endpoints and a midpoint), then $w_1 \equiv w_2$ on the common boundary.

Problem 8.8 Assume that the triangle K has nodes at $\{v^1, v^2, v^3\}$, $v^i = (v^i_1, v^i_2)$, the element nodal basis is the set of functions $\lambda_i \in \mathcal{P}^1(K)$, such that

$$\lambda_i(v^j) = \begin{cases} 1, & i = j \\ 0, & i \neq j. \end{cases} \qquad i = 1, 2, 3$$

Compute the explicit formulas for λ_i.

Problem 8.9 Let K be a triangular element. Show the following identities, for $j,\ k = 1, 2$, and $x \in K$,

$$\sum_{i=1}^{3} \lambda_i(x) = 1, \quad \sum_{i=1}^{3} (v_j^i - x_j)\lambda_i(x) = 0, \tag{8.6.2}$$

$$\sum_{i=1}^{3} \frac{\partial}{\partial x_k}\lambda_i(x) = 0, \quad \sum_{i=1}^{3} (v_j^i - x_j)\frac{\partial \lambda_i}{\partial x_k} = \delta_{jk}, \tag{8.6.3}$$

where $v^i = (v_1^i, v_2^i),\ \ i = 1, 2, 3$ are the vertices of K, $x = (x_1, x_2)$ and $\delta_{jk} = 1$ if $j = k$ and $\delta_{jk} = 0$ otherwise.

Problem 8.10 Using (8.6.2), we obtain a representation for the interpolation error of the form

$$f(x) - \pi_h f(x) = -\sum_{i=1}^{3} r_i(x)\lambda_i(x). \tag{8.6.4}$$

Prove that the remainder term $r_i(x)$ can be estimated as

$$|r_i(x)| \le \frac{1}{2} h_K \|D^2 f\|_{L_\infty(K)}, \qquad i = 1, 2, 3. \tag{8.6.5}$$

Hint: (i) Note that $|v^i - x| \le h_K$. (ii) Start applying Cauchy's inequality to show that

$$\sum_{ij} x_i c_{ij} x_j = \sum_i x_i \sum_j c_{ij} x_j.$$

Problem 8.11 τ_K is the smallest angle of a triangular element K. Show that

$$\max_{x \in K} |\nabla \lambda_i(x)| \le \frac{2}{h_K \sin(\tau_K)}.$$

Problem 8.12 The Euler equation for an incompressible inviscid fluid can be written as

$$u_t + (u \cdot \nabla)u + \nabla p = f, \quad \nabla \cdot u = 0, \tag{8.6.6}$$

where $u(x, t)$ is the velocity and $p(x, t)$ the pressure of the fluid at the point x at time t and f is an applied volume force (e.g. a gravitational force). The second equation $\nabla \cdot u = 0$ expresses the incompressibility. Prove that the first equation follows from the Newton's second law.

Hint: Let $u = (u_1, u_2)$ with $u_i = u_i(x(t), t), \ \ i = 1, 2$ and use the chain rule to derive $\dot{u}_i = \frac{\partial u_i}{\partial x_1} u_1 + \frac{\partial u_i}{\partial x_2} u_2 + \frac{\partial u_i}{\partial t}, \quad i = 1, 2.$

Problem 8.13 Prove that if $u : \mathbb{R}^2 \to \mathbb{R}^2$ satisfies rot $u := \left(\frac{\partial u_2}{\partial x_1}, -\frac{\partial u_1}{\partial x_2}\right) = 0$ in a convex domain $\Omega \subset \mathbb{R}^2$, then there is scalar function φ defined on Ω such that $u = \nabla\varphi$ in Ω.

Problem 8.14 Prove that $\int_\Omega \text{rot } u \, dx = \int_\Gamma \mathbf{n} \times u \, ds$, where Ω is a subset of $\mathbb{R}^3$ with boundary Γ with outward unit normal $\mathbf{n}$.

Problem 8.15 Let V_h be the space of piecewise linears on a partition $\mathcal{T}_h$ and $\mathcal{I}_h^1$ the *piecewise linear* Lagrange interpolation operator, viz.

$$\mathcal{I}_h^1 : C^0(\overline{\Omega}) \ni v \mapsto \sum_{j=0}^{N+1} v(x_j)\varphi_j \in V_h,$$

(a) Show that for all h and $v \in C^0(\overline{\Omega})$,

$$\|\mathcal{I}_h^1 v\|_{C^0(\overline{\Omega})} \leq \|v\|_{C^0(\overline{\Omega})}$$

(b) Show that for all h and $v \in C^1(\overline{\Omega})$,

$$\|v - \mathcal{I}_h^1 v\|_{C^0(\overline{\Omega})} \leq h\|v\|_{C^1(\overline{\Omega})}.$$

Hint: You may use the mean-value theorem.

Problem 8.16 Let K be a rectangle and denote by $\{b_1, b_2, b_3, b_4\}$ the midpoints of its sides. Let $P = \mathbb{Q}_1$ and $\Sigma = \{\sigma_1, \sigma_2, \sigma_3, \sigma_4\}$ such that $\sigma_i(p) = p(b_i)$, $1 \leq i \leq 4$. Show that $\{K, P, \Sigma\}$ is not a finite element, i.e. unisolvence does not hold.

Problem 8.17 Let $\mathcal{I}_h$ be the Lagrange interpolation operator associated with continuous, piecewise linears on triangular elements. Show that

$$\sup_{u \in C^0(\overline{\Omega})} \frac{\|\mathcal{I}_h u\|_{L_\infty}}{\|u\|_{L_\infty}} \leq 1.$$

Is this valid for piecewise quadratics?

9

The Boundary Value Problems in $\mathbb{R}^N$

In this chapter, we rely on the results of the previous chapter and consider the study of finite element approximations for the boundary value problems (BVPs) of the elliptic type in $\Omega \subset \mathbb{R}^d$, $d \geq 2$. Then, the underlying differential operator is the Laplace operator which is the basic, spatial operator in both heat and wave equations. In this regard, in Chapter 8 and Chapter 2, a somewhat general framework was introduced by the L_2- and elliptic-projections with some (semi)-theoretical estimates. Here, first, we outline the weak and strong stabilities and prove mainly, the cG(1) a priori and a posteriori finite element approximations for the Poisson equation (simplest possible form of an elliptic partial differential equation [PDE]). Then, we proceed with the general layout adding both convection and absorption/advection terms and study convection–diffusion equation and convection-dominated problem. In this part, we establish weak formulations considering different boundary conditions (BCs) and show, in distribution sense, that the variational formulations (VFs) are equivalent to underlying PDEs. We also prove corresponding error bounds in both a priori and a posteriori settings. The optimality of the convergence rates as well as the role played by convection versus diffusion are also discussed. The error analysis in the convection-dominating case are based on the use of the duality concept.

9.1 The Poisson Equation

In this section, we shall extend the studies for the one-dimensional problem of the stationary heat conduction in the earlier chapters and solve the Poisson equation, with Dirichlet boundary condition, in higher dimensions, i.e.

$$\begin{cases} -\Delta u = f & \text{in } \Omega \subset \mathbb{R}^d, \quad d = 2, 3 \\ u = 0 & \text{on } \partial\Omega, \end{cases} \tag{9.1.1}$$

where Ω is a bounded domain in $\mathbb{R}^d$ with polygonal boundary $\Gamma := \partial\Omega$.

An Introduction to the Finite Element Method for Differential Equations, First Edition.
Mohammad Asadzadeh.

9.1.1 Weak Stability

To derive stability estimates for (9.1.1), we shall assume an underlying general vector space V (to be specified below) of functions as the solution/trial space. We multiply the equation by $u \in V$ and integrate over Ω to obtain

$$-\int_\Omega (\Delta u)u\,dx = \int_\Omega f\,u\,dx. \tag{9.1.2}$$

Using Green's formula and the boundary condition, $u = 0$ on Γ, we get the concise form

$$(\nabla u, \nabla u) = (f, u), \quad \forall u \in V, \tag{9.1.3}$$

and by Cauchy–Schwarz' inequality

$$\|\nabla u\|^2 = \int_\Omega fu\,dx \le \|f\|\,\|u\|, \tag{9.1.4}$$

where $\|\cdot\|$ denotes the usual $L_2(\Omega)$-norm.

To derive a, so-called, *weak stability estimate* for (9.1.1), we combine the Poincare inequality:

$$\|u\| \le C_\Omega \|\nabla u\|,$$

with the inequality (9.1.4) and get

$$\|\nabla u\| \le C_\Omega \|f\|. \tag{9.1.5}$$

Variational formulation. We start formulating the *weak form* for problem (9.1.1): We multiply the equation by a test function v, integrate over $\Omega \subset \mathbb{R}^2$ (the higher-dimensional case is similar) and use Green's formula. Then, the *VF* reads as follows: find $u \in H_0^1(\Omega)$ such that

$$(\text{VF}): \quad \int_\Omega \nabla u \cdot \nabla v\,dx = \int_\Omega fv\,dx, \quad \forall v \in H_0^1(\Omega). \tag{9.1.6}$$

The minimization problem. As a generalization of the one-dimensional studies, one can show that the variational problem (9.1.6) is equivalent to the following quadratic *minimization problem*: find $u \in H_0^1(\Omega)$ such that

$$F(u) \le F(v), \quad \forall v \in H_0^1(\Omega), \tag{9.1.7}$$

where

$$F(v) = \frac{1}{2}\|\nabla u\|^2 - (f, v),$$

is the energy functional (total energy composed of the inertial energy $\frac{1}{2}\|\nabla u\|^2$ and load potential $-(f, v)$).

9.1.2 Error Estimates for the CG(1) FEM

To prepare for a finite element method, first we assume that the computational domain Ω is a polygon. Then, we approximate the exact solution $u(x)$ by a suitable discrete solution $U(x)$ in a certain finite dimensional subspace of $H_0^1(\Omega)$. To this end, let $\mathcal{T}_h = \{K : \bigcup K = \Omega\}$ be a triangulation of the domain Ω by elements K with the maximal diameter $h = \max_K \text{diam}(K)$.

More specifically, here we consider the cG(1) method describing continuous piecewise linear approximations of the solution $u \in H_0^1(\Omega)$ for Poissons' equation (9.1.1) in a finite dimensional subspace. Thus, we have the (common due to homogeneous Dirichlet BC) test and trial function space given as

$$V_h^0 = \{v(x) : \ v \text{ is continuous, piecewise linear on } \mathcal{T}_h, \ \text{ and } v = 0 \text{ on } \Gamma = \partial\Omega\}.$$

Observe that $V_h^0 \subset H_0^1(\Omega)$ is associated with the chosen partition $\mathcal{T}_h$. We assume n interior partition nodes with $\varphi_j, j = 1, 2, \dots, n$, as the corresponding basis functions with the property that, each $\varphi_j(x)$, $j = 1, 2, \dots, n$, is continuous in Ω, linear on each K, and

$$\varphi_j(N_i) = \begin{cases} 1 & \text{for } i = j \\ 0 & \text{for } i \neq j. \end{cases} \tag{9.1.8}$$

Here, $N_1, N_2, \dots, N_n$ are the interior nodes in the triangulation, and dim $V_h^0 = n$. Now, we set the approximate solution $u_h(x)$ to be a linear combination of the basis functions φ_j, $j = 1, \dots, n$:

$$u_h(x) = u_{h,1}\varphi_1(x) + u_{h,2}\varphi_2(x) + \cdots + u_{h,n}\varphi_n(x), \tag{9.1.9}$$

and seek the coefficients $u_{h,j} = u_h(N_j)$, i.e. the nodal values of $u_h(x)$ (which are not necessarily $= u(N_j)$), at the nodes N_j, $1 \leq j \leq n$, such that

$$\int_\Omega \nabla u_h \cdot \nabla v \, dx = \int_\Omega f \, v \, dx, \quad \forall v \in V_h^0, \tag{9.1.10}$$

or equivalently

$$\text{(FEM)} \quad \int_\Omega \nabla u_h \cdot \nabla \varphi_i \, dx = \int_\Omega f \, \varphi_i \, dx, \quad \text{for } i = 1, 2, \dots, n. \tag{9.1.11}$$

Note that every $v \in V_h^0$ can be represented by a linear combination of $\{\varphi_j\}_{j=1}^n$:

$$v(x) = v(N_1)\varphi_1(x) + v(N_2)\varphi_2(x) + \cdots + v(N_n)\varphi_n(x). \tag{9.1.12}$$

Equations (9.1.11) or (9.1.10), are the finite element formulation for (9.1.1). As in the one-dimensional case, they correspond to linear system of equations with $(\nabla\varphi_i, \nabla\varphi_j)$ being the elements of the corresponding *stiffness matrix*.

The Mass matrix via L_2-projection. Recall the L_2-projection $P_h\xi$ of a function $\xi \in L_2(\Omega)$ into V_h^0 (or V_h) defined by

$$(P_h\xi, \eta) = (\xi, \eta), \quad \forall \eta \in V_h^0.$$

Now, since $P_h\xi \in V_h^0$, there exist scalars (nodal values) χ_j such that $P_h\xi = \sum_{j=1}^{n} \chi_j\varphi_j$ and choosing $\eta = \varphi_i$, $i = 1, 2, \dots, n$, we end up with the linear system of equations

$$\mathbf{M}\chi = b, \tag{9.1.13}$$

with $\mathbf{M}$ being the *Mass matrix* with the coefficients $m_{ij} = (\varphi_j, \varphi_i)$.

Theorem 9.1 (cG(1) a priori error estimate for the gradient $\nabla u - \nabla u_h$). *Let $e = u - u_h$ represent the error in the above continuous, piecewise linear, finite element approximation of the solution for (9.1.1). Then, there is a constant C, independent of h, such that*

$$\|\nabla e\| = \|\nabla(u - u_h)\| \leq C\|h\, D^2u\|. \tag{9.1.14}$$

Proof: For the error $e = u - u_h$, $\nabla e = \nabla(u - u_h)$. Subtracting (9.1.10) from (9.1.6), where we restrict v to V_h^0, we obtain the *Galerkin orthogonality*:

$$\int_\Omega (\nabla u - \nabla u_h) \cdot \nabla v\, dx = \int_\Omega \nabla e \cdot \nabla v\, dx = 0, \quad \forall v \in V_h^0. \tag{9.1.15}$$

Further, we may write

$$\|\nabla e\|^2 = \int_\Omega \nabla e \cdot \nabla e\, dx = \int_\Omega \nabla e \cdot \nabla(u - u_h)\, dx = \int_\Omega \nabla e \cdot \nabla u\, dx - \int_\Omega \nabla e \cdot \nabla u_h\, dx.$$

Now using the Galerkin orthogonality (9.1.15), since $u_h \in V_h^0$ we have the last integral above: $\int_\Omega \nabla e \cdot \nabla u_h\, dx = 0$. Hence, removing the vanishing ∇u_h-term and inserting $\int_\Omega \nabla e \cdot \nabla v\, dx = 0$ for any arbitrary $v \in V_h^0$ we get

$$\|\nabla e\|^2 = \int_\Omega \nabla e \cdot \nabla u\, dx - \int_\Omega \nabla e \cdot \nabla v\, dx = \int_\Omega \nabla e \cdot \nabla(u - v)\, dx \leq \|\nabla e\|\, \|\nabla(u - v)\|.$$

Thus,

$$\|\nabla(u - u_h)\| \leq \|\nabla(u - v)\|, \quad \forall v \in V_h^0. \tag{9.1.16}$$

That is, in the L_2-norm, the gradient of the finite element solution, ∇u_h is closer to the gradient of the exact solution: ∇u, than the gradient ∇v of any other v in V_h^0. In other words, in the H_0^1-norm, the error $u - u_h$ is orthogonal to V_h^0 (interpreting Figure 9.1 for the gradient). One can show that there is a $v \in V_h^0$ (an interpolant), such that

$$\|\nabla(u - v)\| \leq C\|h\, D^2u\|, \tag{9.1.17}$$

where $h = h(x) = \text{diam}(K)$ for $x \in K$ and C is a constant independent of h. This is the case, for example, if v interpolates u at the nodes N_i.

Combining (9.1.16) and (9.1.17) we end up with

$$\|\nabla e\| = \|\nabla(u - u_h)\| \leq C\|h\, D^2u\|, \tag{9.1.18}$$

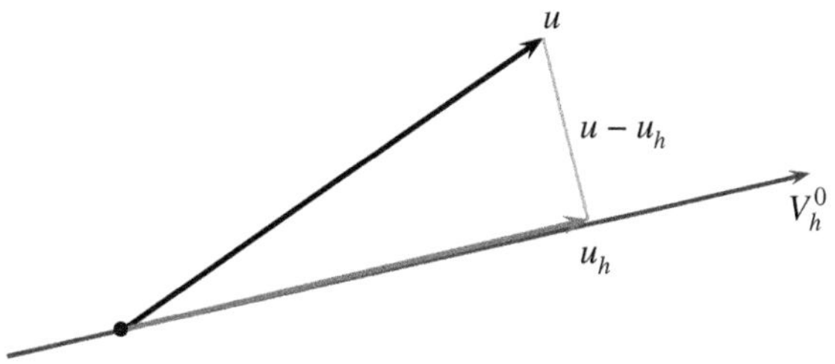

Figure 9.1 The orthogonal (L_2) projection of u on V_h^0.

i.e. the error is small if $h(x)$ is sufficiently small depending on D^2u. □

To prove an a priori error estimate *for the solution* we use the following result:

Lemma 9.1 (Regularity lemma). *Assume that Ω has no re-entrant corners (Ω is convex). We have for $u \in H^2(\Omega)$; with $u = 0$ (or $\frac{\partial u}{\partial n} = 0$) on $\Gamma := \partial\Omega$, that*

$$\|D^2u\| \leq C_\Omega \; \|\Delta u\|, \tag{9.1.19}$$

where

$$D^2u = (u_{xx}^2 + 2u_{xy}^2 + u_{yy}^2)^{1/2}. \tag{9.1.20}$$

We postpone the proof of this lemma and first derive the error estimate.

Theorem 9.2 (cG(1) a priori error estimate for the solution $e = u - u_h$). *For a shape regular mesh, we have the following a priori error estimate for the solution of the Poisson equation (9.1.1),*

$$\|e\| = \|u - u_h\| \leq C^2 \; C_\Omega \; (\max_\Omega h) \; \|h \; D^2u\|, \tag{9.1.21}$$

where C is a constant (generated twice).

Proof: Let φ be the solution of the *dual problem* for (9.1.1):

$$\begin{cases} -\Delta\varphi = e, & \text{in } \Omega \\ \varphi = 0, & \text{on } \Gamma := \partial\Omega. \end{cases} \tag{9.1.22}$$

Then, we have using Green's formula and the Galerkin orthogonality that

$$\begin{aligned} \|e\|^2 &= \int_\Omega e(-\Delta\varphi)\,dx = \int_\Omega \nabla e \cdot \nabla\varphi\,dx = \int_\Omega \nabla e \cdot \nabla(\varphi - v)\,dx \\ &\leq \|\nabla e\|\|\nabla(\varphi - v)\|, \quad \forall v \in V_h^0, \end{aligned} \tag{9.1.23}$$

We now choose v (e.g. as an interpolant of φ, in Figure 9.2) such that

$$\|\nabla(\varphi - v)\| \leq C\|h \; D^2\varphi\| \leq Ch\|D^2\varphi\|. \tag{9.1.24}$$

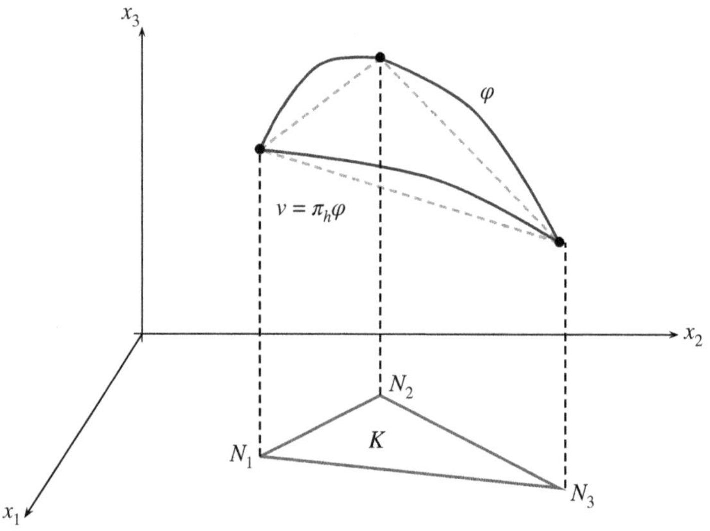

Figure 9.2 The nodal interpolant of φ in $2d$ case.

Applying the lemma to φ, we get

$$\|D^2\varphi\| \leq C_\Omega \, \|\Delta\varphi\| = C_\Omega \|e\|. \tag{9.1.25}$$

Finally, combining (9.1.18)–(9.1.25) yields

$$\begin{aligned} \|e\|^2 &\leq \|\nabla e\| \, \|\nabla(\varphi - v)\| \leq \|\nabla e\| \; Ch \; \|D^2\varphi\| \\ &\leq \|\nabla e\| \; ChC_\Omega \|e\| \leq C\|h \; D^2 u\| \; C \; hC_\Omega \|e\|. \end{aligned} \tag{9.1.26}$$

Thus, we have obtained the desired a priori error estimate

$$\|e\| = \|u - u_h\| \leq C^2 \; C_\Omega \; h \; \|h \; D^2 u\|. \tag{9.1.27}$$

That suggests a refinement strategy, viz. Figure 9.3.

Corollary 9.1 (A priori strong stability estimate). *Using Lemma 9.1, for a piecewise linear approximation, a priori error estimate (9.1.27) can be written as the following* strong stability estimate,

$$\|u - u_h\| \leq C^2 \; C_\Omega \; h^2 \; \|f\|. \tag{9.1.28}$$

Theorem 9.3 (cG(1) a posteriori error estimate). *Let u be the solution of the Poisson equation (9.1.1) and u_h its continuous piecewise linear finite element approximation. Then, there is a constant C, independent of u and h, such that*

$$\|u - u_h\| \leq C \; \|h^2 r\|, \tag{9.1.29}$$

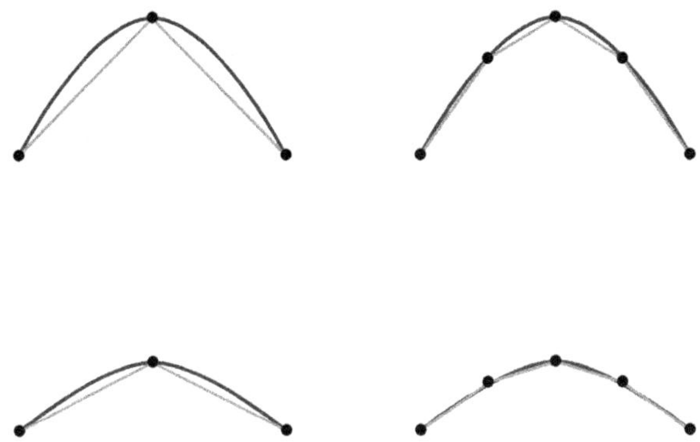

Figure 9.3 The adaptivity principle: to refine mesh for large D^2u.

where $r = f + \Delta_h u_h$ is the residual with Δ_h being the discrete Laplacian operator defined by

$$(\Delta_h u_h, v) := \sum_{K \in \mathcal{T}_h} (\nabla u_h, \nabla v)_K. \tag{9.1.30}$$

Proof: Let $e(x) = u(x) - u_h(x)$ and consider the following dual problem

$$\begin{cases} -\Delta\varphi(x) = e(x), & x \in \Omega, \\ \varphi(x) = 0, & x \in \Gamma. \end{cases} \tag{9.1.31}$$

Then, we have that $e(x) = 0$, $\forall x \in \partial\Omega$, and by the Green's formula

$$\|e\|^2 = \int_\Omega e\, e\, dx = \int_\Omega e(-\Delta\varphi)\, dx = \int_\Omega \nabla e \cdot \nabla\varphi\, dx. \tag{9.1.32}$$

Thus, by the Galerkin orthogonality (9.1.15) and using the boundary data: $v(x) = \varphi(x) = 0$, $\forall x \in \Gamma := \partial\Omega$, we get

$$\begin{aligned} \|e\|^2 &= \int_\Omega \nabla e \cdot \nabla\varphi\, dx - \int_\Omega \nabla e \cdot \nabla v\, dx = \int_\Omega \nabla e \cdot \nabla(\varphi - v)\, dx \\ &= \int_\Omega (-\boldsymbol{\Delta} e)(\varphi - v)\, dx \le \|h^2 r\|\; \|h^{-2}(\varphi - v)\| \\ &\le C\, \|h^2 r\|\; \|\Delta\varphi\| \le C\, \|h^2 r\|\; \|e\|, \end{aligned} \tag{9.1.33}$$

where we use the fact that $-\boldsymbol{\Delta} e := -\Delta u + \Delta_h u_h + [\frac{\partial u_h}{\partial \mathbf{n}}] = f + \Delta_h u_h =: r$ is the residual and v is chosen as an interpolant of φ, so that

$$\|h^{-2}(\varphi - v)\| \le C\|D^2\varphi\| \le C\, C_\Omega \|\Delta\varphi\|.$$

Thus, the final result reads as

$$\|u - u_h\| \le C_\Omega\, \|h^2 r\|. \tag{9.1.34}$$

Corollary 9.2 (A posteriori strong stability estimate) *Observe that for piecewise linear approximations $\Delta u_h = 0$, but $\Delta_h u_h \neq 0$ on each element K. If $\Delta_h u_h = 0$ on each element K then $r \equiv f$ and our a posteriori error estimate above can be viewed as a* strong stability estimate

$$\|e\| \leq C\ \|h^2 f\|. \tag{9.1.35}$$

We now return to the proof of Lemma 9.1. One can show that, in the lemma for a convex Ω the constant $C_\Omega \leq 1$. Otherwise, $C_\Omega > 1$ and increases with the degree of the singularity of a corner point at the boundary of the domain Ω. With Ω having a re-entrant corner, we have that $C_\Omega = \infty$.

9.1.3 Proof of the Regularity Lemma

Proof: Let now Ω be a rectangular domain and set $u = 0$ on $\Gamma := \partial\Omega$, then

$$\|\Delta u\|^2 = \int_\Omega (u_{xx} + u_{yy})^2\, dx\, dy = \int_\Omega (u_{xx}^2 + 2u_{xx}u_{yy} + u_{yy}^2)\, dx\, dy. \tag{9.1.36}$$

Applying Green's formula (in x-direction) for the rectangular domain Ω, we can write (note that partial derivatives of higher order appears),

$$\int_\Omega u_{xx}u_{yy}\, dx\, dy = \int_{\partial\Omega} (u_x \cdot n_x)u_{yy}\, ds - \int_\Omega u_x \underbrace{u_{yyx}}_{=u_{xyy}}\, dx\, dy. \tag{9.1.37}$$

Once again by Green's formula ("with $v = u_x$ and $\Delta w = u_{xyy}$"),

$$\int_\Omega u_x u_{xyy}\, dx\, dy = \int_{\partial\Omega} u_x(u_{yx} \cdot n_y)\, ds - \int_\Omega u_{xy}u_{xy}\, dx\, dy. \tag{9.1.38}$$

Inserting (9.1.38) in (9.1.37) yields

$$\int_\Omega u_{xx}u_{yy}\, dx\, dy = \int_{\partial\Omega} (u_x u_{yy} n_x - u_x u_{yx} n_y)\, ds + \int_\Omega u_{xy}u_{xy}\, dx\, dy. \tag{9.1.39}$$

Now, as we can see from Figure 9.4, we have that $(u_x u_{yy} n_x - u_x u_{yx} n_y) = 0$, on Γ and hence

$$\int_\Omega u_{xx}u_{yy}\, dx\, dy = \int_\Omega u_{xy}u_{xy}\, dx\, dy = \int_\Omega u_{xy}^2\, dx\, dy. \tag{9.1.40}$$

Thus, in this case,

$$\|\Delta u\|^2 = \int_\Omega (u_{xx} + u_{yy})^2\, dx\, dy = \int_\Omega (u_{xx}^2 + 2u_{xy}^2 + u_{yy}^2)\, dx\, dy = \|D^2 u\|^2,$$

and the proof is complete by a constant $C_\Omega \equiv 1$. The general case of a polygonal domain, easily, follows from this proof. □

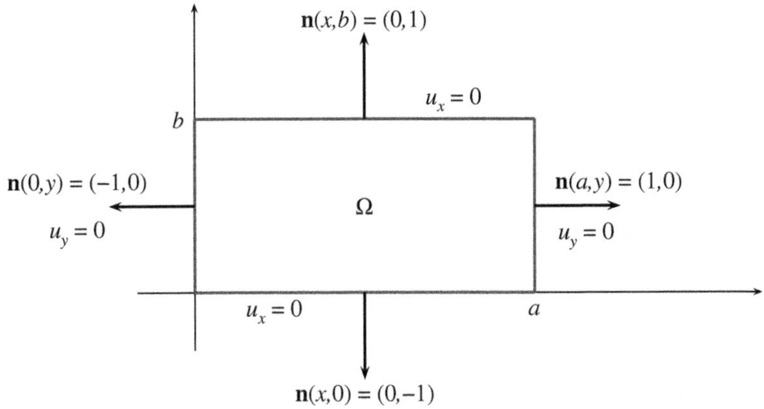

Figure 9.4 Rectangular domain Ω with outward unit normal to its sides.

9.2 Stationary Convection–Diffusion Equation

Recall the general elliptic PDE of *convection–diffusion–absorption* type:

$$\begin{cases} -\nabla \cdot (\varepsilon \cdot \nabla u)(\mathbf{x}) + \beta(\mathbf{x}) \cdot \nabla u(\mathbf{x}) + a(\mathbf{x})u(\mathbf{x}) = f(\mathbf{x}), & \mathbf{x} \in \Omega \subset \mathbb{R}^n, \\ +BCs, \quad e.g. \quad u(\mathbf{x}) = g^-, \quad \mathbf{x} \in \Gamma^-, & \varepsilon\partial_{\mathbf{n}} u(\mathbf{x}) = g^+, \quad \mathbf{x} \in \Gamma^+, \end{cases} \tag{9.2.1}$$

where $\Gamma^- = \{\mathbf{x} \in \Gamma := \partial\Omega \mid \beta(\mathbf{x}) \cdot \mathbf{n} < 0\}$ is the inflow boundary and $\Gamma^+ := \Gamma \backslash \Gamma^-$ represents the outflow. Further, $\mathbf{n} = \mathbf{n}(x)$ is the outward unit normal to Γ at $\mathbf{x} \in \Gamma$. Equation (9.2.1) is modeling, e.g. the *heat transfer* ($a(\mathbf{x}) = 0$) and *advection–diffusion* process.

9.2.1 The Elliptic Case

We write (9.2.1) in concise form as $\mathcal{L}u = f$, and assume that $\mathcal{L}$ is elliptic (see definition below) and $\mathcal{B}$ denotes the boundary operator (on $\Gamma = \Gamma^- \cup \Gamma^+$), then

$$\mathcal{L}u = f \quad \text{in } \Omega, \qquad \mathcal{B}u = g \quad \text{on } \Gamma, \tag{9.2.2}$$

is called an elliptic PDE.

Definition 9.1 The operator $\mathcal{L}$ in (9.2.2) is elliptic if there is a constant $\varepsilon_0 > 0$ such that

$$\forall \eta \in \mathbb{R}^n, \quad \sum_{i,j=1}^{n} \varepsilon_{ij}\eta_i\eta_j \geq \varepsilon_0 \|\eta\|^2 \quad a.e. \text{ in } \Omega. \tag{9.2.3}$$

Here, we study (9.2.1) in the framework of elliptic PDEs considering different boundary conditions. We use *distributions* (see below) for a concise argument that the VFs yield the underlying PDEs. The hyperbolicity and particular features of the inter-relation between convection and diffusion phenomena are discussed at the end of Chapter 7.

9.2.1.1 A Brief Note on Distributions

Definition 9.2 Let $\mathcal{D}(\Omega)$ be the vector space of C^∞ functions with compact support in Ω and $\mathcal{D}'$ be its dual space (not in usual sense). A linear mapping

$$u : \mathcal{D}(\Omega) \ni \varphi \mapsto \langle u, \varphi \rangle_{\mathcal{D}',\mathcal{D}} \in \mathbb{R}, \text{or } \mathbb{C},$$

is a distribution on Ω if and only if for all compact $K \subset \Omega$ (i.e. K is bounded and closed), there is an integer l and a constant C such that

$$\forall \varphi \in \mathcal{D}(\Omega),\ \mathrm{supp}(\varphi) \subset K, \quad |\langle u, \varphi \rangle_{\mathcal{D}',\mathcal{D}}| \le C \sup_{x \in K,\ |\alpha| \le l} |\partial^\alpha \varphi(x)|.$$

Theorem 9.4 ([76] or [120])

(i) $\mathcal{D}(\Omega)$ *is dense in* $L_p(\Omega)$, $1 \le p < \infty$.
(ii) If $f \in L^1_{loc}(\Omega)$ *satisfies* $\int_\Omega f \varphi dx = 0$, $\forall \varphi \in \mathcal{D}(\Omega)$. *Then,* $f = 0$ *a.e. in* Ω.

Definition 9.3 Let $u \in \mathcal{D}'(\Omega)$ be a distribution and $1 \le i \le n$. The distribution derivative $\partial_i u \in \mathcal{D}'(\Omega)$ is defined as

$$\partial_i u : \mathcal{D}(\Omega) \ni \varphi \mapsto \langle \partial_i u, \varphi \rangle_{\mathcal{D}',\mathcal{D}} = -\langle u, \partial_i \varphi \rangle_{\mathcal{D}',\mathcal{D}}.$$

Generally, for a multi-index α, the distribution derivative $\partial^\alpha u$ is defined as

$$\partial^\alpha u : \mathcal{D}(\Omega) \ni \varphi \mapsto \langle \partial^\alpha u, \varphi \rangle_{\mathcal{D}',\mathcal{D}} = (-1)^{|\alpha|} \langle u, \partial^\alpha \varphi \rangle_{\mathcal{D}',\mathcal{D}}.$$

Variational formulations (VFs) "$\Longrightarrow$" BVPs. We establish VF for each of the following boundary conditions for Problem (9.2.1). That the solutions for these VFs satisfy, a.e., in their underlying BVPs are stated in their corresponding propositions. The proofs of these propositions are given using *distribution* theory and hence, in rigorous form, are beyond the scope of this text. Nevertheless, we supply sketchy proofs to some of them and for the rest refer the interested reader to Ern and Guermond [58].

The general framework. (works for all the cases I–III below, except for IB). Consider the Hilbert space V satisfying:

$$H_0^1(\Omega) \subset V \subset H^1(\Omega).$$

The general framework for the weak formulation is as follows: find $u \in V$ such that

$$a(u, v) = (f, v), \quad \forall v \in V, \tag{9.2.4}$$

where $a : V \times V$ is a bilinear form in the weak formulation of the equation and f is the data. (The nonhomogeneous Dirichlet is treated in IB below.)

Assumptions 9.1 Let $\mathcal{H}$ be a Hilbert space associated with a positive continuous symmetric bilinear form $\ell(\cdot,\cdot)$ and the norm $|\cdot|_{\mathcal{H}} := \sqrt{\ell(\cdot,\cdot)}$. Assume that the space V can be continuously embedded in $\mathcal{H}$. Further, we assume that there exists a Banach space $B \subset V$ and a stability constant C_s such that for a given $g \in \mathcal{H}$, the solution of the following **dual problem**

$$\text{Find } \chi_g \in V \text{such that } a(v, \chi_g) = \ell(g, v), \quad \forall v \in V, \tag{9.2.5}$$

satisfies the a priori estimate

$$\|\chi_g\|_B \le C_s |g|_{\mathcal{H}}. \tag{9.2.6}$$

Definition 9.4 (Smoothing property). Problem (9.2.4) is said to have the smoothing property in Ω if the **Assumptions 9.1** are valid for $B = H^2(\Omega) \cap H_0^1(\Omega)$ and $\mathcal{H} = L_2(\Omega)$: More specifically, if there exists a constant C_s such that for all $\varphi \in L_2(\Omega)$, the solution ξ to the dual problem:

$$\text{Find } \xi \in V \text{ such that } a(v, \xi) = \ell(\varphi, v), \quad \forall v \in V, \tag{9.2.7}$$

satisfies the a priori stability estimate

$$\|\xi\|_{H^2(\Omega)} \le C_s \|\varphi\|_{L_2(\Omega)}. \tag{9.2.8}$$

In sketching the proofs, we shall use the concept of distributions which we briefly introduced.

I. *VF for BVP (9.2.1) with Dirichlet boundary condition*:

IA. *Homogeneous Dirichlet*: Let $\mathcal{B}u = 0$ and use the Green's formula with the test function $v = 0$ on Γ to get

$$\int_\Omega -\nabla \cdot (\varepsilon \cdot \nabla u) v dx = \int_\Omega \nabla v \cdot (\varepsilon \cdot \nabla u) dx - \int_{\partial\Omega} v(\mathbf{n} \cdot (\varepsilon \cdot \nabla u))\, d\sigma. \tag{9.2.9}$$

Thus, we have the weak formulation

$$\int_\Omega (\nabla v \cdot (\varepsilon \cdot \nabla u) + v\beta \nabla u + a u v) dx = \int_\Omega f v dx. \tag{9.2.10}$$

Now we define a bilinear form, viz.

$$a(u, v) := \int_\Omega (\nabla v \cdot (\varepsilon \cdot \nabla u) + v\beta \nabla u + a u v)\, dx. \tag{9.2.11}$$

Hence, with the obvious regularity requirement, $u \in H^1(\Omega)$ together with boundary condition $u = 0$ on Γ we can write Eq. (9.2.10) in the compact form as: find $u \in H_0^1(\Omega)$ such that

$$(VF)_{IA} \quad a(u, v) = (\mathcal{L}u, v) \quad \forall v \in H_0^1(\Omega). \tag{9.2.12}$$

Proposition 9.1 (VF$_{\text{IA}}$ "$\Longrightarrow$" BVP) *Assume that u solves Eq. (9.2.10), then $\mathcal{L}u = f$ a.e. in Ω and $u = 0$ a.e. on Γ.*

Proof: See [58]. For $\varphi \in \mathcal{D}(\Omega)$ and the solution u for (9.2.12)

$$\begin{aligned}\langle -\nabla \cdot (\varepsilon \cdot \nabla u), \varphi \rangle_{\mathcal{D}',\mathcal{D}} &= \langle \varepsilon \cdot \nabla u, \nabla \varphi \rangle_{\mathcal{D}',\mathcal{D}} = \int_\Omega \nabla \varphi \cdot \varepsilon \cdot \nabla u \, dx \\ &= \int_\Omega (f - \beta \cdot \nabla u - au)\varphi \, dx,\end{aligned}$$

which implies that $\langle \mathcal{L}u, \varphi \rangle_{\mathcal{D}',\mathcal{D}} = \int_\Omega f\varphi$. Now since $\mathcal{D}(\Omega)$ is dense in $L_2(\Omega)$ we have $\mathcal{L}u = f$ in $L_2(\Omega)$. Hence, $\mathcal{L}u = f$ a.e. in Ω. Further, using trace theorem and the definition of $H_0^1(\Omega)$ we have that $u = 0$ a.e. on $\partial\Omega$. □

IB. *Nonhomogeneous Dirichlet*: Let now $H^{1/2}(\Gamma) \ni g : \Gamma \to \mathbb{R}$ be a given function such that there exists a corresponding *lifting function* $u_g \in H^1(\Omega)$ such that $u_g = g$ on Γ. Then we have the following VF: find $u \in H^1(\Omega)$ such that

$$\begin{cases} u = u_g + w, & \forall w \in H_0^1(\Omega) \\ a(w, v) + a(u_g, v) = (f, v), & \forall v \in H_0^1(\Omega). \end{cases} \tag{9.2.13}$$

Proposition 9.2 *Assume that $g \in H^{1/2}(\Gamma)$ and u solves (9.2.13), then $\mathcal{L}u = f$ a.e. in Ω and $u = g$ a.e. on Γ.*

Proof: As the proof of Proposition 9.1

II. *VF for Neumann boundary condition for* (9.2.1): Assume that $L_2(\Gamma) \ni g : \Gamma \to \mathbb{R}$ is a given function satisfying $\mathbf{n} \cdot \varepsilon \cdot \nabla u = g$ on Γ. Then, $\forall v \in H^1(\Omega)$, the boundary integral in (9.2.9) is nonvanishing, and hence, we have the following VF of the Neumann BVP: find $u \in H^1(\Omega)$ such that

$$a(u, v) = (f, v) + \langle g, v \rangle_\Gamma, \quad \forall v \in H^1(\Omega). \tag{9.2.14}$$

Proposition 9.3 *Assume that $g \in H^{1/2}(\Gamma)$ and u solves (9.2.14), then $\mathcal{L}u = f$ a.e. in Ω and $\mathbf{n} \cdot \varepsilon \cdot \nabla u = g$ a.e. on Γ.*

Proof: Evidently, with all functions in $\mathcal{D}(\Omega)$, we get $\mathcal{L}u = f$ a.e. in Ω. Consequently, $\nabla \cdot (\varepsilon \cdot \nabla u) \in L_2(\Omega)$. Further, as a consequence of Green's formula (see [58], Appendix B.)

$$\mathbf{n} \cdot \varepsilon \cdot \nabla u \in H^{1/2}(\partial\Omega)', \quad \text{and} \quad H^{1/2}(\partial\Omega)' = H^{-1/2}(\partial\Omega).$$

This is due to the fact that for a lifting $u_\chi \in H^1(\Omega)$ of $\chi \in H^1(\Omega)$, we can establish

$$\langle \mathbf{n} \cdot \varepsilon \cdot \nabla u, \chi \rangle_{H^{-1/2}, H^{1/2}} = \int_\Omega -\nabla \cdot (\varepsilon \cdot \nabla u) u_\chi dx + \int_\Omega \nabla u_\chi \cdot \varepsilon \cdot \nabla u dx, \quad \forall \chi \in H^{1/2}(\Omega).$$

Then, (9.2.14) yields

$$\langle \mathbf{n} \cdot \varepsilon \cdot \nabla u, \chi \rangle_{H^{-1/2}, H^{1/2}} = \int_{\partial\Omega} g\chi dx, \quad \forall \chi \in H^{1/2}(\Omega).$$

Hence, $\mathbf{n} \cdot \varepsilon \cdot \nabla u = g$ in $H^{-1/2}(\partial\Omega)$ and, consequently, in $L_2(\partial\Omega)$ since $g \in L_2(\partial\Omega)$. □

I + II *VF for mixed Dirichlet–Neumann boundary conditions*: Here we split the boundary as $\Gamma := \Gamma_D \cup \Gamma_N$. On Γ_D we impose the Dirichlet condition and on Γ_N Neumann. (For nonhomogeneous Dirichlet, we use the above lifting technique to render the boundary condition to homogeneous). Thus, for $g : \Gamma_N \to \mathbb{R}$, the boundary conditions are

$$u = 0 \quad \text{on } \Gamma_D, \quad \text{and} \quad \mathbf{n} \cdot \varepsilon \cdot \nabla u = g \quad \text{on } \Gamma_N. \tag{9.2.15}$$

Further, we define the Dirichlet boundary space

$$H^1_{\Gamma_D}(\Omega) := \{u \in H^1(\Omega); \quad u = 0 \text{ on } \Gamma_D\}. \tag{9.2.16}$$

Then a Neumann type weak formulation is: find $u \in H^1_{\Gamma_D}$ such that

$$a(u, v) = (f, v) + \langle g, v \rangle_\Gamma, \quad \forall v \in H^1_{\Gamma_D}(\Omega). \tag{9.2.17}$$

Proposition 9.4 *Let $\Gamma_D \subset \Gamma$, $\Gamma_N = \Gamma \backslash \Gamma_D$. Assume that meas$(\Gamma_D) > 0$ and $g \in L_2(\Gamma_N)$. Then the solution u of the variational Problem (9.2.17) satisfies $\mathcal{L}u = f$ a.e. in Ω, $u = 0$ a.e. on Γ_D and $(\mathbf{n} \cdot \varepsilon \cdot \nabla u = g)$ a.e. on Γ_N.*

Proof: Follows the same path as in the proof of Proposition 9.3.

III *VF for Robin boundary condition for* (9.2.1): Let g, $\eta : \Gamma \to \mathbb{R}$ be given functions satisfying $\eta u + \mathbf{n} \cdot \varepsilon \cdot \nabla u = g$ on Γ. Inserting a boundary integral term in (9.2.10), we get the following VF for the Robin problem: find $u \in H^1(\Omega)$ such that

$$a(u, v) + \langle \eta u, v \rangle_\Gamma = (f, v) + \langle g, v \rangle_\Gamma, \quad \forall v \in H^1(\Omega). \tag{9.2.18}$$

Proposition 9.5 *Assume that $g \in L_2(\Gamma)$, $\eta \in L_\infty(\Gamma)$. Then, the solution u for Problem (9.2.18) satisfies $\mathcal{L}u = f$ a.e. in Ω and $\eta u + \mathbf{n} \cdot \varepsilon \cdot \nabla u = g$ a.e. on Γ.*

Proof: Is similar to that of the Proposition 9.3.

Remark 9.1 Here, for the sake of completeness, we state the well-posedness conditions for the BVPs of the weak formulations I–III:

$$\min\left(\operatorname*{ess\,inf}_{\Omega}\left(a-\frac{1}{2}\nabla\cdot\beta\right),\ \operatorname*{ess\,inf}_{\Gamma}\left(\eta+\frac{1}{2}\mathbf{n}\cdot\beta\right)\right)>0. \tag{9.2.19}$$

This is a rather rough and strong estimate. For more specific conditions, see [37] or [58].

9.2.2 Error Estimates

In this section, we derive some of the basic error estimates for the finite element approximations of the problem setting in variational formulations I–III (with a natural modification for the nonhomogeneous Dirichlet data case IB). Here, we shall consider the study of the general form (9.2.4). To this end, let $\{\mathcal{T}_h\}$ be a family of partitions of Ω, consider the finite element $\{K, P_K, \Sigma\}$ of order k and define the approximation spaces

$$V_h = \{v_h \in C^0(\overline{\Omega}); \forall K \in \mathcal{T}_h, v_h \in P_k(K)\}, \tag{9.2.20}$$

$$V_h^0 = \{v_h \in V_h; v_h = 0, \text{ on } \Gamma\}. \tag{9.2.21}$$

and

$$V_{h,D}^0 = \{v_h \in V_h; v_h = 0, \text{ on } \Gamma_D\}. \tag{9.2.22}$$

The General Finite Element for the BVPs of I–III. We shall put a common finite element formulation for the BVPs with the boundary conditions of type I–III, viz.

$$\text{Find } u_h \in \tilde{V}_h \text{ such that } \quad a(u_h, v_h) = (f, v_h), \quad \forall v_h \in \tilde{V}_h, \tag{9.2.23}$$

where

$$\tilde{V}_h := \begin{cases} V_h^0, & \text{(homogeneous Dirichlet)} \\ V_h, & \text{(Neumann and Robin)} \\ V_{h,D}^0, & \text{(mixed Dirichlet–Neumann)}, \end{cases} \tag{9.2.24}$$

with a straightforward reformulation to nonhomogeneous Dirichlet IB.

Now, we are ready to state and prove the finite element error estimates:

Theorem 9.5 (The gradient/H^1- estimate). *Let $\{\mathcal{T}_h\}$ be a family of shape-regular meshes of a polyhedron $\Omega \subset \mathbb{R}^n$. Let further $\tilde{V}_h$ be defined as in (9.2.24). Then, there is a constant C such that for the solution of problem (9.2.4) and with $\frac{n}{2} < s < k+1$, we have that*

$$\|u - h_h\|_{H^1(\Omega)} \le Ch^{s-1}|u|_s. \tag{9.2.25}$$

Proof: We have using the interpolation estimate for the Hilbert space (8.4.8) that

$$||u - u_h||_{H^1(\Omega)} \leq C||u - \mathcal{I}_h^k u||_{H^1(\Omega)} \leq Ch^{s-1}|u|_s, \tag{9.2.26}$$

and the proof is complete.

Theorem 9.6 (The L_2 error estimate). *Under the assumptions of theorem 9.5 and with $V = H_0^1(\Omega)$ in (9.2.20)–(9.2.24), and for problem (9.2.4), with the smoothing property above, there is a constant C such that*

$$\|u - h_h\|_{L_2(\Omega)} \leq Ch^1|u - u_h|_{H^1(\Omega)}. \tag{9.2.27}$$

Proof: See [37]. □

Finally, formulating a dual problem, one can also derive the following negative norm estimate:

Theorem 9.7 *Under the assumption of theorem 9.5, with $V_h \subset H_0^1(\Omega)$, $1 \leq s \leq k - 1$ and the dual solution ξ, for problem (9.2.7), satisfying*

$$\|\xi\|_{H^{s+2}(\Omega)} \leq C\|\varphi\|_{H^s(\Omega)} \tag{9.2.28}$$

we have that

$$\|u - u_h\|_{H^{-s}(\Omega)} \leq Ch^{s+1}\|u - u_h\|_{H^1(\Omega)}. \tag{9.2.29}$$

Proof: Let $\xi \in H^{s+2}(\Omega)$ be the solution of the dual Problem (9.2.7) with the data $\varphi \in H^s(\Omega) \cap H_0^1(\Omega)$. Then, by the Galerkin orthogonality and with $\pi_h\xi$ being the interpolant of ξ in V_h, we can write

$$\begin{aligned}(u - u_h, \varphi) &= a(u - u_h, \xi) = a(u - u_h, \xi - \pi_h\xi)\\ &\leq C\|u - u_h\|_{H^1(\Omega)}\|\xi - \pi_h\xi\|_{H^1(\Omega)}\\ &\leq C\|u - u_h\|_{H^1(\Omega)}h^{s+1}|\xi|_{H^{s+2}(\Omega)}\\ &\leq Ch^{s+1}\|u - u_h\|_{H^1(\Omega)}\|\varphi\|_{H^s(\Omega)}.\end{aligned} \tag{9.2.30}$$

Now recalling the definition of the negative norm:

$$\|w\|_{H^{-s}(\Omega)} := \sup_{\varphi \in H^s(\Omega) \cap H_0^1(\Omega)} \frac{(w, \varphi)}{\|\varphi\|_{H^s(\Omega)}}, \tag{9.2.31}$$

and with $w = u - u_h$, we get the desired result and the proof is complete. □

9.3 Hyperbolicity Features

Beyond the general framework in the Sections 9.1 and 9.2, special features of (9.2.1) require extra care in designing finite element methods for it. In particular, in the convection- dominating problems, i.e. when $\varepsilon/|\beta|$ is small problem (9.2.1) has hyperbolic character where conventional Galerkin's schemes for elliptic equations yield spurious oscillations (see, e.g. [57], Chapter 18).

To see this let us first consider the purely hyperbolic case of the coupled convection–absorption, i.e. when $\varepsilon \equiv 0$ and problem (9.2.1) is reduced to

$$\begin{cases} \beta(\mathbf{x}) \cdot \nabla u + a(\mathbf{x})u = f, & \text{in } \Omega \\ u = g^-, & \text{on } \Gamma^-. \end{cases} \tag{9.3.1}$$

The convection velocity field $\beta(\mathbf{x})$ is associated with the streamline curves $y(s)$, parametrized by $s \geq 0$ and satisfying the ordinary differential equation

$$\begin{cases} \dfrac{dy}{ds} = \beta(y(s)), \text{ for } s > 0 \\ y(0) = \bar{y}. \end{cases} \tag{9.3.2}$$

Equation (9.3.2) describes the path followed by a particle started at $\bar{y}$ and convected with velocity $\beta(y)$. Here s plays the role of time and dy/ds is the particle velocity. For a particle returning to $\bar{y}$, the streamline is called *closed*. Assume there are no closed streamlines. Then, by the chain rule (9.3.2) is written as

$$\frac{d}{ds}u(y(s)) + a(y(s))u(y(s)) = (\beta \cdot \nabla u + au)(y(s)) = f(y(s)), \quad s > 0,$$

with the initial data $u(y(0))$ given by the inflow boundary condition: $g^-(\bar{y})$ at $\bar{y} \in \Gamma^-$. Thus, Problem (9.3.1) is equivalent to the initial value problem of the form studied in Chapter 6:

$$\begin{cases} \dot{v}(s) + a(s)v(s) = \bar{f}(s), & s > 0 \\ v(0) = g^-(y(0)), \end{cases}$$

with $v(s) = u(y(s)),\ a(s) = a(y(s))$, and $\bar{f}(s) = f(y(s))$. Hence, the parabolic case in Chapter 6, i.e. $a(s) \geq 0$, corresponds to the nonnegative absorption $a(\mathbf{x}) \geq 0$ in Eq. (9.2.1).

As a consequence of the above observation, we conclude that in the absence of the diffusion, the information propagates along the streamlines drown from the inflow toward the outflow boundary. In particular, a discontinuity at $\bar{y} \in \Gamma^-$ continues to cross the entire streamline.

9.3.1 Convection Dominating Case

Assume that in Problem (9.2.1), $\varepsilon \neq 0$, but $\varepsilon/|\beta| > 0$ is small. Then the presence of positive diffusion smears out a discontinuity over a layer of width $\mathcal{O}(\sqrt{\varepsilon})$, otherwise, it yields a continuous solution. Further, in case of Dirichlet boundary condition, on the outflow boundary Γ^+, the solution exhibits an outflow layer of width $\mathcal{O}(\varepsilon)$ in the vicinity of Γ^+. Thus, as expected, the diffusion has a regularizing effect and that is why, in the numerical study of the hyperbolic type PDEs, it is popular to supply the problem/approximation procedure with an artificial diffusion. This, however, is made more natural (than artificial) using the *streamline-diffusion (SD) method* in finite element procedure which supplies the desired amount of diffusion, say δ, to the equation through adding to the test function a multiple of the streaming term in terms of test function by δ (see Chapter 7).

9.3.2 The SD Method for Convection Diffusion Problem

We shall consider Problem (9.2.1) with constant ε and homogeneous Dirichlet boundary condition. Let $\mathcal{T}_h$ be a quasi-uniform triangulation of Ω and $V_h \subset H_0^1(\Omega)$ be the space of continuous piecewise linear functions. Then, the SD method for (9.2.1) reads as:

$$(\mathcal{A}u_h, v + \delta \mathcal{A}v) + (\tilde{\varepsilon}\nabla u_h, \nabla v) = (f, v + \delta \mathcal{L}v), \quad \forall v \in V_h. \tag{9.3.3}$$

Here $(\cdot, \cdot)$ denotes the usual inner product,

$$\begin{aligned} \mathcal{A}\chi &= \beta \cdot \nabla \chi + a\chi, \quad \delta = \frac{h}{2|\beta|}, \quad \text{and} \\ (\tilde{\varepsilon}\nabla u_h, \nabla v) &= (\varepsilon \nabla u_h, \nabla(v + av)) + \delta(\varepsilon \Delta_h u_h, \beta \cdot \nabla v), \end{aligned}$$

where Δ_h is the discrete Laplacian.

Proposition 9.6 *One can show that (see details in [57])*

$$\tilde{\varepsilon}(u_h, h) = \max\{\varepsilon, c_1 h^2 |\mathcal{A}u_h - f|, c_2 h^{3/2}\} \approx \begin{cases} h^{3/2} & \textit{smooth } u \\ h & \textit{nonsmooth } u, \end{cases}$$

where c_k, $k = 1, 2$ are certain constants to be specified.

Proof: By the definition $\tilde{\varepsilon}$ is proportional to the residual of u_h:

$$\mathcal{R}(u_h) := \beta \cdot \nabla u_h + a u_h - f = \mathcal{A}u_h - f.$$

and

$$|\mathcal{A}u_h - f| \approx \begin{cases} h^{-1} & \text{in general} \\ h^{-3/2} & \text{in the vicinity of characteristic layers} \end{cases}$$

we deduce the desired result. □

Note that $\tilde{\varepsilon}$ depends on u_h and, hence, Problem (9.3.3) is nonlinear. This, however, exhibits a minor effect in computations when solving the discrete equation (9.3.3). Further, as in the smooth case $\tilde{\varepsilon} \approx h^{3/2}$ in the characteristic layers, while on outflow layer, it behaves as in nonsmooth case: $\tilde{\varepsilon} \approx h$ in the vicinity of outflow layer.

9.3.3 Stability Estimates

Theorem 9.8 *Assume that there is constant $c > 0$ such that*

$$a(\mathbf{x}) - \frac{1}{2}\nabla \cdot \beta(\mathbf{x}) \geq c. \tag{9.3.4}$$

Then the following stability estimates hold true

(i) *Weak stability for the continuous problem (9.2.1)*

$$\|\sqrt{\varepsilon}\nabla u\| + \|u\| \leq \left(\frac{\sqrt{c}+2}{2c}\right)\|f\|. \tag{9.3.5}$$

(ii) *Strong stability for continuous problem (9.2.1) with Neumann outflow*

$$\|\beta \cdot \nabla u + au\| + \|\varepsilon D^2 u\| \leq C\|f\|. \tag{9.3.6}$$

(iii) *Weak stability for discrete problem (9.3.3)*

$$\|\sqrt{\tilde{\varepsilon}}\nabla u_h\| + \|\sqrt{\delta}(\beta \cdot \nabla u_h + au_h)\| + \|u_h\| \leq C\|f\|. \tag{9.3.7}$$

Proof:

(i) Multiply Eq. (9.2.1) by u and integrate over Ω to get

$$(\varepsilon \cdot \nabla u, \nabla u) + (\beta \cdot \nabla u, u) + (au, u) = (f, u).$$

By the Green's formula and the boundary condition in (9.2.1)

$$(\beta \cdot \nabla u, u) = -\frac{1}{2}(\nabla \cdot \beta u, u) + \frac{1}{2}\int_\Gamma u^2 \beta \cdot \mathbf{n}\, ds = -\frac{1}{2}(\nabla \cdot \beta u, u).$$

Combining these two relations with (9.3.4), we get the desired estimate.

(ii) See the proof in the two-dimensional dual problem below.

(iii) Follows immediately choosing $v = u$ in (9.3.3). □

Remark 9.2 Due to the presence of the second derivative D^2u, the analog of the strong stability (9.3.6) for the discrete problem is not achievable. Neither a full control of the pure streaming term is possible. The best we can reach is through the weak stability estimate (9.3.7), where instead an $\tilde{\varepsilon}$-weighted L_2-norm of ∇u_h in $\|\sqrt{\tilde{\varepsilon}} u_h\|$ and a δ-weighted L_2 norm of the streaming term: $\|\sqrt{\delta}(\beta \cdot \nabla u_h + au_h)\|$ are controlled by the norm of the right-hand side.

9.3.4 Error Estimates for Convention Dominating in 2*d*

Here, to skip the tedious detail in the path of a duality procedure, we shall consider a two-dimensional model problem where in (9.2.1) we take $\Omega{:}(0,1) \times (0,1)$ and assume constant coefficients, viz. $a(\mathbf{x}) \equiv 1$, $\beta = (1,0)$ and ε a *small* constant. Then, obviously

$$\beta \cdot \nabla u = (1,0) \cdot (u_x, u_y) = u_x,$$

and Problem (9.2.1) is reduced to

$$-\varepsilon \Delta u + u_x + u = f, \quad \text{in } \Omega, \quad u = 0, \quad \text{on } \Gamma. \tag{9.3.8}$$

The solution to this problem may have characteristic layers of width $\mathcal{O}(\sqrt{\varepsilon})$ along characteristics $\{(x,y) : y = y_0\}$ and outflow layers of width $\mathcal{O}(\varepsilon)$ along the outflow boundary (for the dual problem), i.e. $\Gamma^- = \{(0,y) : 0 \le y \le 1\}$.

The dual problem, strong stability. We consider the following dual problem associated with (9.3.8):

$$\begin{cases} -\varepsilon \Delta \varphi - \varphi_x + \varphi = f, & \text{in } \Omega \\ \varphi = 0, & \text{on } \Gamma^+ \\ \varphi_x = 0, & \text{on } \Gamma^-. \end{cases} \tag{9.3.9}$$

Proposition 9.7 *Problem (9.3.9) satisfies the stability estimates*

$$\left(\|\varphi\| + 2\|\sqrt{\varepsilon}\nabla\varphi\|^2 + \int_{\Gamma^-} \varphi^2 \, ds \right)^{1/2} \le \|f\|, \tag{9.3.10}$$

$$\|\varepsilon D^2 \varphi\| \le \|f\|. \tag{9.3.11}$$

Proof: Multiply (9.3.9) by 2φ and integrate over Ω to get

$$-(2\varepsilon\Delta\varphi, \varphi) - 2(\varphi_x, \varphi) + 2(\varphi, \varphi) = 2(f, \varphi). \tag{9.3.12}$$

Now, using Green's formula and the boundary data, we have

$$-(2\varepsilon\Delta\varphi, \varphi) = 2(\sqrt{\varepsilon}\nabla\varphi, \sqrt{\varepsilon}\nabla\varphi) + \int_\Gamma \varphi \, (\mathbf{n} \cdot \nabla\varphi) \, ds = 2\|\sqrt{\varepsilon}\nabla\varphi\|^2,$$

and

$$-2(\varphi_x, \varphi) = \int_{\Gamma^-} \varphi^2 \, ds.$$

Inserting there two relations in (9.3.12), we end up with

$$2\|\varphi\|^2 + 2\|\sqrt{\varepsilon}\nabla\varphi\|^2 + \int_{\Gamma^-} \varphi^2 \, ds = 2(f, \varphi) \le \|f\|^2 + \|\varphi\|^2,$$

which gives the stability estimate (9.3.10).

To prove (9.3.11), we multiply (9.3.9) by $-\varepsilon\Delta\varphi$, integrate over Ω and use Green's formula to obtain

$$\|\varepsilon\Delta\varphi\|^2 + (\varepsilon\Delta\varphi, \varphi_x) + (\varepsilon\nabla\varphi, \nabla\varphi) = (f, -\varepsilon\Delta\varphi). \tag{9.3.13}$$

Now we have that

$$(\varepsilon\Delta\varphi, \varphi_x) = (\varphi_{xx} + \varphi_{yy}, \varphi_x) = (\varphi_{xx}, \varphi_x) - (\varphi_y, \varphi_{xy})$$
$$= \frac{1}{2}\left(\int_{\Gamma^+} \varphi_{x'}^2 \, ds + \int_{\Gamma^-} \varphi_y^2 \, ds\right),$$

where the first boundary integral is due to the fact that $\varphi_x = 0$ on Γ^-. As for the second boundary integral, we note that

$$-(\varphi_y, \varphi_{xy}) = (\varphi_{xy}, \varphi_y) + \int_{\Gamma} \varphi_y^2 \, ds = \int_{\Gamma^-} \varphi_y^2 \, ds,$$

where we use that $\varphi = 0$ on Γ^+ implies that $\varphi_y = 0$ on Γ^+. Hence, we conclude that

$$\|\varepsilon\Delta\varphi\|^2 \le (f, -\varepsilon\Delta\varphi) \le \|f\|\|\varepsilon\Delta\varphi\|.$$

Now recalling the regularity lemma, we have for the rectangular region Ω that $\|D^2\varphi\| = \|\Delta\varphi\|$ and the proof is complete.

A priori error estimates for pure convection. We derive a priori error estimate for (9.3.3) when $\tilde{\varepsilon} \equiv 0$. □

Theorem 9.9 *Let* $a(\mathbf{x}) - \frac{1}{2}\nabla \cdot \beta \ge c > 0$, *then we have the following a priori error estimate for (9.3.3)*

$$\|u - u_h\| \le CC_i\|h^{3/2}D^2u\|.$$

Proof: We let $\varepsilon \equiv 0$ and consider the problem with vanishing inflow boundary condition $u = 0$ on Γ^-. Then for the piecewise continuous functions χ with $\chi = 0$ on Γ^- we can derive

$$\|\sqrt{\delta}\mathcal{A}\chi\|^2 + c\|\chi\|^2 \le (\mathcal{A}\chi, \chi + \delta\mathcal{A}\chi). \tag{9.3.14}$$

Now let $e = u - u_h$ and set $\mathcal{A} = \frac{\partial}{\partial x} + I$. Then using Galerkin orthogonality, an analog of (9.3.14) and combining the strong stability of the continuous problem (9.3.6) with the weak stability (9.3.7) for the discrete problem yield

$$\begin{aligned}\|\sqrt{\delta}\mathcal{A}e\|^2 + c\|e\|^2 &\le (\mathcal{A}e, e + \delta\mathcal{A}e) = (\mathcal{A}e, u - \pi_h u + \delta\mathcal{A}(u - \pi_h u))\\ &\le \tfrac{1}{2}\|\sqrt{\delta}\mathcal{A}e\|^2 + 2\|\delta^{-1/2}(u - \pi_h u)\|^2 + \|\delta^{1/2}\mathcal{A}(u - \pi_h u)\|^2,\end{aligned} \tag{9.3.15}$$

where π_{hu} is the usual interpolant of u. Hence, choosing $\delta = h$, by the standard interpolation estimates, we get

$$\frac{1}{2}\left\|\sqrt{h}\mathcal{A}e\right\|^2 + c\|e\|^2 \le C_i^2\|h^{3/2}D^2u\|^2, \tag{9.3.16}$$

where C_i denotes the interpolation constant. This yields the desired result for $\varepsilon < h$ and the proof is complete. □

A posteriori error estimate for standard Galerkin.

Theorem 9.10 *Let $\tilde{\varepsilon} = \varepsilon$ and set $\delta = 0$ in the SD procedure for the two-dimensional problem (9.3.8). Then, there is a constant C such that*

$$\|e\| \le C\left(\left\|\frac{h^2}{\varepsilon^\star}\mathcal{R}(u_h)\right\| + \|\varepsilon\partial_{\mathbf{n}}u\|_{\Gamma^-} + \|\varepsilon\partial_{\mathbf{n}}u_h\|_{\Gamma^-}\right), \tag{9.3.17}$$

where $\mathcal{R}(u_h) = \mathcal{R}_1(u_h) + \mathcal{R}_2(u_h)$ with

$$\mathcal{R}_1(u_h) = |u_{h,x} + u_h - f|,$$

and

$$\mathcal{R}_2(u_h) = \frac{1}{2}\varepsilon \max_{S\subset\partial K} h_K^{-1}|[\partial_S u_h]|, \quad \text{on } K \in \mathcal{T}_h,$$

where $[\partial_S w]$ denotes the jump of w across the side $S \subset \partial K$ in the normal derivative of the function $w \in V_h$ and

$$\varepsilon^\star = \begin{cases} \varepsilon h^{1/2} \text{ on } K & \text{if } K \cap \Gamma^- \ne \emptyset \\ \varepsilon, & \text{else} \end{cases}$$

Proof: Let φ denote the solution of the dual problem in two dimensions: (9.3.9) with $g = e = u - u_h$. Then, using Galerkin orthogonality and the equations for u and u_h, we derive the following error representation formula

$$\begin{aligned}\|e\|^2 &= (e, -\varepsilon\Delta\varphi - \varphi_x + \varphi) = (\varepsilon\nabla e, \nabla\varphi) + (e_x + e, \varphi)\\ &= (\varepsilon\nabla u, \nabla\varphi) + (u_x + u, \varphi) - (\varepsilon\nabla u_h, \nabla\varphi) - (u_{h,x} + u_h, \varphi)\\ &= (f, \varphi) + \int_{\Gamma^-} \varepsilon\partial_{\mathbf{n}}u\ \varphi\ ds - (u_{h,x} + u_h, \varphi) - (\varepsilon\nabla u_h, \nabla\varphi)\\ &= (f - u_{h,x} - u_h, \varphi - \pi_h\varphi) - (\varepsilon\nabla u_h, \nabla(\varphi - \pi_h\varphi)) + \textstyle\int_{\Gamma^-}\varepsilon\partial_{\mathbf{n}}u\ \varphi\ ds.\end{aligned}$$

Then desired estimates follows by interpolation error estimates and (9.3.10).

Remark 9.3 The modified $\varepsilon^\star$ is due to compatibility of the boundary condition. A more elaborate motivation is given in [57]. □

9.4 Exercises

Problem 9.1 Derive stability estimate for the Neumann BVP:

$$\begin{cases} -\Delta u + u = f, & \text{in } \Omega\\ \frac{\partial u}{\partial n} = 0, & \text{on } \Gamma = \partial\Omega. \end{cases} \tag{9.4.1}$$

Problem 9.2 Consider Eq. (9.1.1) and derive the a posteriori error estimate

$$\|\nabla(u - u_h)\| \le C\|hr\|.$$

Problem 9.3 Verify that for v being the interpolant of φ, we have

$$\|e\| \leq C\, \|h^2 f\| \times \begin{cases} \|h^{-2}\,(\varphi - v)\| \leq C\, \|\Delta\varphi\|, & \text{and} \\ \|h^{-1}(\varphi - v)\| \leq C\, \|\nabla\varphi\|. \end{cases} \tag{9.4.2}$$

Problem 9.4 Derive the estimate corresponding to (9.1.34) for the one-dimensional case.

Problem 9.5 Consider the following two-dimensional problem:

$$\begin{cases} -\Delta u = 1, & \text{in } \Omega \\ u = 0, & \text{on } \Gamma_D \\ \frac{\partial u}{\partial n} = 0, & \text{on } \Gamma_N \end{cases} \tag{9.4.3}$$

See Figure 9.5 below

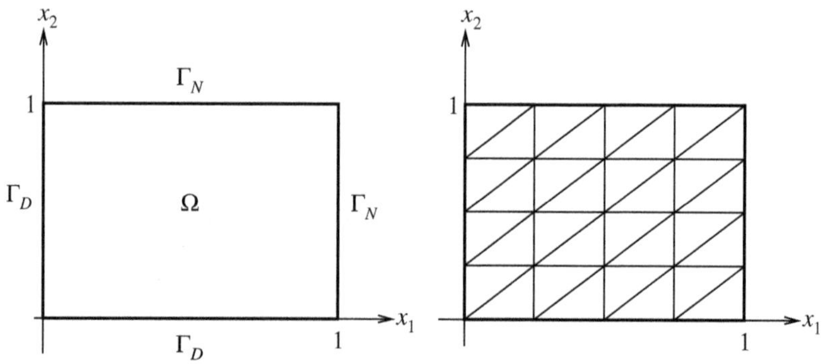

Figure 9.5 A uniform triangulation of Ω.

Triangulate Ω as in the figure and let

$$U(x) = U_1\varphi_1(x) + \cdots + U_{16}\varphi_{16}(x),$$

where $x = (x_1, x_2)$ and φ_j, $j = 1, \dots, 16$ are the basis functions, see Figure 9.5, and determine $U_1, \dots, U_{16}$ so that

$$\int_\Omega \nabla U \cdot \nabla\varphi_j\, dx = \int_\Omega \varphi_j\, dx, \quad j = 1, 2, \dots, 16.$$

Problem 9.6 Generalize the procedure in the previous problem to the following case with mesh-size h.

$$\begin{cases} -\nabla(a\nabla u) = f, & \text{in } \Omega \\ u = 0, & \text{on } \Gamma_D \\ a\frac{\partial u}{\partial n} = 7, & \text{on } \Gamma_N \end{cases}, \text{where } \begin{cases} a = 1 & \text{for } x_1 < \frac{1}{2} \\ a = 2 & \text{for } x_1 > \frac{1}{2} \\ f = x_2. \end{cases}$$

Problem 9.7 Consider the Dirichlet problem

$$-\nabla \cdot (a(x)\nabla u(x)) = f(x), \quad x \in \Omega \subset \mathbb{R}^2, \quad u(x) = 0, \quad \text{for } x \in \partial\Omega.$$

Assume that c_0 and c_1 are constants such that $0 < c_0 \le a(x) \le c_1$, $\forall x \in \Omega$ and let U be a Galerkin approximation of u in a subspace M of $H_0^1(\Omega)$. Prove the a priori error estimate

$$||u - U||_{H_0^1(\Omega)} \le \frac{c_1}{c_0} \inf_{\chi \in M} ||u - \chi||_{H_0^1(\Omega)}.$$

Problem 9.8 Consider the following *Schrödinger* equation

$$\dot{u} + i\Delta u = 0, \quad \text{in } \Omega, \quad u = 0, \quad \text{on } \partial\Omega,$$

where $i = \sqrt{-1}$ and $u = u_1 + iu_2$. (a) Show that the L_2 norm of the solution, i.e. $\int_\Omega |u|^2 dx$ is time-independent.
Hint: Multiply the equation by $\bar{u} = u_1 - iu_2$, integrate over Ω and consider the real part.
(b) Consider the corresponding *eigenvalue problem*, of finding $(\lambda, u \neq 0)$, such that

$$-\Delta u = \lambda u \quad \text{in } \Omega, \quad u = 0, \quad \text{on } \partial\Omega.$$

Show that $\lambda > 0$, and give the relation between $||u||$ and $||\nabla u||$ for the corresponding eigenfunction u.
(c) What is the optimal constant C (expressed in terms of smallest eigenvalue λ_1), for which the inequality $||u|| \le C||\nabla u||$ can fulfill for all functions u, such that $u = 0$ on $\partial\Omega$?

Problem 9.9 Determine the stiffness matrix and load vector if the cG(1) finite element method is applied to the Poisson's equation on a triangulation with triangles of side length 1/2 in both x_1- and x_2-directions:

$$\begin{cases} -\Delta u = 1, & \text{in} \quad \Omega = \{(x_1, x_2) : 0 < x_1 < 2,\ 0 < x_2 < 1\}, \\ u = 0, & \text{on} \quad \Gamma_1 = \{(0, x_2)\} \cup \{(x_1, 0)\} \cup \{(x_1, 1)\}, \\ \frac{\partial u}{\partial n} = 0, & \text{on} \quad \Gamma_2 = \{(2, x_2) : 0 \le x_2 \le 1\}. \end{cases}$$

Problem 9.10 Let $\Omega = (0, 2) \times (0, 2)$, $B_1 = \partial\Omega \backslash B_2$ and $B_2 = \{2\} \times (0, 2)$. Determine the stiffness matrix and load vector in the cG(1) solution for the problem

$$\begin{cases} -\frac{\partial^2 u}{\partial x_1^2} - 2\frac{\partial^2 u}{\partial x_2^2} = 1, \ \text{in } \Omega = (0, 2) \times (0, 2), \\ u = 0, \quad \text{on } B_1, \quad \frac{\partial u}{\partial x_1} = 0, \quad \text{on } B_2, \end{cases}$$

with piecewise linear approximation applied on the triangulation in Figure 9.6:

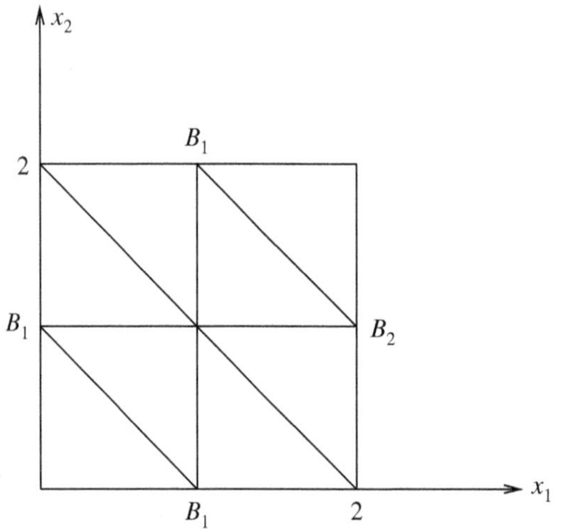

Figure 9.6 The considered triangulation for Ω.

Problem 9.11 Determine the stiffness matrix and load vector if the cG(1) finite element method, with piecewise linear approximation, is applied to the following Poisson's equation with mixed boundary conditions:

$$\begin{cases} -\Delta u = 1, \text{ on } & \Omega = (0,1) \times (0,1), \\ \frac{\partial u}{\partial n} = 0, \text{ for } & x_1 = 1, \\ u = 0, \text{ for } & x \in \partial\Omega \backslash \{x_1 = 1\}, \end{cases}$$

on a triangulation with triangles of side length $1/4$ in the x_1-direction and $1/2$ in the x_2-direction.

Problem 9.12 Formulate the cG(1) method for the boundary value problem

$$-\Delta u + u = f, \quad x \in \Omega; \quad u = 0, \quad x \in \partial\Omega.$$

Write down the matrix form of the resulting system of equations using the uniform mesh as in Figure 9.7 below:

Problem 9.13 Write a code to solve the system $\mathbf{A}\chi = \mathbf{b}$, where $\mathbf{A}$ is a band structure matrix.

Problem 9.14 Implement an *error estimate* routine for a code that approximates the Poisson problem using cG(1) finite element method. Given f, construct a model problem with a known solution $u(x)$ for the Poisson problem

$$-\Delta u = f \text{ in } \Omega := (0,1)^2, \quad u = 0, \text{on } \partial\Omega.$$

Compare the error estimates to the true error in a few meshes.

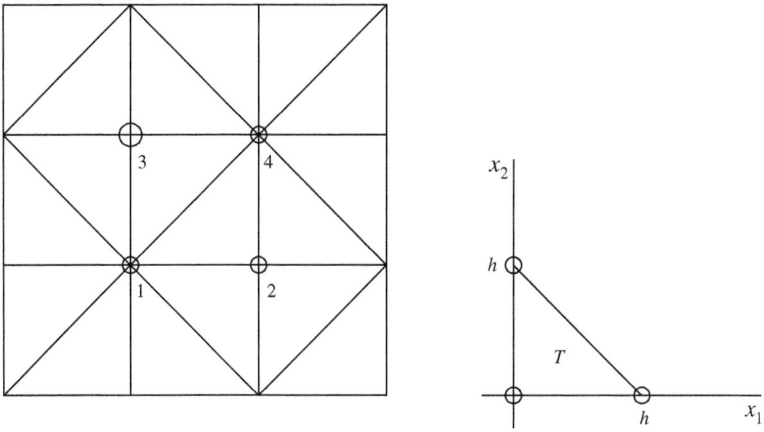

Figure 9.7 A uniform mesh of square Ω (a) and its standard element T (b).

Problem 9.15 Write a Matlab code to approximate the solution of the Poisson equation on an L-shaped domain

$$\Omega = \{(x,y) : -1 < x < 0, -1 < y < 1\} \cup \{(x,y) : 0 < x, y < 1\}.$$

with homogeneous Dirichlet boundary data on

$$\Gamma^{\perp} := \{(0,y) : -1 \le y \le 0\} \cup \{(x,0) : 0 \le x \le 1\}$$

and nonhomogeneous conditions on $\partial\Omega \setminus \Gamma^{\perp}$, so that the exact solution in polar coordinates becomes $u(r,\theta) = r^{2/3}\sin(2\theta/3)$. Start with a coarse triangulation and a given small tolerance (TOL). Plot the value of h versus r and compare it with $h \approx r^{1/3}\mathrm{TOL}$.

Problem 9.16 Prove the equivalence between (9.1.6) and (9.1.7).

Problem 9.17 Prove that the mass matrix $\mathbf{M}$ in (9.1.13) is symmetric and positive definite.

Problem 9.18

(a) Use a standard uniform triangulation and derive the discrete system of equations for the finite element approximation of the BVP

$$\begin{cases} -\Delta u = 1, & \text{in } \Omega := (0,1)^2 \\ u = 0, & \text{on } \Gamma_1 = \partial\Omega \backslash \{y = 0\} \\ \partial_n u + u = 1 & \text{on } \Gamma_2 = \partial\Omega \backslash \Gamma_1, \end{cases}$$

(b) Implement a cG(1) approximation of this problem by, e.g. a Matlab code.

Problem 9.19 **(Discrete maximum principle)** Consider Problem (9.1.1). Let $\Omega \subset \mathbb{R}^2$ be a polygonal domain and $\mathcal{T}_h$ a quasi-uniform, shape regular triangulation of Ω. Consider the continuous piecewise linear approximation space V_h on $\mathcal{T}_h$ with the basis functions $\{\varphi_j\}_{j=1}^n$ and let S be the stiffness matrix: $S_{ij} = \int_\Omega \nabla\varphi_i \cdot \nabla\varphi_j dx,\ 1 \le i,j \le N$.

(a) Show that all off-diagonal entries of S are nonpositive and its row-wise sums are nonnegative.

(b) Prove the maximum principle:

$$f \in L_2(\Omega), \quad f \le 0 \text{ in } \Omega \Longrightarrow u_h \le 0 \text{ in } \Omega,$$

where u_h is the cG(1) finite element approximation for the solution of (9.1.1).

10

The Initial Boundary Value Problems in $\mathbb{R}^N$

This chapter is devoted to the study of time-dependent problems formulated in spatial domain in $\mathbb{R}^N$, $N \geq 2$. More specifically, we shall consider the finite element methods for two fundamental equations: the heat equation and the wave equation. In this setting, we have already studied time discretizations for initial value problem as well as spatial discretization of boundary value problems, (specifically, finite element analysis for the Laplace's equation), in Chapters 6 and 9, respectively. Here, we shall derive certain stability estimates for the heat and wave equations and construct semidiscrete schemes for spatial discretizations (which is time-parameterized version of the schemes of Chapter 9) and fully discrete schemes for both temporal and spatial discretizations. We also formulate convergence results in the a priori setting. These can be viewed as a combination of error estimates in Chapters 6 and 9. More detailed studies on a priori error estimates can be found in, e.g. Thomée [136]. The a posteriori error estimates are direct generalization of the results of Chapter 7 for one-space dimension and can be found in, e.g. Eriksson et al. [57].

10.1 The Heat Equation in $\mathbb{R}^N$

In this section, we shall study the stability of the heat equation in $\mathbb{R}^N$, $N \geq 2$. These estimates in the higher dimensions are direct generalizations of the ones for the one-dimensional problem addressed in Chapter 7.

The initial value problem for the heat conduction can be formulated as

$$\begin{cases} \dot{u}(x,t) - \Delta u(x,t) = f(x,t), & \text{in } \mathbb{R}^d, d = 1,2,3 \quad \text{(DE)} \\ u(x,0) = u_0(x), & \text{for } x \in \mathbb{R}^d. \quad \text{(IC)} \end{cases} \tag{10.1.1}$$

Equation (10.1.1) is of parabolic type with significant *smoothing* and *stability* properties. It can also be used as a model for a variety of physical phenomena involving *diffusion processes*. Here, we shall focus on the stability concept and construction

An Introduction to the Finite Element Method for Differential Equations, First Edition.
Mohammad Asadzadeh.

of the fully discrete schemes, which are a combination of the discretization procedures in Chapters 7 and 9. For details in a priori and a posteriori error analysis, we refer the reader to Chapter 16 in [57].

10.1.1 The Fundamental Solution

We consider the homogeneous heat conduction in the general N-dimensional Euclidean space:

$$\begin{cases} \dot{u}(x,t) - \Delta u(x,t) = 0, & \text{for } (x,t) \in \mathbb{R}^N \times (0,\infty) \\ u(x,0) = u_0(x), & \text{for } x \in \mathbb{R}^N. \end{cases} \tag{10.1.2}$$

First, we seek the *fundamental solution* of (10.1.2) in the two-dimensional case and when the initial data is assumed to be the Dirac δ function: $u_0(x) = \delta_0(x)$. We denote this solution by $E(x,t)$. One may easily verify that

$$E(x,t) = \frac{1}{(4\pi t)} \exp\left(-\frac{|x|^2}{4t}\right). \tag{10.1.3}$$

Then, for an arbitrary initial data u_0, the solution for (10.1.2), with $N = 2$, can be expressed in terms of fundamental solution and is obtained through convolution between u_0 and E:

$$u(x,t) = \frac{1}{4\pi t} \int_{\mathbb{R}^2} u_0(y) \exp\left(-\frac{|x-y|^2}{4t}\right) dy. \tag{10.1.4}$$

For the general Problem (10.1.2), using the N-dimensional Fourier transform, see [92], we have the N-dimensional versions of (10.1.3) and (10.1.4), viz.

$$E(x,t) = \frac{1}{(4\pi t)^{N/2}} \exp\left(-\frac{|x|^2}{4t}\right), \quad x \in \mathbb{R}^N,\ t > 0, \tag{10.1.5}$$

and

$$u(x,t) =: (E(t)u_0)(x) = \frac{1}{(4\pi t)^{N/2}} \int_{\mathbb{R}^N} u_0(y) \exp\left(-\frac{|x-y|^2}{4t}\right) dy. \tag{10.1.6}$$

Here, if we let $\eta = \dfrac{y-x}{\sqrt{4t}}$, we get

$$\frac{1}{(4\pi t)^{N/2}} \int_{\mathbb{R}^N} \exp\left(-\frac{|x-y|^2}{4t}\right) dy = \frac{1}{\pi^{N/2}} \int_{\mathbb{R}^N} e^{-|\eta|^2}\, d\eta = 1, \tag{10.1.7}$$

consequently, we have that

$$|u(x,t)| \le \|u_0\|_{L_\infty(\mathbb{R}^N)}, \quad \text{for } t > 0. \tag{10.1.8}$$

Now, we return to the original, nonhomogeneous, heat equation (10.1.1) with $f \neq 0$ and recall that E, defined as in (10.1.6), is interpreted as an operator acting on initial data u_0. This can be extended to f and using the *Duhamel's principle*, see [92], the solution to problem (10.1.1) can be expressed as

$$u(t) = E(t)u_0 + \int_0^t E(t-s)f(x,s)\, ds. \tag{10.1.9}$$

Then as usual, it is easy to verify (see below), that

$$||u(t)|| \leq ||u_0|| + \int_0^t ||f(s)||\, ds, \tag{10.1.10}$$

which yields the uniqueness in the standard way.

10.1.2 Stability

Here, we consider the following initial boundary value problem for the homogeneous heat equation in a bounded domain $\Omega \subset \mathbb{R}^2$,

$$\begin{cases} \dot{u}(x,t) - \Delta u(x,t) = 0, & \text{for } (x,t) \in \Omega \times (0,\infty) \\ u(x,t) = 0, & \text{for } x \in \partial\Omega,\ t > 0, \\ u(x,0) = u_0(x), & \text{for } x \in \Omega. \end{cases} \tag{10.1.11}$$

The stability estimates for Eq. (10.1.11) are summarized in following theorem. It roughly states a control of certain energy norms of the solution u by the L_2 norm of the initial data u_0, where some of the estimates below up in the vicinity of the initial time $t = 0$.

Theorem 10.1 (Energy estimates). *The solution u of problem* (10.1.11) *satisfies the following stability estimates*

$$\max\left(||u(\cdot,t)||^2,\ 2\int_0^t ||\nabla u||^2(s)\, ds \right) \leq ||u_0||^2 \tag{10.1.12}$$

$$||\nabla u(\cdot,t)|| \leq \frac{1}{\sqrt{2\,t}}||u_0|| \tag{10.1.13}$$

$$\left(\int_0^t s||\Delta u(\cdot,s)||^2\, ds \right)^{1/2} \leq \frac{1}{2}||u_0|| \tag{10.1.14}$$

$$||\Delta u(\cdot,t)|| \leq \frac{1}{\sqrt{2}\,t}||u_0|| \tag{10.1.15}$$

$$\text{For } \varepsilon > 0 \text{ small,} \quad \int_\varepsilon^t ||\dot{u}(\cdot,s)||\, ds \leq \frac{1}{2}\sqrt{\ln\frac{t}{\varepsilon}}\,||u_0||. \tag{10.1.16}$$

Note that (10.1.16) indicates strong instabilities for ε (as well as t) close to 0. This is the reason behind the splitting strategy used in the analysis below.

Proof: To derive the first two estimates in (10.1.12), we multiply (10.1.1) by u and integrate over Ω, to get

$$\int_\Omega \dot{u}u\, dx - \int_\Omega (\Delta u)u\, dx = 0. \tag{10.1.17}$$

Note that $\dot{u}u = \frac{1}{2}\frac{d}{dt}u^2$. Further, using Green's formula with the Dirichlet boundary data, $u = 0$ on Γ, we have that

$$-\int_\Omega (\Delta u)u\, dx = -\int_\Gamma (\nabla u \cdot \mathbf{n})\, u\, ds + \int_\Omega \nabla u \cdot \nabla u\, dx = \int_\Omega |\nabla u|^2\, dx. \tag{10.1.18}$$

Thus, Eq. (10.1.17) can be written in the following two equivalent forms:

$$\frac{1}{2}\frac{d}{dt}\int_\Omega u^2\, dx + \int_\Omega |\nabla u|^2\, dx = 0 \iff \frac{1}{2}\frac{d}{dt}\|u\|^2 + \|\nabla u\|^2 = 0, \tag{10.1.19}$$

where $\|\cdot\|$ denotes the $L_2(\Omega)$ norm. We substitute t by s and integrate Eq. (10.1.19) over $s \in (0, t)$ to get

$$\frac{1}{2}\int_0^t \frac{d}{ds}\|u\|^2(s)\, ds + \int_0^t \|\nabla u\|^2(s)\, ds = \frac{1}{2}\|u\|^2(t) - \frac{1}{2}\|u\|^2(0) + \int_0^t \|\nabla u\|^2\, ds = 0.$$

Hence, inserting the initial data $u(0) = u_0$, we have

$$\|u\|^2(t) + 2\int_0^t \|\nabla u\|^2(s)\, ds = \|u_0\|^2. \tag{10.1.20}$$

In particular, we have our first two stability estimates

$$\|u\|(t) \le \|u_0\|, \quad \text{and} \quad \int_0^t \|\nabla u\|^2(s)\, ds \le \frac{1}{2}\|u_0\|^2. \tag{10.1.21}$$

To derive (10.1.13) and (10.1.14), since $\dot{u} = \Delta u$, and we need to estimate the L_2-norm of $\dot{u}$, so we multiply the equation, $\dot{u} - \Delta u = 0$, by $-t\,\Delta u$ and integrate over Ω to get

$$-t\int_\Omega \dot{u}\,\Delta u\, dx + t\int_\Omega (\Delta u)^2\, dx = 0. \tag{10.1.22}$$

Using Green's formula ($u = 0$ on $\Gamma \iff \dot{u} = 0$ on Γ) yields

$$\int_\Omega \dot{u}\,\Delta u\, dx = -\int_\Omega \nabla\dot{u} \cdot \nabla u\, dx = -\frac{1}{2}\frac{d}{dt}\|\nabla u\|^2, \tag{10.1.23}$$

so that (10.1.22) can be rewritten as

$$t\frac{1}{2}\frac{d}{dt}\|\nabla u\|^2 + t\|\Delta u\|^2 = 0. \tag{10.1.24}$$

Now, using the relation $t\frac{d}{dt}\|\nabla u\|^2 = \frac{d}{dt}(t\|\nabla u\|^2) - \|\nabla u\|^2$, (10.1.24) can be rewritten as

$$\frac{d}{dt}(t\|\nabla u\|^2) + 2t\|\Delta u\|^2 = \|\nabla u\|^2. \tag{10.1.25}$$

Once again, we substitute t by s and integrate over $(0, t)$ to get

$$\int_0^t \frac{d}{ds}(s\|\nabla u\|^2(s))\, ds + 2\int_0^t s\|\Delta u\|^2(s)\, ds = \int_0^t \|\nabla u\|^2(s)\, ds.$$

Using (10.1.21), the last equality is estimated as

$$t\,\|\nabla u\|^2(t) + 2\int_0^t s\,\|\Delta u\|^2(s)\, ds \leq \frac{1}{2}\|u_0\|^2. \tag{10.1.26}$$

In particular, we have

$$\|\nabla u\|(t) \leq \frac{1}{\sqrt{2t}}\|u_0\| \quad \text{and} \quad \left(\int_0^t s\|\Delta u\|^2(s)\, ds\right)^{1/2} \leq \frac{1}{2}\|u_0\|,$$

which are our third and fourth stability estimates (10.1.13) and (10.1.14). The stability estimate (10.1.15) is proved analogously. Now, using (10.1.1): $(\dot{u} = \Delta u)$ and (10.1.15), we may write

$$\int_\varepsilon^t \|\dot{u}\|(s)\, ds \leq \frac{1}{\sqrt{2}}\|u_0\| \int_\varepsilon^t \frac{1}{s}\, ds = \frac{1}{\sqrt{2}} \ln \frac{t}{\varepsilon}\|u_0\| \tag{10.1.27}$$

or more carefully

$$\begin{aligned}
\int_\varepsilon^t \|\dot{u}\|(s)\, ds &= \int_\varepsilon^t \|\Delta u\|(s)\, ds = \int_\varepsilon^t 1\ \|\Delta u\|(s)\, ds \\
&= \int_\varepsilon^t \frac{1}{\sqrt{s}}\ \sqrt{s}\|\Delta u\|(s)\, ds \\
&\leq \left(\int_\varepsilon^t s^{-1}\, ds\right)^{1/2} \left(\int_\varepsilon^t s\|\Delta u\|^2(s)\, ds\right)^{1/2} \\
&\leq \frac{1}{2}\sqrt{\ln \frac{t}{\varepsilon}}\|u_0\|,
\end{aligned}$$

where in the last two inequalities, we use Cauchy–Schwarz' inequality and (10.1.14), respectively. This yields (10.1.16), and the proof is complete. □

10.1.3 The Finite Element for Heat Equation

10.1.3.1 The Semidiscrete Problem

Here, we consider only the space discretization; hence, the problem is continuous in time and is referred to as a *spatially semidiscrete* problem. We focus on a two-dimensional heat conduction formulated for a bounded convex domain

$\Omega \subset \mathbb{R}^2$ with smooth boundary Γ. Hence, we consider the *nonhomogeneous* initial-boundary value problem (10.1.11):

$$\begin{cases} \dot{u}(x,t) - \Delta u(x,t) = f(x,t), & \text{for } x \in \Omega \subset \mathbb{R}^2, \quad \text{(DE)} \\ u(x,t) = 0, & x \in \Gamma,\ 0 < t \le T, \quad \text{(BC)} \\ u(x,0) = u_0(x), & \text{for } x \in \Omega. \quad \text{(IC)} \end{cases} \tag{10.1.28}$$

Using Green's formula, we have the following weak formulation:

$$(u_t, \varphi) + (\nabla u, \nabla \varphi) = (f, \varphi), \quad \forall \varphi \in H_0^1(\Omega), \quad t > 0. \tag{10.1.29}$$

Let now $\mathcal{T}_h$ be a triangulation of Ω of the types introduced in Chapter 8 (see Tables 8.1–8.3), with the interior nodes $\{x_j\}_{j=1}^{M_h}$. Further, let V_h be the continuous piecewise linear functions on $\mathcal{T}_h$ vanishing on $\partial\Omega$. Finally, let $\{\psi_j\}_{j=1}^{M_h}$ be the standard basis of V_h corresponding to the nodes $\{x_j\}_{j=1}^{M_h}$.

Then, we formulate an approximate problem as for each t find the finite element solution $u_h(t) = u_h(\cdot, t) \in V_h$ such that

$$\begin{aligned} (u_{h,t}, \chi) + (\nabla u_h, \nabla \chi) = (f, \chi), \quad \forall \chi \in V_h, \quad t > 0, \\ u_h(0) = u_{0,h}, \end{aligned} \tag{10.1.30}$$

where $u_{0,h}$ in an approximation of u_0. Problem (10.1.30) is equivalent to find the coefficients $\xi_j(t)$ in the representation

$$u_h(x,t) = \sum_{j=1}^{M_h} \xi_j(t)\psi_j(x),$$

such that

$$\sum_{j=1}^{M_h} \xi'(t)(\psi_j, \psi_i) + \sum_{j=1}^{M_h} \xi(t)(\nabla\psi_j, \nabla\psi_i) = (f(t), \psi_i), \quad i = 1, \dots, M_h,$$

and, with η_j denoting the nodal values of the initial approximation $u_{0,h}$:

$$\xi_j(0) = \eta_j, \quad j = 1, \dots, M_h.$$

In the matrix form, this can be written as

$$\mathbf{M}\dot{\xi}(t) + \mathbf{S}\xi(t) = b(t), \quad t > 0, \text{ and } \xi(0) = \eta, \tag{10.1.31}$$

where $\mathbf{M} = (m_{ij}) = ((\psi_j, \psi_i))$ is the mass matrix, $\mathbf{S} = (s_{ij}) = ((\nabla\psi_j, \nabla\psi_i))$ is the stiffness matrix, and $b = (b_i) = ((f, \psi_i))$ is the load vector. Recall that both mass and stiffness matrices are symmetric and positive definite, in particular, $\mathbf{M}$ is invertible and hence, (10.1.31) can be written as a system of ordinary differential equations (ODEs), viz.

$$\dot{\xi}(t) + \mathbf{M}^{-1}\mathbf{S}\xi(t) = \mathbf{M}^{-1}b(t), \quad t > 0, \text{ and } \xi(0) = \eta, \tag{10.1.32}$$

which has a unique solution for $t > 0$.

We can easily derive the counterpart of the stability estimates for the semidiscrete problem (10.1.30) with χ replaced by u_h:

$$\begin{aligned}(u_{h,t}, u_h) + (\nabla u_h, \nabla u_h) &= (f, u_h), \quad t > 0,\\ u_h(0) &= u_{0,h},\end{aligned} \tag{10.1.33}$$

and end up, e.g. with the $L_2(\Omega)$ estimate

$$\|u_h(t)\| \leq \|u_{0,h}\| + \int_0^t \|f\| \, ds. \tag{10.1.34}$$

To proceed, we introduce the *discrete Laplacian* operator Δ_h: $V_h \to V_h$, as in discrete analogue of Green's formula (note! test space has zero b.c.),

$$(-\Delta_h \psi, \chi) := (\nabla \psi, \nabla \chi), \quad \forall \psi, \ \chi \in V_h. \tag{10.1.35}$$

One can show that the operator Δ_h is self-adjoint and $-\Delta_h$ is positive definite with respect to the L_2-inner product. Let now P_h denote the L_2-projection onto V_h, then Eq. (10.1.30) can be written as

$$(u_{h,t} - \Delta_h u_h - P_h f, \chi) = 0, \quad \forall \chi \in V_h.$$

Now, since $u_{h,t} - \Delta_h u_h - P_h f \in V_h$ taking $\chi = u_{h,t} - \Delta_h u_h - P_h f$, we thus obtain

$$u_{h,t} - \Delta_h u_h = P_h f, \quad t > 0, \quad u_h(0) = u_{0,h}. \tag{10.1.36}$$

Now, considering $E_h(t)$ as the solution operator of homogeneous (10.1.36), with $f = 0$, we have that $u_h(t) = E_h(t)u_{0,h}$. Recall that the Duhamel's principle reads as

$$u_h(t) = E_h(t)u_{0,h} + \int_0^t E_h(t-s)P_h f(s) \, ds. \tag{10.1.37}$$

Recalling the interpolant $\mathcal{I}_h$: $C_0(\overline{\Omega}) \to V_h$ and the $L_2(\Omega)$ version of the error bound stated in Theorem 8.2, we can now prove the following error estimate:

Theorem 10.2 *Let u_h and u be the solutions of (10.1.30) and (10.1.28), respectively. Then, we have the following $L_2(\Omega)$-error estimate*

$$\|u_h(t) - u(t)\| \leq \|u_{0,h} - u_0\| + Ch^2 \left(\|u_0\|_2 + \int_0^t \|u_t\|_2 \, ds \right).$$

Proof: We recall the common estimate for interpolation $\mathcal{I}_h$, L_2-projection P_h and elliptic (Ritz) projection R_h,

$$\|v - v_h\| \leq Ch^2 \|v\|_2, \quad \text{where } v_h = \mathcal{I}_h v, \ v_h = P_h v, \ \text{or } v_h = R_h v. \tag{10.1.38}$$

Below we shall use the Ritz projection and rewrite the error as

$$u - u_h = (u - R_h u) + (R_h u - u_h) = \rho + \theta. \tag{10.1.39}$$

Now, recalling (8.5.16), the first term is estimated as

$$\begin{aligned}\|\rho(t)\| \le Ch^2\|u(t)\|_2 &= Ch^2\|u_0 + \int_0^t u_t\, ds\|_2 \\ &\le Ch^2\left(\|u_0\|_2 + \int_0^t \|u_t\|_2\, ds\right).\end{aligned} \tag{10.1.40}$$

As for a bound for θ, using the notation $a(w,\xi) := (\nabla w, \nabla \xi)$ and the definition of R_h, we have

$$\begin{aligned}(\theta_t, \varphi) + a(\theta, \varphi) &= -(u_{h,t}, \varphi) - a(u_h, \varphi) + (R_h u_{h,t}, \varphi) + a(R_h u_h, \varphi) \\ &= -(f, \varphi) + (R_h u_t, \varphi) + a(u, \varphi) = (-u_t + R_h u_t, \varphi),\end{aligned} \tag{10.1.41}$$

which, using the fact that R_h commutes with time differentiation, i.e. $R_h u_t = (R_h u)_t$, yields

$$(\theta_t, \varphi) + a(\theta, \varphi) = -(\rho_t, \varphi), \quad \forall \varphi \in V_h. \tag{10.1.42}$$

Now, using the variational formulation (10.1.29) and the finite element formulation for the discrete problem (10.1.33) together with the estimate (10.1.34), we get

$$\|\theta(t)\| \le \|\theta(0)\| + \int_0^t \|\rho_t\|\, ds. \tag{10.1.43}$$

For the $\theta(0)$, we have that

$$\begin{aligned}\|\theta(0)\| &= \|R_h u_0 - u_{0,h}\| \le \|R_h u_0 - u_0\| + \|u_0 - u_{h,0}\| \\ &\le \|u_0 - u_{h,0}\| + Ch^2\|u_0\|_2.\end{aligned} \tag{10.1.44}$$

Further,

$$\|\rho_t\| = \|u_t - R_h u_t\| \le Ch^r\|u_t\|_r, \quad r = 1, 2. \tag{10.1.45}$$

Combining the above estimates, we get the desired result. □

The error estimate for the gradient has a similar form, viz.

Theorem 10.3 *Let u_h and u be the solutions of (10.1.30) and (10.1.28), respectively. Then, we have the following $L_2(\Omega)$-estimate for the gradient of the solution*

$$|u_h(t) - u(t)|_1 \le |u_{0,h} - u_0|_1 + Ch\left[\|u_0\|_2 + \|u(t)\|_2 + \left(\int_0^t \|u_t\|_1^2\, ds\right)^{1/2}\right].$$

Proof: Let $\rho(t)$ be defined as in (10.1.39). Then, using (8.5.16) we have

$$|\rho(t)|_1 = |u - R_h u|_1 \le Ch\|u(t)\|_2. \tag{10.1.46}$$

To estimate $\nabla\theta$, we use (10.1.42) with $\varphi = \theta_t$. Then

$$\|\theta_t\|^2 + \frac{1}{2}\frac{d}{dt}|\theta|_1^2 = -(\rho_t, \theta_t) \leq \frac{1}{2}(\|\rho_t\|^2 + \|\theta_t\|^2), \tag{10.1.47}$$

and hence

$$\frac{d}{dt}|\theta|_1^2 \leq \|\rho_t\|^2, \tag{10.1.48}$$

which integrating in t, yields

$$|\theta(t)|_1^2 \leq |\theta(0)|_1^2 + \int_0^t \|\rho_t\|^2\, ds \leq (|R_h u_0 - u_0|_1^2 + |u_0 - u_{0,h}|_1^2) + \int_0^t \|\rho_t\|^2\, ds.$$

Now, recalling the elliptic projection estimate:

$$\|R_h w - w\| + h|R_h w - w|_1 \leq Ch^s\|w\|_s, \quad s = 1, 2, \tag{10.1.49}$$

we get using (10.1.45) that

$$|\theta(t)|_1 \leq |u_0 - u_{0,h}|_1 + Ch\left[\|u_0\|_2 + \left(\int_0^t \|u_t\|^2\, ds\right)^{1/2}\right], \tag{10.1.50}$$

which, combining with (10.1.46), completes the proof. □

10.1.4 A Fully Discrete Algorithm

For the time discretization of (10.1.1), we shall consider a partition of the time interval $I = [0, T]$: $0 = t_0 < t_1 < \cdots < t_N = T$ into subintervals $I_n = (t_{n-1}, t_n]$ of length $k_n = t_n - t_{n-1}$. Then, we divide the *space–time slab* $\mathbf{S}_n = \Omega \times I_n$ into prisms $K \times I_n$, where $K \in \mathcal{T}_n$ and $\mathcal{T}_n$ is a triangulation of Ω with the mesh function h_n and minimal angle β_n. In the error estimates, we shall use $\beta = \min_n \beta_n$. We construct a finite element mesh based on approximations by continuous piecewise linear functions in space and discontinuous polynomials of degree r in time. We shall refer to this method as cG(1)dG(r) method. We define the trial space as,

$$W_k^{(r)} = \{v(x,t)\colon v|_{\mathbf{S}_n} \in W_{kn}^{(r)}\}, \quad n = 1, 2, \ldots, N,$$

where

$$W_{kn}^{(r)} = \left\{ v(x,t)\colon v(x,t) = \sum_{j=0}^{r} t^j \psi_j(x), \quad \psi_j \in V_n,\ (x,t) \in \mathbf{S}_n \right\},$$

with $V_n = V_{h_n}$ as the space of continuous piecewise linear functions vanishing on Γ associated to $\mathcal{T}_n$. Generally, the functions in $W_k^{(r)}$ are discontinuous across the discrete time levels t_n with jump discontinuities $[w_n] = w_n^+ - w_n^-$ and $w_m^\pm = \lim_{s\to 0^\pm} w(t_n + s)$.

The cG(1)dG(r) method is based on a variational formulation for, nonhomogeneous version of, (10.1.11), viz. find $U \in W_k^{(r)}$ such that for $n = 1, 2, \dots, N$,

$$\int_{I_n} \left((\dot{U}, v) + (\nabla U, \nabla v) \right) \, dt + ([U_{n-1}], v_{n-1}^+) = \int_{I_n} (f, v) \, dt, \quad \forall v \in W_{kn}^{(r)}, \tag{10.1.51}$$

where $U_0^- = u_0$ and $(\cdot, \cdot)$ is the $L_2(\Omega)$ inner product.

Using the discrete Laplacian Δ_{h_n}, in the case of $r = 0$, we may write (10.1.51) as follows: find $U_n \in V_n$ such that

$$(I - k_n \Delta_n) U_n = P_n U_{n-1} + \int_{I_n} P_n f \, dt, \tag{10.1.52}$$

where $U_n = U_n^- = U|_{I_n} \in V_n$, and P_n is the $L_2(\Omega)$ projection onto V_n. Note that the *initial data*, $U_{n-1} \in V_{n-1}$ from the previous time interval I_{n-1}, is projected into the space V_n. If $V_{n-1} \subset V_n$, then $P_n U_{n-1} = U_{n-1}$. For the case $r = 1$, writing $U(t) = F_n + (t - t_{n-1}) G_n$ on I_n with $F_n, \ G_n \in V_n$, the variational formulation (10.1.51) takes the following form

$$\begin{cases} (I - k_n \Delta_n) F_n + \left(I - \dfrac{k_n}{2} \Delta_n \right) G_n = P_n U_{n-1} + \displaystyle\int_{I_n} P_n f \, dt, \\ \left(\dfrac{1}{2} I - \dfrac{k_n}{3} \Delta_n \right) G_n - \dfrac{k_n}{2} \Delta_n F_n = \displaystyle\int_{I_n} \dfrac{t - t_{n-1}}{k_n} P_n f \, dt, \end{cases} \tag{10.1.53}$$

and gives a system of equations for F_n and G_n.

10.1.5 The Discrete Equations

To derive the matrix equation for U_n in the case $r = 0$, we use (10.1.52). To this end, let $\varphi_{n,j}$ be the nodal basis for V_n associated to J_n interior nodes of $\mathcal{T}_n$, then we can represent U_n as

$$U_n = \sum_{j=1}^{J_n} \xi_{n,j} \varphi_{n,j},$$

with $\xi_{n,j}$ being the nodal values of U_n. We denote the vector coefficients by $\xi_n = (\xi_{n,j})$, the $J_n \times J_n$ mass matrix by $\mathbf{M}_n$ and the stiffness matrix by $\mathbf{S}_n$. Further, we let $\mathbf{b}_n$ denote the $J_n \times 1$ data vector. Then, the associated elements are given by

$$(\mathbf{M}_n)_{i,j} = (\varphi_{n,j}, \varphi_{n,i}), \quad (\mathbf{S}_n)_{i,j} = (\nabla \varphi_{n,j}, \nabla \varphi_{n,i}), \quad (\mathbf{b}_n)_i = \int_{t_{n-1}}^{t_n} (f, \varphi_{n,i}) dt,$$

for $1 \le i,\ j \le J_n$. Finally, we define the $J_n \times J_{n-1}$ matrix $\mathbf{M}_{n-1,n}$ with entries

$$(\mathbf{M}_{n-1,n})_{ij} = (\varphi_{n,j}, \varphi_{n-1,i}), \quad 1 \le i \le J_n,\ 1 \le j \le J_{n-1}. \tag{10.1.54}$$

Now, the discrete equation for cG(1)dG(0) approximation on I_n reads as follows: find ξ_n such that

$$(\mathbf{M}_n + k_n\mathbf{S}_n)\xi_n = \mathbf{M}_{n-1,n}\xi_{n-1} + \mathbf{b}_n. \tag{10.1.55}$$

One can easily verify that the coefficient matrix of the system (10.1.55) is sparse, symmetric, and positive-definite. Hence, (10.1.55) can be solved by a direct or an iterative algorithm presented in Chapter 4.

10.1.6 A Priori Error Estimate: Fully Discrete Problem

To derive a priori (or a posteriori) error estimates for the heat equation in higher dimensions is a rather involved procedure. Below we outline the prerequisites and formulate an a priori error estimate theorem without a proof. Detailed studies of the error analysis, for the heat equation can be found in [57].

We assume that there are positive constants α_i, $i = 1, 2$, with α_2 sufficiently small, such that for $n = 1, \dots, N$,

$$\alpha_1 k_n \le k_{n+1} \le \alpha_1^{-1} k_n, \tag{10.1.56}$$

$$\alpha_1 h_n(x) \le h_{n+1}(x) \le \alpha_1^{-1} h_n, \quad x \in \Omega \tag{10.1.57}$$

$$\left(\max_{x\in\overline{\Omega}} h(x)\right)^2 \le \alpha_2 k_n, \tag{10.1.58}$$

Relation (10.1.58) enters if $V_{n-1} \nsubseteq V_n$. An a priori error estimate is then given as:

Theorem 10.4 *For a convex Ω and with α_2 sufficiently small, there exists a constant C_i depending only on β (which is the minimum angle in the triangulation of Ω) and α_i, $i = 1, 2$, such that for $N \ge 1$, and under Assumptions (10.1.56)–(10.1.58), we have*

$$\|u(t_N) - U_N\| \le C_i L_N \max_{1\le n\le N} (k_n \|\dot{u}\|_{I_n} + \|h_n^2 D^2 u\|_{I_n}), \tag{10.1.59}$$

where

$$L_N = 2 + \max_{1\le n\le N} \max \left\{ \left(\log\left(\frac{t_n}{k_n}\right)\right)^{1/2}, \log\left(\frac{t_n}{k_n}\right) \right\}.$$

An approach (technical) to a proof as well as error estimates for an a posteriori version can be found in [57]. We skip this proof.

10.2 The Wave Equation in $\mathbb{R}^d$

The study of the wave equation in $\mathbb{R}^d$, $d \geq 2$ is an extension of the results in the one-dimensional case introduced in Chapter 7. For a full finite element study of wave propagation in higher dimensions, we refer to [84]. In this section, first, we shall prove the *law of conservation of energy*, derive spatial and temporal discretizations, and give a stability estimate for the wave equation in $\mathbb{R}^d$ $d \geq 2$. Next, we prove a priori error estimates for the semidiscreet (time-continuous) problem. Finally, we conclude with a space–time discretization and derive a priori error estimates for the fully discrete problem.

Theorem 10.5 (Conservation of energy) *For the wave equation given by*

$$\begin{cases} \ddot{u} - \Delta u = 0, & \text{in } \Omega \subset \mathbb{R}^d \quad (DE) \\ u = 0, & \text{on } \partial\Omega = \Gamma \quad (BC) \\ (u = u_0) \text{ and } (\dot{u} = v_0) & \text{in } \Omega, \text{ for } t = 0, \quad (IC) \end{cases} \tag{10.2.1}$$

where $\ddot{u} = \partial^2 u/\partial t^2$, *with* $t \in [0, T]$, *or* $t \in (0, \infty)$, *we have that*

$$\frac{1}{2}\|\dot{u}(\cdot, t)\|^2 + \frac{1}{2}\|\nabla u(\cdot, t)\|^2 = \text{constant, independent of } t, \tag{10.2.2}$$

i.e. the total energy is conserved, where $\frac{1}{2}\|\dot{u}\|^2$ *is the kinetic energy, and* $\frac{1}{2}\|\nabla u\|^2$ *is the potential (elastic) energy.*

Proof: We multiply the equation by $\dot{u}$ and integrate over Ω to obtain

$$\int_\Omega \ddot{u}\dot{u}\, dx - \int_\Omega \Delta u \cdot \dot{u}\, dx = 0. \tag{10.2.3}$$

Using Green's formula:

$$-\int_\Omega (\Delta u)\dot{u}\, dx = -\int_\Gamma (\nabla u \cdot n)\dot{u}\, ds + \int_\Omega \nabla u \cdot \nabla \dot{u}\, dx, \tag{10.2.4}$$

and the boundary condition $u = 0$ on Γ (which implies $\dot{u} = 0$ on Γ), we get

$$\int_\Omega \ddot{u} \cdot \dot{u}\, dx + \int_\Omega \nabla u \cdot \nabla \dot{u}\, dx = 0. \tag{10.2.5}$$

Consequently, we have that

$$\int_\Omega \frac{1}{2}\frac{d}{dt}(\dot{u}^2)\, dx + \int_\Omega \frac{1}{2}\frac{d}{dt}(|\nabla u|^2)\, dx = 0 \Longleftrightarrow \frac{1}{2}\frac{d}{dt}(\|\dot{u}\|^2 + \|\nabla u\|^2) = 0,$$

and hence,

$$\frac{1}{2}\|\dot{u}\|^2 + \frac{1}{2}\|\nabla u\|^2 = \frac{1}{2}\|v_0\|^2 + \frac{1}{2}\|\nabla u_0\|^2,$$

which is the desired result, and the proof is complete. □

10.2.1 The Weak Formulation

We multiply the equation $\ddot{u} - \Delta u = f$ by a test function $v \in H_0^1(\Omega)$ and integrate over Ω to obtain

$$\begin{aligned} \int_\Omega fv\,dx &= \int_\Omega \ddot{u}v\,dx - \int_\Omega \Delta u\, v\,dx \\ &= \int_\Omega \ddot{u}v\,dx + \int_\Omega \nabla u \cdot \nabla v\,dx - \int_{\partial\Omega} (n\cdot\nabla u)v\,ds \\ &= \int_\Omega \ddot{u}v\,dv + \int_\Omega \nabla u\cdot\nabla v\,dx. \end{aligned} \tag{10.2.6}$$

where we used Green's formula and boundary data on v. Thus, we have the following *variational formulation* for the wave equation (10.2.1): find $u \in H_0^1(\Omega)$ such that for every fixed $t \in [0, T]$,

$$\int_\Omega \ddot{u}v\,dv + \int_\Omega \nabla u\cdot\nabla v\,dx = \int_\Omega fv\,dx, \quad \forall v \in H_0^1(\Omega). \tag{10.2.7}$$

10.2.2 The Semidiscrete Problem

Let V_h^0 be a subspace of $H_0^1(\Omega)$ consisting of continuous piecewise linear functions on a partition $\mathcal{T}_h$ of Ω that vanish on $\partial\Omega$. The semidiscrete (spatial discretization) counterpart of (10.2.7) reads as follows: find $u_h \in V_h^0$, such that for every fixed $t \in (0, T]$,

$$\int_\Omega \ddot{u}_h v\,dx + \int_\Omega \nabla u_h\cdot\nabla v\,dx = \int_\Omega fv\,dx, \quad \forall v \in V_h^0. \tag{10.2.8}$$

Now, let $\{\varphi_j\}_{j=1}^N$ be a set of basis functions for V_h^0, Thus, in (10.2.8), we may choose $v = \varphi_i$, $i = 1, \dots, N$. Next, we make the ansatz

$$u_h(x,t) = \sum_{j=1}^N \xi_j(t)\varphi_j(x), \tag{10.2.9}$$

where ξ_j, $j = 1, \dots, N$ are N time-dependent unknown coefficients. Hence, with $i = 1, \dots, N$ and $t \in I = (0, T]$, (10.2.8) can be written as

$$\sum_{j=1}^N \ddot{\xi}_j(t)\int_\Omega \varphi_j\varphi_i\,dx + \sum_{j=1}^N \xi_j(t)\int_\Omega \nabla\varphi_j\cdot\nabla\varphi_i\,dx = \int_\Omega f\varphi_i\,dx. \tag{10.2.10}$$

In the matrix form, (10.2.10) is written as the $N \times N$ linear system of ODEs,

$$\mathbf{M}\ddot{\xi}(t) + \mathbf{S}\xi(t) = b(t), \quad t \in I = (0, T], \tag{10.2.11}$$

where $\mathbf{M}$ and $\mathbf{S}$ are the usual mass and stiffness matrices, respectively, and b is the load vector.

Recall $\mathbf{M}$ is a positive-definite, invertible sparse matrix and (10.2.11) can be written, in compact form, as a system of, second-order ODEs, viz.

$$\ddot{\xi}(t) + \mathbf{M}^{-1}\mathbf{S}\xi(t) = \mathbf{M}^{-1}b(t), \quad t \in I = (0, T]. \tag{10.2.12}$$

10.2.2.1 A Priori Error Estimates for the Semidiscrete Problem

Theorem 10.6 *The spatial discrete solution u_h given by (10.2.9) satisfies the a priori error estimate*

$$\|u(t) - u_h(t)\| \leq Ch^2 \left(\|\Delta u(\cdot, t)\| + \int_0^t \|\ddot{u}(\cdot, s)\| \, ds \right). \tag{10.2.13}$$

Proof: One approach to the proof follows the general path of a priori error estimates (see the proof of Theorem 10.2) combined with conservation of energy. □

Below, for the sake of completeness, we include a general semidiscrete error estimate, due to [92], that contains the proof of Theorem 10.6 as well.

Theorem 10.7 *Let u and u_h be the solution of (10.2.1) and (10.2.8), respectively. Then, there is nondecreasing Time-dependent constant $C(t)$ such that for $t > 0$,*

$$\begin{aligned} &\|u(t) - u_h(t)\| + |u(t) - u_h(t)|_1 + \|u_t(t) - u_{h,t}(t)\| \\ &\leq C(|u_{0,h} - R_h u_0|_1 + \|v_h - R_h v\|) \\ &+ C(t)h^2 \left[\|u(t)\|_2 + \|u_t(t)\|_2 + \left(\int_0^t \|u_{tt}\|_2^2 \, ds \right)^{1/2} \right]. \end{aligned} \tag{10.2.14}$$

Proof: We start with the, Ritz projection, split

$$u - u_h = (u - R_h u) + (R_h u - u_h) = \rho + \theta.$$

We may bound ρ and ρ_t as in the case of heat equation:

$$\|\rho(t)\| + h|\rho(t)|_1 \leq Ch^2\|u(t)\|_2, \quad \|\rho_t(t)\| \leq Ch^2\|u_t(t)\|_2.$$

As for $\theta(t)$, we note that

$$(\theta_{tt}, \psi) + a(\theta, \psi) = -(\rho_{tt}, \psi), \quad \forall \psi \in V_h^0, \quad t > 0. \tag{10.2.15}$$

To separate the effects from the initial values and the data, we let $\theta = \eta + \zeta$, where

$$\begin{cases} (\eta_{tt}, \psi) + a(\eta, \psi) = 0, & \forall \psi \in V_h^0, \quad t > 0 \\ \eta(0) = \theta(0), & \eta_t(0) = \theta_t(0). \end{cases}$$

Then, η and η_t are bounded (as in conservation of energy).

As for the contribution from the ζ, we note that by the initial data for η, ζ satisfies (10.2.15) with the data $\zeta(0) = \zeta_t(0) = 0$, where for $\psi = \zeta_t$, we end up with

$$\frac{1}{2}\frac{d}{dt}(\|\zeta_t\|^2 + |\zeta|_1^2) = -(\rho_{tt}, \zeta_t) \leq \frac{1}{2}\|\rho_{tt}\|^2 + \leq \frac{1}{2}\|\zeta_t\|^2.$$

Then, integrating with respect to t, we have that

$$\|\zeta_t(t)\|^2 + |\zeta(t)|_1^2 \le \int_0^t \|\rho_{tt}\|^2\, ds + \int_0^t \|\zeta_t\|^2\, ds.$$

Finally, using Grönwall's lemma (with $C(t) = e^t$), we get

$$\|\zeta_t(t)\|^2 + |\zeta(t)|_1^2 \le C(t)\int_0^t \|\rho_{tt}\|^2\, ds \le C(t)h^4 \int_0^t \|u_{tt}\|_2^2\, ds,$$

where we used the bound for ρ_{tt}. This bounds the first two terms on the left-hand side in theorem. The third term is estimated in a similar way. □

10.2.3 The Fully Discrete Problem

Here, we discretize the ODE system (10.2.11) in the time variable. To this end, we let $\eta = \dot{\xi}$ and rewrite (10.2.11) as two first-order systems:

$$\begin{aligned} \mathbf{M}\dot{\xi}(t) &= \mathbf{M}\eta(t) \\ \mathbf{M}\dot{\eta}(t) + \mathbf{S}\xi(t) &= b(t). \end{aligned} \tag{10.2.16}$$

We apply the Crank–Nicolson method to each system, as in Chapter 6, and end up with

$$\mathbf{M}\frac{\xi_n - \xi_{n-1}}{k_n} = \mathbf{M}\frac{\eta_n + \eta_{n-1}}{2}, \tag{10.2.17}$$

$$\mathbf{M}\frac{\eta_n - \eta_{n-1}}{k_n} + \mathbf{S}\frac{\xi_n + \xi_{n-1}}{2} = \frac{b_n + b_{n-1}}{2}. \tag{10.2.18}$$

In the block matrix form, this can be written as

$$\begin{bmatrix} \mathbf{M} & -\frac{k_n}{2}\mathbf{M} \\ \frac{k_n}{2}\mathbf{S} & \mathbf{M} \end{bmatrix} \begin{bmatrix} \xi_n \\ \eta_n \end{bmatrix} = \begin{bmatrix} \mathbf{M} & \frac{k_n}{2}\mathbf{M} \\ -\frac{k_n}{2}\mathbf{S} & \mathbf{M} \end{bmatrix} \begin{bmatrix} \xi_{n-1} \\ \eta_{n-1} \end{bmatrix} + \begin{bmatrix} 0 \\ \frac{k_n}{2}(b_n + b_{n-1}) \end{bmatrix}.$$

Here we may, e.g. choose the starting iterates ξ_0 and η_0 as the nodal interpolations of u_0 and v_0, respectively.

Note that, as we mentioned earlier in Chapter 6, the Crank–Nicolson method is more accurate than the Euler methods, and it has a certain kind of *norm preserving property*. Therefore, it is a more suitable method to the numerical study of the wave equation. Below, for the sake of completeness, we include a fully discrete scheme analog to that of the heat equation and derive the corresponding a priori error estimates.

10.2.3.1 Finite Elements for the Fully Discrete Problem

We shall follow the procedure for the heat equation and hence need to rewrite the second-order wave equation (10.2.1) in a system form as two first-order equations in time. To this end, we let $v := u_t$ and reformulate the wave equation, viz. Find (u, v) such that

$$\begin{cases} v_t - \Delta u = f & \text{in } \Omega \times (0, T) \\ -\Delta u_t + \Delta v = 0 & \text{in } \Omega \times (0, T) \\ u = v = 0 & \text{on } \Gamma \times (0, T) \\ u(\cdot, 0) = u_0, \quad v(\cdot, 0) = \dot{u}_0, & \text{in } \Omega. \end{cases} \tag{10.2.19}$$

As for the finite element approximation for (10.2.19), we consider continuous piecewise linear functions that might have jump discontinuities across the discrete time levels. Let now $0 = t_0 < t_1 < \cdots < t_L = T$ be a partition of the time interval $[0, T]$ into subintervals $I_l = (t_{l-1}, t_l]$ of length $k_l = t_l - t_{l-1}$. Then, we divide the *space–time slab* $\mathbf{S}_l = \Omega \times I_l$ into prisms $K \times I_l$, where $K \in \mathcal{T}_l$ and $\mathcal{T}_l$ is a triangulation of Ω with the mesh function h_l and the corresponding finite element space W_l of continuous piecewise linear vector functions in Ω which are vanishing on Γ. Then, for $r = 0, 1$, we define the space

$$W_{k_l}^{(r)} = \left\{ (w_1, w_2) \colon w_j(x, t) = \sum_{m=0}^{r} t^m \psi_j^{(m)}(x), \quad \psi_j^{(m)} \in W_l, \ j = 1, 2 \right\}$$

on the l-th slab $\mathbf{S}_l := \Omega \times I_l$ and

$$W_k^{(r)} := \{(w_1, w_2) \colon (w_1, w_2)|_{\mathbf{S}_l} \in W_{k_l}^{(r)}, \quad l = 1, 2, \dots, L\},$$

as the space of continuous piecewise polynomials in space (with possible time-discontinuities).

Let now $\pi_l \chi$ be the Galerkin approximation of the solution χ for the homogeneous Dirichlet boundary value problem for the Poisson's equation on Ω denoted as the elliptic projection into V_l and defined by

$$(\nabla \pi_l \chi, \nabla \zeta) = (\nabla \chi, \nabla \zeta), \quad \forall \zeta \in V_l.$$

A variational formulation of (10.2.19) for the cG(1) method for (10.2.1) in $W_k^{(1)}$ reads as follows: find $(v_h, u_h) \in W_k^{(1)}$ such that for all $l = 1, 2, \dots, L$, and all $w = (w_1, w_2) \in W_{kl}^{(0)}$,

$$\begin{cases} \displaystyle\int_{I_l} ((\dot{v}_h, w_1) + (\nabla u, \nabla w_1))\, dt = \int_{I_l} (f, \ w_1)\, dt, \\ \displaystyle\int_{I_l} ((\nabla \dot{u}_h, \nabla w_2) - (\nabla v_h, \nabla w_2))\, dt = 0, \\ u_{h,l-1}^{+} = \pi_l u_{h,l-1}^{-}, \quad v_{h,l-1}^{+} = P_n v_{h,l-1}^{-}, \end{cases} \tag{10.2.20}$$

where $u_{h,0}^- = u_0$, $v_{h,0}^- = \dot{u}_0$ and with $\chi_1 = u_h$ and $\chi_2 = v_h$,

$$\chi_j(x,t)|_{S_l} = \chi_{j,l}^-(x)\frac{t - t_{l-1}}{k_l} + \chi_{j,l-1}^+(x)\frac{t - t_l}{-k_l}. \quad j = 1,\ 2.$$

Remark 10.1 For an unchanged mesh, $\mathcal{T}_l = \mathcal{T}_{l-1}$ across the time level t_{l-1}, both π_l and P_l are reduced to identity and the approximate solution (u_h, v_h) is continuous across t_{l-1}. Whereas for the refined mesh across t_{l-1}, i.e. for $V_{l-1} \subset V_l$, the coefficients of $(v_{h,l-1}^+, u_{h,l-1}^+)$ are found through interpolation of $(v_{h,l-1}^-, u_{h,l-1}^-)$ at the nodes $\{N_{l,k}\}$ in $\mathcal{T}_l$.

At each time level, $l = 1, 2, \ldots, L$, let $\varphi_{l,i}$, $i = 1, 2, \ldots, N_l$ be the slab-wise nodal basis functions for V_l with dimension N_l and introduce the associated, slabwise, $N_l \times N_l$ mass- and stiffness matrices:

$$\mathbf{M}_l := (\varphi_{l,i}, \varphi_{l,j})_{i,j} \quad \text{and} \quad \mathbf{S}_l := (\nabla\varphi_{l,i}, \nabla\varphi_{l,j})_{i,j}, \quad i,j = 1, 2, \ldots, N_l,$$

and the cross time-layer mass and stiffness matrices

$$\begin{cases} (\mathbf{M}_{l-1,l})_{i,j} := (\varphi_{l,i}, \varphi_{l-1,j}), & 1 \le i \le N_l,\ 1 \le j \le N_{l-1}, \\ (\mathbf{S}_{l-1,l})_{i,j} := (\nabla\varphi_{l,i}, \nabla\varphi_{l-1,j}), & 1 \le i \le N_l,\ 1 \le j \le N_{l-1}. \end{cases}$$

Further, with (ξ_l^-, η_l^-) and $(\xi_{l-1}^+, \eta_{l-1}^+)$, denoting the vectors of coefficients in the basis $\{\varphi_{l,i}\}$ of $(v_{h,l}^-, u_{h,l}^-)$ and $(v_{h,l-1}^+, u_{h,l-1}^+)$, we can now write (10.2.20) in the matrix form as

$$\begin{cases} \mathbf{M}_l(\xi_l^- - \xi_{l-1}^-) + \frac{k_l}{2}\mathbf{S}_l(\eta_l^- + \eta_{l-1}^+) = F_l \\ \mathbf{S}_l(\eta_l^- - \eta_{l-1}^+) - \frac{k_l}{2}\mathbf{S}_l(\xi_l^- + \xi_{l-1}^+) = 0 \\ \mathbf{M}_l\xi_{l-1}^+ = \mathbf{M}_{l-1,l}\xi_{l-1}^-, \quad \mathbf{S}_l\eta_{l-1}^+ = S_{l-1,l}\eta_{l-1}^-. \end{cases} \tag{10.2.21}$$

Finally, the load vector F has its slabwise components

$$(F_l)_i = \int_{I_l} (f, \varphi_{l,i})\, dt, \quad 1 \le i \le N_l,\ 1 \le l \le L.$$

An energy conservation. One of the advantages of the cG(1) scheme (10.2.20) for the wave equation is the conservation of energy for the homogeneous data ($f \equiv 0$) and time continuous mesh (i.e. when $V_{l-1} \subset V_l$, $\forall l$). To see this, let in homogeneous (10.2.20)

$$\begin{cases} w_1 = (v_{h,l-1}^+ + v_{h,l}^+)/2, \\ w_2 = (u_{h,l-1}^+ + u_{h,l}^+)/2, \end{cases}$$

add the two equations and use cancellation effects to obtain

$$\int_{I_l} ((\dot{v}_h, v_h) + (\nabla\dot{u}_h, \nabla u_h))\, dt = 0,$$

which yields the *discrete version of the conservation of energy*:

$$\|\nabla u_{h,l}^-\| + \|v_{h,l}^-\| = \|\nabla u_{h,l-1}^-\| + \|v_{h,l-1}^-\|. \tag{10.2.22}$$

10.2.4 Error Estimate for cG(1)

Theorem 10.8 (An a priori error estimate) *Let (v_h, u_h) be the cG(1) approximation of the wave equation (10.2.19) that satisfies (10.2.20). Then, there is a constant C such that for $L = 1, 2, \dots$, we have*

$$\|u - u_h\| \le C \int_0^{t_L} (\|k^2 \nabla \ddot{u}\| + \|k^2 \ddot{v}\| + \|h^2 D^2 \dot{u}\|)\, dt. \tag{10.2.23}$$

Proof: For simplicity, we give a sketchy proof cf. [57] and consider only the temporal discretization. On each time level l, $1 \le l \le L$, we define

$$V_l := \{w\colon \|\nabla w\| < \infty, \quad w = 0 \text{ on } \Gamma = \partial\Omega\},$$

and let $\pi_k u$ be the piecewise linear time-interpolant of u and for $\zeta := (v, u)$ write

$$e = (v, u) - (v_k, u_k) = \zeta - \zeta_k = (\zeta - \pi_k\zeta) + (\pi_k\zeta - \zeta_k) \equiv \sigma + \tilde{\theta}.$$

Since we have already interpolation error estimates for σ, it remains to estimate the contributions form $\tilde{\theta} = \pi_k\zeta - \zeta_k$. To proceed, we use the discrete dual problem, viz. Find $(\phi_1, \phi_2) \in W_k^{(1)}$ such that for $l = L, L-1, \dots, 1$,

$$\begin{cases} -(\xi, \dot{\phi}_1)_l - (\nabla\xi, \nabla\phi_2)_l = 0, & \\ -(\nabla\eta, \nabla\dot{\phi}_2)_l + (\nabla\eta, \nabla\phi_1)_l = 0, & \forall(\xi, \eta) \in W_{kl}^{(0)}, \\ \phi_{1,L} = 0, -\Delta\phi_{2,L} = e_{2,L} := \pi_k u_L - u_{h,L}. & \end{cases} \tag{10.2.24}$$

In the second terms above, we make the replacements $\tilde{\phi}_{j,l} := (\phi_{j,l-1} + \phi_{j,l})/2$ on I_l. This is justified since ξ and η are piecewise constant. Then, we may raise ξ and η to piecewise linears due to the fact that now both $\dot{\phi}_j$ and $\tilde{\phi}_j$ are piecewise constants. In particular, we let $\xi = e_1 := \pi_k v - v_h$ and $\eta = e_2 = \pi_k u - u_h$. Then, adding up the resulting equations, summing over l, replacing (from second equality) ϕ_j by $\tilde{\phi}_j$, we end up with

$$\begin{aligned} \|e_{2,L}\|^2 &= \sum_{l=1}^{L} ((\nabla\dot{e}_2, \nabla\phi_2)_l + (\nabla e_2, \nabla\tilde{\phi}_1)_l + (\dot{e}_1, \phi_1)_l - (\nabla e_1, \nabla\tilde{\phi}_2)_l) \\ &= \sum_{l=1}^{L} ((\dot{e}_1, \tilde{\phi}_1)_l + (\nabla e_2, \nabla\tilde{\phi}_1)_l + (\nabla\dot{e}_2, \nabla\tilde{\phi}_2)_l - (\nabla e_1, \nabla\tilde{\phi}_2)_l) \\ &= -\sum_{l=1}^{L} ((\dot{\sigma}_1, \tilde{\phi}_1)_l + (\nabla\sigma_2, \nabla\tilde{\phi}_1)_l + (\nabla\dot{\sigma}_2, \nabla\tilde{\phi}_2)_l - (\nabla\sigma_1, \nabla\tilde{\phi}_2)_l), \end{aligned}$$

where we used Galerkin orthogonality to replace u_h by u. Further, since $\dot{\sigma}_j = 0, j = 1, 2$, we get the error representation formula:

$$\|e_{2,L}\|^2 = -\sum_{l=1}^{L}((\nabla\sigma_2, \nabla\tilde{\phi}_1)_l - (\sigma_1, \Delta\tilde{\phi}_2)_l). \tag{10.2.25}$$

Finally, combining the error representation formula with the interpolation error estimate for σ, we get the desired a priori error estimate. □

Stability. If we let in (10.2.24) $\xi = -\Delta\tilde{\phi}_1$ and $\eta = -\Delta\tilde{\phi}_2$, we get the stability estimate:

$$\|\nabla\phi_{1,l}\|^2 + \|\Delta\phi_{2,l}\|^2 = \|e_{2,L}\|^2, \quad \forall l \leq L. \tag{10.2.26}$$

10.3 Exercises

Problem 10.1 Consider the heat equation. Show that $\|\nabla u(t)\| \leq \|\nabla u_0\|$ (the stability estimate for the gradient).
Hint: Assume $f = 0$, multiply (10.1.1) by $-\Delta u$ and integrate over Ω.
Is this inequality valid for $u_0 =$ constant?

Problem 10.2 Derive the corresponding estimate for Neumann boundary condition:

$$\frac{\partial u}{\partial n} = 0. \tag{10.3.1}$$

Problem 10.3 Consider the following general form of the heat equation

$$\begin{cases} u_t(x,t) - \Delta u(x,t) = f(x,t), & \text{for } x \in \Omega,\ 0 < t \leq T, \\ u(x,t) = 0, & \text{for } x \in \Gamma,\ 0 < t \leq T, \\ u(x,0) = u_0(x), & \text{for } x \in \Omega, \end{cases} \tag{10.3.2}$$

where $\Omega \in \mathbb{R}^2$ with boundary Γ. Let $\tilde{u}$ be the solution of (10.3.2) with a modified initial data $\tilde{u}_0(x) = u_0(x)\varepsilon(x)$.
(a) Show that $w := \tilde{u} - u$ solves (10.3.2) with initial data $w_0(x) = \varepsilon(x)$.
(b) Give estimates for the difference between u and $\tilde{u}$.
(c) Prove that the solution of (10.3.2) is unique.

Problem 10.4 Formulate the equation for cG(1)dG(1) for the two-dimensional heat equation using the discrete Laplacian.

Problem 10.5 In two dimensions the heat equation, in the case of radial symmetry, can be formulated as $r\dot{u} - (ru'_r)'_r = rf$, where $r = |x|$ and $w'_r = \frac{\partial w}{\partial r}$.

(a) Verify that $u = \frac{1}{4\pi t}\exp(-\frac{r^2}{4t})$ is a solution for the homogeneous equation ($f = 0$) with the initial data being the Dirac δ function $u(r, 0) = \delta(r)$.
(b) Sketching $u(r, t)$ for $t = 1$ and $t = 0.01$, deduce that $u(r, t) \to 0$ as $t \to 0$ for $r > 0$.
(c) Show that $\int_{\mathbb{R}^2} u(x, t)\, dx = 2\pi \int_0^\infty u(r, t)\, r\, dr = 1$ for all t.
(d) Determine a stationary solution to the heat equation with data

$$f = \begin{cases} 1/(\pi\varepsilon)^2, & \text{for } r < \varepsilon, \\ 0, & \text{otherwise.} \end{cases}$$

(e) Determine the fundamental solution corresponding to $f = \delta$, letting $\varepsilon \to 0$.

Problem 10.6 Consider the Schrödinger equation

$$iu_t - \Delta u = 0, \quad \text{in } \Omega, \qquad u = 0, \quad \text{on } \partial\Omega.$$

where $i = \sqrt{-1}$ and $u = u_1 + iu_2$.

(a) Show that the *total probability* $\int_\Omega |u|^2 dx$ is independent of the time.
Hint: Multiplying by $\bar{u} = u_1 - iu_2$, and consider the imaginary part
(b) Consider the corresponding eigenvalue problem, i.e. find the eigenvalue λ and the corresponding eigenfunction $u \neq 0$ such that

$$-\Delta u = \lambda u \quad \text{in } \Omega, \qquad u = 0, \quad \text{on } \partial\Omega.$$

Show that $\lambda > 0$ and give the relationship between the norms $\|u\|$ and $\|\nabla u\|$ for the corresponding eigenfunction u.
(c) Determine (in terms of the smallest eigenvalue λ_1), the smallest possible value for the constant C in the Poincare estimate

$$\|u\| \le C\|\nabla u\|,$$

derived for all solutions u vanishing at the boundary ($u = 0$, on $\partial\Omega$).

Problem 10.7 Consider the initial-boundary value problem

$$\begin{cases} u_t(x, t) - \Delta u(x, t) = f(x, t), & \text{for } x \in \Omega,\ t > 0, \\ u(x, t) = 0, & \text{for } x \in \Gamma,\ t > 0, \\ u(x, 0) = u_0(x), & \text{for } x \in \Omega. \end{cases} \tag{10.3.3}$$

(a) Prove (with $\|u\| = (\int_\Omega u^2\, dx)^{1/2}$) that

$$\|u(t)\|^2 + \int_0^t \|\nabla u(s)\|^2\, ds \le \|u_0\|^2 + \int_0^t \|f(s)\|^2\, ds$$

$$\|\nabla u(t)\|^2 + \int_0^t \|\Delta u(s)\|^2\, ds \le \|\nabla u_0\|^2 + \int_0^t \|f(s)\|^2\, ds$$

(b) Formulate $dG(0) - cG(1)$ method for this problem.

Problem 10.8 Formulate and prove $dG(0) - cG(1)$ a priori and a posteriori error estimates for the two-dimensional heat equation (cf. the previous problem) that uses lumped mass and midpoint quadrature rule.

Problem 10.9 Consider the convection problem

$$\beta \cdot \nabla u + \alpha u = f, \quad x \in \Omega, \quad u = g, \quad x \in \Gamma_-, \tag{10.3.4}$$

Define the outflow Γ_+ and inflow Γ_- boundaries. Assume that $\alpha - \frac{1}{2}\nabla \cdot \beta \ge c > 0$. Show the following stability estimate

$$c\|u\|^2 \int_{\Gamma_+} n \cdot \beta u^2\, ds\, dt \le \|u_0\|^2 + \frac{1}{c}\|f\|^2 + \int_{\Gamma_-} |n \cdot \beta| g^2\, ds. \tag{10.3.5}$$

Hint: Show first that

$$2(\beta \cdot \nabla u, u) = \int_{\Gamma_+} n \cdot \beta\, u^2\, ds - \int_{\Gamma_-} \|n \cdot \beta\|\, u^2\, ds - ((\nabla \cdot \beta)u, u).$$

Formulate the streamline diffusion for this problem.

Problem 10.10 Consider the convection problem

$$\begin{aligned} &\dot{u} + \beta \cdot \nabla u + \alpha u = f, \; x \in \Omega, && t > 0, \\ &u = g, \qquad\qquad x \in \Gamma_-, && t > 0, \\ &u(x,0) = u_0(x), \qquad x \in \Omega, && \end{aligned} \tag{10.3.6}$$

with the obvious definitions for Γ_+ and Γ_-. Assume that $\alpha - \frac{1}{2}\nabla \cdot \beta \ge c > 0$. Show the following stability estimate

$$\begin{aligned} &\|u(\cdot,T)\|^2 + c\int_0^T \|u(\cdot,t)\|^2\, dt + \int_0^T \int_{\Gamma_+} n \cdot \beta u^2\, ds\, dt \\ &\le \|u_0\|^2 + \frac{1}{c}\int_0^T \|f(\cdot,t)\|^2\, dt + \int_0^T \int_{\Gamma_-} |n \cdot \beta| g^2\, ds\, dt, \end{aligned} \tag{10.3.7}$$

where $\|u(\cdot,T)\|^2 = \int_\Omega u(x,T)^2\, dx$.

Problem 10.11 Compute the coefficient matrices in (10.1.55), for $\Omega = (0,1) \times (0,1]$ and the uniform triangulation on each time step.

Problem 10.12 Use the system (10.1.53) to specify the matrix equations for the cG(1)cG(1) method for heat equation.

Problem 10.13 Derive problem (10.1.53) for the modified heat conduction

$$\dot{u} - \Delta u + u = f, \quad \text{in } \Omega \subset \mathbb{R}^2, \quad \frac{\partial u}{\partial n} = 0, \quad \text{on } \partial\Omega.$$

Problem 10.14 Derive a total conservation of energy relation using the Robin type boundary condition: $\dfrac{\partial u}{\partial n} + u = 0$.

Problem 10.15 Determine a solution for the following equation

$$\ddot{u} - \Delta u = e^{it}\delta(x),$$

where $\Delta = \frac{\partial^2}{\partial x_1^2} + \frac{\partial^2}{\partial x_2^2} + \frac{\partial^2}{\partial x_3^2}$, $i = \sqrt{-1}$, $x = (x_1, x_2, x_3)$ and δ is the Dirac δ function. *Hint*: Let $u = e^{it}v(x)$, $v(x) = w(r)/r$ where $r = |x|$. Further, $rv = w \to \frac{1}{4\pi}$ as $r \to 0$. Use spherical coordinates.

Problem 10.16 Consider the initial boundary value problem

$$\begin{cases} \ddot{u} - \Delta u + u = 0, & x \in \Omega, \quad t > 0, \\ u = 0, & x \in \partial\Omega, \quad t > 0, \\ u(x,0) = u_0(x), \quad \dot{u}(x,0) = u_1(x), & x \in \Omega. \end{cases} \tag{10.3.8}$$

Rewrite the problem as a system of two equations with a time derivative of order at most 1. Why this modification?

Problem 10.17 Consider the initial boundary value problem

$$\begin{cases} \ddot{u} - \Delta u = 0, & x \in \Omega, \quad t > 0, \\ u = 0, & x \in \partial\Omega, \quad t > 0, \\ u(x,0) = u_0(x), \quad \dot{u}(x,0) = u_1(x), & x \in \Omega. \end{cases} \tag{10.3.9}$$

Formulate the cG(1)cG(1) method for this problem. Show that the energy is conserved.

Problem 10.18 Let ω be a given frequency and consider the following wave equation:

$$\ddot{u} - \Delta u = g(x)e^{i\omega t}, \quad \omega > 0.$$

(a) Show that the solution to this equation is given by

$$u(x,t) = e^{i\omega t} v(x).$$

(b) Show that the amplitude $v(x)$ is the solution of the stationary Helmholtz's equation:

$$-\Delta v - \omega^2 v = g. \tag{10.3.10}$$

(c) Write a Matlab code for a cG(1) solution of Eq. (10.3.10) in $2d$, e.g. $\Omega := (0,1)^2$, with a physically appropriate boundary data.

Problem 10.19

(a) Show that in spherical coordinates, the spherically symmetric version of the homogeneous wave equation (10.2.1) in $\mathbb{R}^3$, through origin and with $u_0 = 0$ satisfies the $1d$ equation

$$\ddot{u} - u_{rr} + \frac{2}{r} u_r = 0, \quad r > 0, \quad t > 0, \tag{10.3.11}$$

where r is the distance to the origin.

(b) Show that using $w = ru$ (10.3.11) is transformed to the $1d$ wave equation

$$\ddot{w} - w_{rr} = 0, \quad r > 0, \quad t > 0. \tag{10.3.12}$$

(c) We associate (10.3.12) with data:

$$w(0,t) = 0, \quad w(r,0) = ru_0(r), \quad \dot{w}(r,0) = 0,$$

Show that a generalized version of d'Alembert formula yields the solution of (10.3.11):

$$u(r,t) = \begin{cases} (u_0(r+t) + u_0(r-t)) + \frac{t}{r}(u_0(r+t) - u_0(r-t)), & r \geq t \\ (u_0(t+r) + u_0(t-r)) + \frac{t}{r}(u_0(t+r) - u_0(t-r)), & r < t. \end{cases} \tag{10.3.13}$$

Problem 10.20 Compute, both explicit and within a Matlab code, the spherically symmetric solution above for $\dot{u}_0 = 0$ and

$$u_0 = \begin{cases} 1 & |x| \leq 1 \\ 0, & \text{else.} \end{cases}$$

Appendix A

Answers to Some Exercises

Chapter 1. Exercise Section 1.4.1

1.1 The solution is unique only for $\ell \neq n\pi$, where n is an integer.

1.2 (a) No!
(b) $\int_0^\ell f(x)\,dx = 0$.

1.3 $\sum_{i=1}^{N} c_i = 1$.

1.4 (b) $c_1 = 0, \; c_2 = 1$.

1.5 (a) $a + 3b = 0$. (b) $3a - \pi b = 0$. (c) $2a + eb = 0$.

1.6 (a) $u_{ss} = 0; \quad v(s,t) = u(x,y)$
(b) $u(x,y) = f(x-y) + xg(x-y)$.
(c) $v_{ss} + v_{tt} = 0, \quad v(s,t) = u(x,y)$.

1.7 (c) $\dfrac{\sin 2\pi x \sinh 2\pi y}{\sinh 2\pi} - \dfrac{\sin 3\pi x \sinh 3\pi y}{\sinh 3\pi} + \dfrac{2\sin \pi x \sinh \pi(1-y)}{\sinh \pi}$.

Chapter 1. Exercise Section 1.5.4

1.11 $u =$ constant gives circles with center $\left(1, \dfrac{-1}{u}\right)$ and radius $1/|u|$.
$v =$ constant gives circles with center $\left(\dfrac{-v}{v-1}, 0\right)$ and radius $1/|v-1|$.

An Introduction to the Finite Element Method for Differential Equations, First Edition.
Mohammad Asadzadeh.

1.15 The term represents heat loss at a rate proportional to the excess temperature over θ_0.

1.18 $a^2 = b^2c^2$.

1.19 $\alpha = \pm c$.

Chapter 2. Exercise Section 2.11

2.4 No.

2.5 Linearly dependent. Possible answer

$$-x - 3 = (2x^2 + 3x + 1) - 2(x^2 + 2x + 2).$$

2.6 Possible answer: $\{(1, 0, 1, 0),\ (0, 1, -2, 1)\}$; dimension is 2.

2.7 Possible answer:

$$\{(0, -1, 0, 0, 1),\ (7, -4, 0, 1, 0),\ (-5, 3, 1, 0, 0)\};$$

Dimension is 3.

2.8 $\alpha = \pm 2$.

2.9 $\alpha = 1$.

Chapter 3. Exercise Section 3.5

3.3

$$Pu(t) = 0.9991 + 1.0183t + 0.4212t^2 + 0.2786t^3.$$

3.5 (a) $u(x) = \frac{1}{2}x(1 - x)$
(b) $R(x) = \pi^2 A \sin\ \pi x + 4\pi^2 B \sin\ 2\pi x - 1$
(c) $A = 4/\pi^3$ and $B = 0$.
(d) –

3.6 (a) –

(b) $R(x) = (\pi^2 + 1)A \sin\ \pi x + (4\pi^2 + 1)B \sin\ 2\pi x + (9\pi^2 + 1)C \sin\ 3\pi x - x$

(c) $A = \dfrac{2}{\pi(\pi^2 + 1)}$, $B = -\dfrac{1}{\pi(4\pi^2 + 1)}$, and $C = \dfrac{2}{3\pi(9\pi^2 + 1)}$.

3.7 (a) $u(x) = \dfrac{1}{6}(\pi^3 - x^3) + \dfrac{1}{2}(x^2 - \pi^2)$

(b) $R(x) = -U''(x) - x + 1 = \dfrac{1}{4}\xi_0 \cos\ \dfrac{x}{2} + \dfrac{9}{4}\xi_1 \cos\ \dfrac{3x}{2} - x + 1$

(c) $\xi_0 = 8(2\pi - 6)/\pi$ and $\xi_1 = \dfrac{8}{9}(\dfrac{2}{9} - \dfrac{2}{3}\pi)/\pi$.

3.8 $U(x) = (16 \sin x + \dfrac{16}{27} \sin 3x)/\pi^3 + 2x^2/\pi^2$.

Chapter 3. Exercise Section 3.8

3.11 (a) x, (b) 0.

3.12

$$\Pi_1 f(x) = \begin{cases} 4 - 11(x + \pi)/(2\pi), & -\pi \le x \le -\dfrac{\pi}{2}, \\ 5/4 - (x + \dfrac{\pi}{2})/(2\pi), & -\dfrac{\pi}{2} \le x \le 0, \\ 1 - 7x/(2\pi), & 0 \le x \le \dfrac{\pi}{2}, \\ 3(x - \pi)/(2\pi), & \dfrac{\pi}{2} \le x \le \pi. \end{cases}$$

3.17 Check the conditions required for a vector space.

3.18

$$\Pi_1 f(x) = f(a)\frac{2x - a - b}{a - b} + f\left(\frac{a + b}{2}\right)\frac{2(x - a)}{b - a}.$$

3.19 *Hint*: Use the procedure in the proof of Theorem 3.1, with somewhat careful estimates at the end.

3.21

$$\pi_4(e^{-8x^2}) \approx 0.25x^4 - 1.25x^2 + 1.$$

3.22 For example we may choose the following basis:

$$\varphi_{i,j}(x) = \begin{cases} 0, & x \in [x_{i-1}, x_i], \\ \lambda_{i,j}(x), & i = 1, \dots, m + 1,\ j = 0, 1, 2. \end{cases}$$

$$\lambda_{i,0}(x) = \frac{(x-\xi_i)(x-x_i)}{(x_{i-1}-\xi_i)(x_{i-1}-x_i)}, \quad \lambda_{i,1}(x) = \frac{(x-x_{i-1})(x-x_i)}{(\xi_i-x_{i-1})(\xi_i-x_i)},$$
$$\lambda_{i,2}(x) = \frac{(x-x_{i-1})(x-\xi_i)}{(x_i-x_{i-1})(x_i-\xi_i)}, \quad \xi_i \in (x_{i-1}, x_i).$$

3.23 This is a special case of Problem 3.22.

3.25 Hint: Use Taylor expansion of f about $x = \dfrac{x_1+x_2}{2}$.

Chapter 4. Exercise Section 4.3

4.2

$$LU = \begin{bmatrix} 1 & 0 & 0 \\ 2 & 1 & 0 \\ -2 & 0 & 1 \end{bmatrix}\begin{bmatrix} 1 & 3 & 2 \\ 0 & -1 & 3 \\ 0 & 0 & 5 \end{bmatrix}.$$

4.3

$$x = \begin{bmatrix} 0 \\ -1 \end{bmatrix}.$$

4.4

$$LDU = \begin{bmatrix} 1 & 0 & 0 \\ 0 & 1 & 0 \\ 3 & -4 & 1 \end{bmatrix}\begin{bmatrix} 1 & 0 & 0 \\ 0 & 1 & 0 \\ 0 & 0 & 14 \end{bmatrix}\begin{bmatrix} 1 & 1 & -3 \\ 0 & 1 & 1 \\ 0 & 0 & 1 \end{bmatrix}.$$

4.7 The exact solution is $(1/15, -11/15) = (0.066\,666, -0.733\,333)$.

(a) $(u_1^3, u_2^3) = (5/64, -47/64)$, $\rho(J) = 1/4$ and $||e_3||_\infty = 0.011$.

(b) $(u_1^3, u_2^3) = (0.067\,382\,8, -0.733\,154\,3)$, $\rho(G) = 1/16$ and $||e_3||_\infty = 7 \times 10^{-4}$.

(c) $(u_1^3, u_2^3) = (0.066\,789, -0.733\,317)$, $\rho(\omega_0) = 0.017$ and $||e_3||_\infty = 1 \times 10^{-4}$.

4.13

$$A^T = (BB^T)^T = BB^T = A \cdot x^T A.$$

and

$$x^T \mathbf{A} x = x^T(BB^T)x = (B^T x)^T(B^T x) > 0.$$

4.14 (a)

$$A^{-1} = \begin{pmatrix} -1 & 1 & 0 \\ -2/a & 1/a & 1/a \\ 2+2/a & -1-1/a & -1/a \end{pmatrix}. \quad \kappa(A) = \begin{cases} 12+6/a, & a \le 2 \\ 10+3a+8/a, & a \ge 2. \end{cases}$$

(b) The condition number is smallest for $a = 2$ (from Matlab code in (b)).

Chapter 5. Exercise Section 5.4

5.1 (c) $\sin\ \pi x, x \ln\ x$ and $x(1-x)$ are test functions of this problem. x^2 and $e^x - 1$ are not test functions.

5.3 (a) U is the solution for

$$AU = f \iff 1/h \begin{pmatrix} 2 & -1 & 0 \\ -1 & 2 & -1 \\ 0 & -1 & 2 \end{pmatrix} \begin{pmatrix} \xi_1 \\ \xi_2 \\ \xi_3 \end{pmatrix} = h \begin{pmatrix} 1 \\ 1 \\ 1 \end{pmatrix}$$

with $h = 1/4$.

(b) A is invertible; therefore, U is unique.

5.6 (a) ξ is the solution for

$$2 \begin{pmatrix} 2 & -1 \\ -1 & 2 \end{pmatrix} \begin{pmatrix} \xi_1 \\ \xi_2 \end{pmatrix} = \begin{pmatrix} 0 \\ 7 \end{pmatrix}$$

(b) $(\xi_1, \xi_2) = 7(1/2, 1)$ and $U(x) = 7x$ (same as the exact solution).

5.7 (a) In case of $N = 3$, ξ is the solution for

$$A\xi = f \iff 1/h \begin{pmatrix} 2 & -1 & 0 \\ -1 & 2 & -1 \\ 0 & -1 & 2 \end{pmatrix} \begin{pmatrix} \xi_0 \\ \xi_1 \\ \xi_2 \end{pmatrix} = \begin{pmatrix} -5 \\ 0 \\ 0 \end{pmatrix}$$

with $h = 1/3$. That is: $(\xi_0, \xi_1, \xi_2) = -\frac{1}{3}(15, 10, 5)$.

(b) $U(x) = 5x - 5$ (same as the exact solution).

5.8 (a) No solution!

(b) Trying to get a finite element approximation ends up with the matrix equation

$$A\xi = f \Longleftrightarrow \begin{pmatrix} 2 & -2 & 0 \\ -2 & 4 & -2 \\ 0 & -2 & 2 \end{pmatrix}\begin{pmatrix} \xi_0 \\ \xi_1 \\ \xi_2 \end{pmatrix} = \frac{1}{4}\begin{pmatrix} 1 \\ 2 \\ 1 \end{pmatrix}$$

where the coefficient matrix is singular ($det\ A = 0$). There is no finite element solution.

5.9 (d) $||U||_E^2 = \xi^T A\xi$ (check spectral theorem, linear algebra!)

5.10 For an $M+1$ partition (here $M = 2$), we get $a_{ii} = 2/h$, $a_{i,i+1} = -1/h$ except $a_{M+1,M+1} = 1/h - 1$, $b_i = 0$, $i = 1, \dots, M$ and $b_{M+1} = -1$:
(a) $U = (0, 1/2, 1, 3/2)$.
(b) e.g. $U_3 = U(1) \to 1$, as $k \to \infty$.

5.11 (c) Set $\alpha = 2$ and $\beta = 3$ in the general FEM solution:$\xi = \frac{\alpha}{3}(-1,1,1)^T + \beta(0,0,2)^T$:

$$\begin{pmatrix} \xi_1 \\ \xi_2 \\ \xi_3 \end{pmatrix} = 2/3\begin{pmatrix} -1 \\ 1 \\ 1 \end{pmatrix} + 3\begin{pmatrix} 0 \\ 0 \\ 2 \end{pmatrix}.$$

5.12

$$3\begin{bmatrix} 2 & -1 \\ -1 & 2 \end{bmatrix}\begin{bmatrix} \xi_1 \\ \xi_2 \end{bmatrix} + \frac{1}{18}\begin{bmatrix} 4 & 1 \\ 1 & 4 \end{bmatrix}\begin{bmatrix} \xi_1 \\ \xi_2 \end{bmatrix} = \frac{1}{3}\begin{bmatrix} 1 \\ 1 \end{bmatrix}$$

$$\Longleftrightarrow \text{(MATLAB)} \quad \xi_1 = \xi_2 = 0.102.$$

5.13 Just follow the procedure in the theory.

5.15 A priori: with $\alpha = 1$, and $||u||_E^2 = ||u||^2 + ||u'||^2$ we have
$||u - U||_E \le C(h + h^2)||u''||_{L_2(I)}$
A posteriori: $||u - U||_E \le C_i||hR(U)||_{L_2(I)}$.

5.16 (a) $||e'||_a \le C_i||h(aU')'||_{1/a}$.

(b) The matrix equation:

$$\begin{pmatrix} 1 & -1 & 0 & 0 \\ -1 & 2 & -1 & 0 \\ 0 & -1 & 3 & -2 \\ 0 & 0 & 2 & 4 \end{pmatrix} \begin{pmatrix} \xi_0 \\ \xi_1 \\ \xi_2 \\ \xi_3 \end{pmatrix} = \begin{pmatrix} -3 \\ 0 \\ 0 \\ 0 \end{pmatrix},$$

which yields the approximate solution $U = -3(1/2, 1, 2, 3)^T$.

(c) Since a is constant and U is linear on each subinterval, we have that

$$(aU')' = a'U' + aU'' = 0.$$

By the a posteriori error estimate, we have that $||e'||_a = 0$, i.e. $e' = 0$. Combining with the fact that $e(x)$ is continuous and $e(1) = 0$, we get that $e \equiv 0$, which means that the finite element solution, in this case, coincides with the exact solution.

5.17 A priori: $||e||_{H^1} \le C_i(||hu''|| + ||h^2u''||)$.
A posteriori: $||e||_{H^1} \le C_i||hR(U)||_{L_2(I)}$.

5.18 (a) A priori: $||e||_E \le (1 + c)C(h + h^2)||u''||$,
and a posteriori: $||e||_E \le C_i||hR(U)||_{L_2(I)}$.
(b) Since $c \ge 0$, the a priori error estimate in (a) yields optimality for $c \equiv 0$, i.e. in the case of no convection (does this tell anything to you?).

5.19 A priori: $||e||_{H^1} \le C_i(||hu''|| + 4||h^2u''||)$, a posteriori estimate is similar (see [5.17.]).

Chapter 6. Exercise Section 6.7

6.3 (a) $a_{ij} = \dfrac{j}{j+i} - \dfrac{1}{j+i+1}, \quad b_i = \dfrac{1}{i+1}, \quad i,j = 1, 2, \dots,$
(b) $q = 1 : \ U(t) = 1 + 3t$. $q = 2 : \ U(t) = 1 + \dfrac{8}{11}t + \dfrac{10}{11}t^2$.

6.5 (a) $u(t) = e^{-4t} + \dfrac{1}{32}(8t^2 - 4t + 1)$.
(b) $u(t) = e^{\frac{1}{2}t^2} - t + \dfrac{\sqrt{\pi}}{\sqrt{2}} e^{\frac{1}{2}t^2} erf\left(\dfrac{t}{\sqrt{2}}\right), \quad erf(x) = \dfrac{2}{\sqrt{\pi}} \displaystyle\int_0^x e^{-y^2}\, dy.$

6.6 $U_i(x_i) = \dfrac{[(x_i^3 - x_{i-1}^3)/3] - U_i(x_{i-1}) \cdot (2(x_i - x_{i-1}) - 1)}{1 + 2(x_i - x_{i-1})}$

Chapter 7. Exercise Section 7.2.3

7.14

$$u(x,t) = \begin{cases} \frac{1}{2}(u_0(x+2t) + u_0(x-2t)), & x \geq 2t \\ \frac{1}{2}(u_0(2t+x) + u_0(2t-x)), & x < 2t \end{cases}$$

7.17 (a) $u(x,t) = \frac{1}{2}[u_0(x+ct) + u_0(ct-x)] + \frac{1}{2c}\left(\int_0^{x+ct} v_0 dx + \int_0^{ct-x} v_0 dx\right).$

(b) $u(x,t) = \frac{1}{2c}\int_0^t 2c(t-s)\, ds = t^2/2.$

Chapter 7. Exercise Section 7.3.3

7.20 A priori: $||e||_{H^1} \leq C_i(||hu''|| + ||h^2u''||).$
A posteriori: $||e||_{H^1} \leq C_i||hR(U)||.$

7.21 A priori: $||e||_E \leq C_i(||hu''|| + ||h^2u''||).$
A posteriori: $||e||_E \leq C_i||hR(U)||.$

Chapter 9. Poisson Equation. Exercise Section 9.4

9.9

$$A = \begin{pmatrix} 4 & -1 & 0 & 0 \\ -1 & 4 & -1 & 0 \\ 0 & -1 & 4 & -1 \\ 0 & 0 & -1 & 2 \end{pmatrix} \quad b = \frac{1}{8}\begin{pmatrix} 2 \\ 2 \\ 2 \\ 1 \end{pmatrix}.$$

9.10

$$A = \begin{bmatrix} 6 & -1 \\ -1 & 3 \end{bmatrix} \quad b = \begin{bmatrix} 1 \\ 1/2 \end{bmatrix}.$$

9.11

$$A = \begin{pmatrix} 5 & -2 & 0 & 0 \\ -2 & 5 & -2 & 0 \\ 0 & -2 & 5 & -2 \\ 0 & 0 & -2 & 5/2 \end{pmatrix} \quad b = \frac{1}{16}\begin{pmatrix} 2 \\ 2 \\ 2 \\ 1 \end{pmatrix}.$$

9.12

$$\mathbf{M} = \frac{h^2}{12}\begin{bmatrix} 8 & 1 & 1 & 1 \\ 1 & 4 & 0 & 1 \\ 1 & 0 & 4 & 1 \\ 1 & 1 & 1 & 8 \end{bmatrix}, \qquad \mathbf{S} = \begin{bmatrix} 4 & -1 & -1 & 0 \\ -1 & 4 & 0 & -1 \\ -1 & 0 & 4 & -1 \\ 0 & -1 & -1 & 4 \end{bmatrix}.$$

Chapter 10. IBVPs: Exercise Section 10.3

10.3

$$\|w(T)\|^2 + 2\int_0^T \|\nabla w\|^2 \, dt \le \|\varepsilon\|^2.$$

10.6 $1/\sqrt{\lambda_1}$.

10.15 $v = \frac{1}{4\pi}\frac{\cos(r)}{r}$ and the corresponding solution $u = e^{it}\frac{1}{4\pi}\frac{\cos(r)}{r}$.

Appendix B

Algorithms and Matlab Codes

To streamline the computational aspects, we have gathered suggestions for some algorithms and Matlab codes that can be used in implementations. These are simple specific Matlab codes on the concepts such as

- The L_2-projection.
- Numerical integration rules: Midpoint, Trapezoidal, Simpson.
- Finite difference methods: Forward Euler, Backward Euler, Crank–Nicolson.
- Matrices/vectors: Stiffness- Mass-, and Convection Matrices. Load vector.

The Matlab codes are not optimized for speed, but rather intended to be easy to read.

An algorithm for L_2-projection:

1. Choose a partition $\mathcal{T}_h$ of the interval I into N subintervals, $N+1$ nodes, and define the corresponding space of piecewise linear functions V_h.
2. Compute the $(N+1) \times (N+1)$ mass matrix **M** and the $(N+1) \times 1$ load vector **b**, viz.

$$m_i j = \int_I \varphi_j \varphi_i \, dx, \qquad b_i = \int_I f \varphi_i \, dx.$$

3. Solve the linear system of equations

$$\mathbf{M}\xi = \mathbf{b}.$$

4. Set

$$P_h f = \sum_{j=0}^{N} \xi_j \varphi_j.$$

An Introduction to the Finite Element Method for Differential Equations, First Edition.
Mohammad Asadzadeh.

Below are two versions of Matlab codes for computing the mass matrix **M**:

```
function M = MassMatrix(p, phi0, phiN)

%------------------------------------------------------------------------
% Syntax:    M = MassMatrix(p, phi0, phiN)
% Purpose:   To compute mass matrix M of partition p of an interval
% Data:      p -     vector containing nodes in the partition
%            phi0 - if 1: include basis function at the left endpoint
%                   if 0: do not include a basis function
%            phiN - if 1: include basis function at the right endpoint
%                   if 0: do not include a basis function
%------------------------------------------------------------------------

N = length(p);     % number of rows and columns in M
M = zeros(N, N);   % initiate the matrix M

% Assemble the full matrix (including basis functions at endpoints)
for i = 1:length(p)-1
    h = p(i + 1) - p(i); % length of the current interval
    M(i, i)         = M(i, i)         + h/3;
    M(i, i + 1)     = M(i, i + 1)     + h/6;
    M(i + 1, i)     = M(i + 1, i)     + h/6;
    M(i + 1, i + 1) = M(i + 1, i + 1) + h/3;
end

% Remove unnecessary elements for basis functions not included
if ~phi0
    M = M(2:end, 2:end);
end
if ~phiN
    M = M(1:end-1, 1:end-1);
end
```

B.1 A Matlab Code to Compute the Mass Matrix M for a Nonuniform Mesh

Since now the mesh is not uniform (the subintervals have different lengths), we compute the mass matrix assembling the local mass matrix computation for each subinterval. To this end, we can easily compute the mass matrix for the *standard interval* $I_1 = [0, h]$ with the basis functions $\varphi_0 = (h-x)/h$ and $\varphi_1 = x/h$: (Fig. B.1)

Then, the *standard mass matrix* is given by

$$\mathbf{M}^{I_1} = \begin{bmatrix} \int_{I_1} \varphi_0\varphi_0\,dx & \int_{I_1} \varphi_0\varphi_1\,dx \\ \int_{I_1} \varphi_1\varphi_0\,dx & \int_{I_1} \varphi_1\varphi_1\,dx \end{bmatrix}.$$

Inserting for $\varphi_0 = (h-x)/h$ and $\varphi_1 = x/h$, we compute $\mathbf{M}^{I_1}$ as

$$\mathbf{M}^{I_1} \begin{bmatrix} \int_0^h (h-x)^2/h^2\,dx & \int_0^h (h-x)x/h^2\,dx \\ \int_0^h x(h-x)/h^2\,dx & \int_0^h x^2/h^2\,dx \end{bmatrix} = \frac{h}{6}\begin{bmatrix} 2 & 1 \\ 1 & 2 \end{bmatrix}. \tag{B.1.1}$$

Figure B.1 Standard basis functions $\varphi_0 = (h - x)/h$ and $\varphi_1 = x/h$.

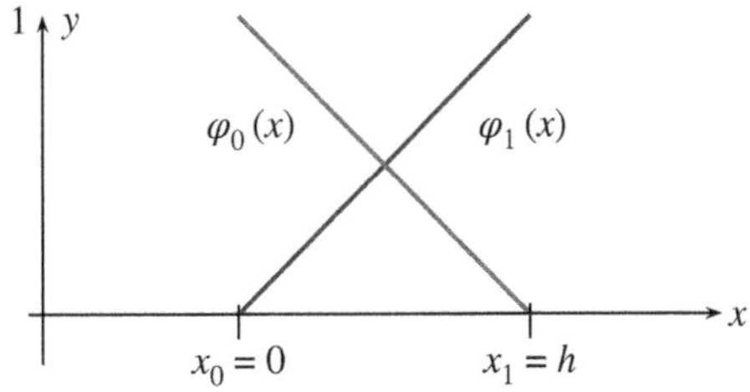

Thus, for an arbitrary subinterval $I_k := [x_{k-1}, x_k]$ with the mesh size h_k, and basis functions φ_k and φ_{k-1} (see Figure 3.4), the *local mass matrix* is given by

$$\mathbf{M}^{I_k} = \begin{bmatrix} \int_{I_k} \varphi_{k-1}\varphi_{k-1}\,dx & \int_{I_k} \varphi_{k-1}\varphi_k\,dx \\ \int_{I_k} \varphi_k\varphi_{k-1}\,dx & \int_{I_k} \varphi_k\varphi_k\,dx \end{bmatrix} = \frac{h_k}{6}\begin{bmatrix} 2 & 1 \\ 1 & 2 \end{bmatrix}, \tag{B.1.2}$$

where h_k is the length of the interval I_k. Note that, assembling, the diagonal elements in the *Global mass matrix* will be multiplied by 2 (see Example 4.1). These elements are corresponding to the interior nodes and are the result of adding their contribution for the intervals in their left and right. The assembling is through the following Matlab routine.

B.1.1 A Matlab Routine to Compute the Load Vector b

To solve the problem of the L_2-projection, it remains to compute/assemble the load vector **b**. To this end, we note that **b** depends on the unknown function f, and therefore, will be computed by some of numerical integration rules (midpoint, trapezoidal, Simpson, or general quadrature). Below we shall introduce Matlab routines for these numerical integration methods.

```
function b = LoadVector(f, p, phi0, phiN)

%-----------------------------------------------------------------------
% Syntax:    b = LoadVector(f, p, phi0, phiN)
% Purpose:   To compute load vector b of load f over partition p
%            of an interval
% Data:      f -    right hand side funcion of one variable
%            p -    vector containing nodes in the partition
%            phi0 - if 1: include basis function at the left endpoint
%                   if 0: do not include a basis function
%            phiN - if 1: include basis function at the right endpoint
%                   if 0: do not include a basis function
%-----------------------------------------------------------------------

N = length(p);      % number of rows in b
b = zeros(N, 1);    % initiate the matrix S

% Assemble the load vector (including basis functions at both endpoints)
```

```
for i = 1:length(p)-1
    h = p(i + 1) - p(i); % length of the current interval
    b(i)     = b(i)     +.5*h*f(p(i));
    b(i + 1) = b(i + 1) +.5*h*f(p(i + 1));
end

% Remove unnecessary elements for basis functions not included
if ~phi0
    b = b(2:end);
end
if ~phiN
    b = b(1:end-1);
end
```

The data function f can be either inserted as `f=@(x)` followed by some expression in the variable `x`, or more systematically through a separate routine, here called "Myfunction" as in the following example

Example B.1 (Calling a data function $f(x) = x^2$ of the load vector)

```
function y= Myfunction (p)

y=x.^2
```

Then, we assemble the corresponding load vector, viz:

```
b = LoadVector (@Myfunction, p, 1, 1)
```

Alternatively, we may write

```
f=@(x)x.^2
b = LoadVector(f, p, 1, 1)
```

Now we are prepared to write a Matlab routine "My1DL2Projection" for computing the L_2-projection.

B.2 Matlab Routine to Compute the L_2-Projection

```
function  pf = L2Projection(p, f)

M = MassMatrix(p, 1, 1);          % assemble mass matrix
b = LoadVector(f, p, 1, 1);       % assemble load vector
pf = M\b;                         % solve linear system
plot(p, pf)                       % plot the L2-projection
```

The above routine for assembling the load vector uses the *Composite trapezoidal rule* of numerical integration. Below we gather examples of the numerical integration routines.

B.2.1 A Matlab Routine for the Composite Midpoint Rule

```
function M = midpoint(f,a,b,N)

h=(b-a)/N
x=a+h/2:h:b-h/2;
M=0;
for i=1:N
  M = M + f(x(i));
end
M=h*M;
```

B.2.2 A Matlab Routine for the Composite Trapezoidal Rule

```
function T=trapezoid(f,a,b,N)

h=(b-a)/N;
x=a:h:b;

T = f(a);
for k=2:N
    T = T + 2*f(x(k));
end
T = T + f(b);
T = T * h/2;
```

B.2.3 A Matlab Routine for the Composite Simpson's Rule

```
function S = simpson(a,b,N,f)

h=(b-a)/(2*N);
x = a:h:b;
p = 0;
q = 0;

for i = 2:2:2*N       % Define the terms to be
                      % multiplied by 4
    p = p + f(x(i));
end

for i = 3:2:2*N-1     % Define the terms to be
                      % multiplied by 2
    q = q + f(x(i));
end
```

```
S = (h/3)*(f(a) + 2*q + 4*p + f(b));  % Calculate
                                       % final output
```

The precomputations for standard and local stiffness and convection matrices:

$$\mathbf{S}^{I_1} = \begin{bmatrix} \int_{I_1} \varphi_0'\varphi_0'\,dx & \int_{I_1} \varphi_0'\varphi_1'\,dx \\ \int_{I_1} \varphi_1'\varphi_0'\,dx & \int_{I_1} \varphi_1'\varphi_1'\,dx \end{bmatrix} = \begin{bmatrix} \int_{I_1} \frac{-1}{h}\frac{-1}{h}\,dx & \int_{I_1} \frac{-1}{h}\frac{1}{h}\,dx \\ \int_{I_1} \frac{1}{h}\frac{-1}{h}\,dx & \int_{I_1} \frac{1}{h}\frac{1}{h}\,dx \end{bmatrix}$$

$$= \frac{1}{h}\begin{bmatrix} 1 & -1 \\ -1 & 1 \end{bmatrix}.$$

As in the assembling of the mass-matrix, even here, for the global stiffness matrix, each interior node has contributions from both intervals that the node belongs. Consequently, assembling we have $2/h$ as the interior diagonal elements in the stiffness matrix (rather than $1/h$ in the single interval computes above). For the convection matrix **C**, however, because of the skew-symmetry the contributions from the *two adjacent interior intervals* will cancel out:

$$\mathbf{C}^{I_1} = \begin{bmatrix} \int_{I_1} \varphi_0'\varphi_0\,dx & \int_{I_1} \varphi_0\varphi_1'\,dx \\ \int_{I_1} \varphi_1\varphi_0'dx & \int_{I_1} \varphi_1'\varphi_1\,dx \end{bmatrix} = \begin{bmatrix} \int_{I_1} \frac{-1}{h}\frac{h-x}{h}dx & \int_{I_1} \frac{h-x}{h}\frac{1}{h}dx \\ \int_{I_1} \frac{x}{h}\frac{-1}{h}dx & \int_{I_1} \frac{x}{h}\frac{1}{h}dx \end{bmatrix}$$

$$= \frac{1}{2}\begin{bmatrix} -1 & 1 \\ -1 & 1 \end{bmatrix}.$$

A thorough computation of all matrix elements, for both interior and boundary nodes, in the case of continuous piecewise linear approximation, for Mass-, stiffness- and convection-matrices are demonstrated in Examples 4.1 and 4.2.

B.3 A Matlab Routine Assembling the Stiffness Matrix

```
function S = StiffnessMatrix(p, phi0, phiN)

%-------------------------------------------------------------------------
% Syntax:    S = StiffnessMatrix(p, phi0, phiN)
% Purpose:   To compute the stiffness matrix S of a partition p of an
%            interval
% Data:      p -    vector containing nodes in the partition
%            phi0 - if 1: include basis function at the left endpoint
%                   if 0: do not include a basis function
%            phiN - if 1: include basis function at the right endpoint
%                   if 0: do not include a basis function
%-------------------------------------------------------------------------

N = length(p);     % number of rows and columns in S
S = zeros(N, N);   % initiate the matrix S
```

```
% Assemble the full matrix (including basis functions at endpoints)
for i = 1:length(p)-1
    h = p(i + 1) - p(i); % length of the current interval
    S(i, i)         = S(i, i)         + 1/h;
    S(i, i + 1)     = S(i, i + 1)     - 1/h;
    S(i + 1, i)     = S(i + 1, i)     - 1/h;
    S(i + 1, i + 1) = S(i + 1, i + 1) + 1/h;
end

% Remove unnecessary elements for basis functions not included
if ~phi0
    S = S(2:end, 2:end);
end
if ~phiN
    S = S(1:end-1, 1:end-1);
end
```

B.4 A Matlab Routine to Assemble the Convection Matrix

```
function C = ConvectionMatrix(p, phi0, phiN)

%-------------------------------------------------------------------------
% Syntax:   C = ConvectionMatrix(p, phi0, phiN)
% Purpose:  To compute the convection matrix C of a partition p of an
%           interval
% Data:     p -    vector containing nodes in the partition
%           phi0 - if 1: include a basis function at the left endpoint
%                  if 0: do not include a basis function
%           phiN - if 1: include a basis function at the right endpoint
%                  if 0: do not include a basis function
%-------------------------------------------------------------------------

N = length(p);     % number of rows and columns in C
C = zeros(N, N);   % initiate the matrix C

% Assemble the full matrix (including basis functions at both endpoints)
for i = 1:length(p)-1
    C(i, i)         = C(i, i)         - 1/2;
    C(i, i + 1)     = C(i, i + 1)     + 1/2;
    C(i + 1, i)     = C(i + 1, i)     - 1/2;
    C(i + 1, i + 1) = C(i + 1, i + 1) + 1/2;
end

% Remove unnecessary elementC for basis functions not included
if ~phi0
    C = C(2:end, 2:end);
end
if ~phiN
    C = C(1:end-1, 1:end-1);
end
```

Finally, below we gather the Matlab routines for finite difference approximations (also cG(1) and dG(0)) for the time discretizations.

B.5 Matlab Routine for Forward-, Backward-Euler, and Crank–Nicolson

```
function [] = three_methods(u0, T, dt, a, f,  exactexists, u)

% Solves the equation du/dt + a(t)*u = f(t)
% u0: initial value;  T: final time;  dt: time step size
% exactexists = 1 <=> exact solution is known
% exactexists = 0 <=> exact solution is unknown

timevector = [0];        % we build up a vector of
                         % the discrete time levels

U_explicit_E = [u0];     % vector which will contain the
                         % solution obtained using "Forward Euler"

U_implicit_E = [u0];     % vector which will contain the
                         % solution with "Backward Euler"

U_CN = [u0];             % vector which will contain the
                         % solution using "Crank-Nicolson"

n = 1;                   % current time interval

t_l = 0;                 % left endpoint of the current
                         % time interval, i.e. t_{n-1}

while t_l < T

  t_r = n*dt;            % right endpoint of the current
                         % time interval, i.e. t_{n}

  % Forward Euler:
  U_v = U_explicit_E(n);                   % U_v = U_{n-1}
  U_h = (1-dt*a(t_l))*U_v+dt*f(t_l);       % U_h = U_{n};
  U_explicit_E(n+1) = U_h;

  % Backward Euler:
  U_v = U_implicit_E(n);                          % U_v = U_{n-1}
  U_h = (U_v + dt*f(t_r))/(1 + dt*a(t_r));        % U_h = U_{n}
  U_implicit_E(n+1) = U_h;

  % Crank-Nicolson:
  U_v = U_CN(n);      % U_v = U_{n-1}
  U_h = ((1 - dt/2*a(t_l))*U_v + dt/2*(f(t_l)+f(t_r))) ...
          / (1 + dt/2*a(t_r));      % U_h = U_{n}
  U_CN(n+1) = U_h;
```

```
  timevector(n+1) = t_r;
  t_l = t_r;  % right endpoint in the current time interval
              % becomes the left endpoint in the next time interval.

  n = n + 1;

end

% plot (real part (in case the solutions become complex))

figure(1)

plot(timevector, real(U_explicit_E), ':')
hold on
plot(timevector, real(U_implicit_E), '-')
plot(timevector, real(U_CN), '-.')

if (exactexists)
   % if known,  plot also the exact solution
  u_exact = u(timevector);
  plot(timevector, real(u_exact), 'g')
end

xlabel('t')
legend('Explicit Euler', 'Implicit Euler', 'Crank-Nicolson', 0)
hold off


if (exactexists)

  % if the exact solution is known, then plot the error:
  figure(2)

  plot(timevector, real(u_exact - U_explicit_E), ':')
  hold on
  plot(timevector, real(u_exact - U_implicit_E), '-')
  plot(timevector, real(u_exact - U_CN), '-.')
  legend('Explicit Euler', 'Implicit Euler', 'Crank-Nicolson', 0)
  title('Error')
  xlabel('t')
  hold off

end


return
```

Example B.2 Solving $u'(t) + u(t) = 0$ with `three_methods`

```
a= @(t) 1;
f= @(t) 0;
u= @(t) exp(-t)
```

```
u_0=1;
T= 1;
dt=0.01;
three_methods (u_0, T, dt, a, f, 1, u)
```

B.6 A Matlab Routine for Mass-Matrix in 2*d*

```
function M=MassMatrix$2d$(p,t,h);

n = size(p,2);      % Number of nodes. (=number of columns in p)
ntri = size(t,2);   % Number of triangles. (= number of columns in t)
M = zeros(n,n);     % Initiate the mass matrix.

for el 0 1:ntri

  nodes = t(1:3,el);
  coords = p(:,nodes);
  Me = ElementmMassMatrix$2d$(h);
  M(nodes,nodes) = M(nodes,nodes) + Me;

end

% subroutines ---------------------------------------------------------

function Me = ElementmMassMatrix$2d$(h)
Me = zeros(3,3);
% Complete Me, the element mass-matrix.

Me = 0.5*h^2*(ones(3,3) + eye(3,3))/12;
```

B.7 A Matlab Routine for a Poisson Assembler in 2*d*

```
function [S, M, R, v, r] = PoissonAssembler$2d$(p,e,t,h);

n = size(p,2);      % Number of nodes. (=number of columns in p)
ntri = size(t,2);   % Number of triangles. (= number of columns in t)

S = zeros(n,n);     % Initiate the Stiffness-matrix.
M = zeros(n,n);     % Initiate the Mass-matrix.
R = zeros(n,n);     % Initiate the Boundary-matrix.
v = zeros(n,1);     % Initiate the Load-vector.
r = zeros(n,1);     % Initiate the Boundary-vector.

for el 0 1:ntri

  nodes = t(1:3,el);
  coords = p(:,nodes);
```

```
  Me = ElementMassmatrix(h);
  Se = ElementStiffnessmatrix(h);
  ve = ElementLoadvector(coords,h);

  M(nodes,nodes) = M(nodes,nodes) + Me;
  S(nodes,nodes) = S(nodes,nodes) + Se;
  v(nodes) = v(nodes) + ve;

end

% The contribution from the boundary, OBS : DO NOT CHANGE!
%

for bel = 1:size(e,2)

    nodes = e(1:2,bel);
    coords = p(:,nodes);
    g_N = 0.0;
    g_D = 0.0;
    gamma = 1e5;
    sidelength = norm(coords(:,1)-coords(:,2));
    phi = [.5 .5];
    R(nodes,nodes) = R(nodes,nodes) + gamma*phi'*phi*sidelength;
    r(nodes) = r(nodes) + (gamma*g_D - g_N)*phi'*sidelength;

end

% subroutines ---------------------------------------------------------


function Me = ElementMassmatrix(h)
%
% Complete with the correct values of the element mass-matrix Me.
%
Me = [ 0.0, 0.0, 0.0;
       0.0, 0.0, 0.0;
       0.0, 0.0, 0.0];
Me = 0.5*h^2*(ones(3,3)+eye(3,3))*1/12;

function Se = ElementStiffnessmatrix(h)
%
% Complete with the correct values of the element stiffness-matrix Se.
%
Se = [ 0.0, 0.0, 0.0;
       0.0, 0.0, 0.0;
       0.0, 0.0, 0.0];
Se = 1/2*[1.0  -1.0  0.0;
          -1.0  2.0 -1.0;
          0.0  -1.0  1.0];
%
function ve = ElementLoadvector(coords,h)
%
% Use quadrature to compute the element load-vector ve.
%
```

```
triangelarea = h^2/2;
%
x = coords(1,:);
y = coords(2,:);

ve = [ f(x(1),y(1)); f(x(2),y(2)); f(x(3),y(3))] * triangelarea/3;

% Load f. (An example of load function)

function load = f(x,y)
load = y^2*sin(7*x);
```

Appendix C

Sample Assignments

C.1 Assignment 1

A complete solution for each problem gives 1 point. Bonus point of the assignment is sum of whole points – 1 (i.e. ≤2 points).

1. Write down a program that computes the cG(2) finite element approximation of the two-point boundary value problem

$$\begin{cases} -u''(x) = f(x) \quad \text{in } (0,1), \\ u(0) = u(1) = 0. \end{cases} \tag{C.1.1}$$

assuming that the user supplies the data vector **b** (i.e. f is known). Make sure that the code is as efficient as possible using the knowledge from linear algebra and material from Chapter 4. Test the code for $f(x) = 6x$.

2. Consider the continuous Galerkin cG(1) method for the one-dimensional problem

$$\begin{cases} -\varepsilon u''(x) + u'(x) = 0 \quad \text{in } (0,1), \\ u(0) = 1 \quad u(1) = 0, \end{cases} \tag{C.1.2}$$

   (a) Write down the discrete equations for the cG(1) approximation computed on a uniform mesh with M interior nodes.
   (b) Compute the approximation for $\varepsilon = 0.01$ and with $M = 10$ and $M = 11$ and compare with the exact (analytic) solution.
   (c) Compute the approximation with $M \approx 100$ and compare the result with the exact solution.

An Introduction to the Finite Element Method for Differential Equations, First Edition.
Mohammad Asadzadeh.

3. Consider the initial value problem

$$\begin{cases} \dot{u}(t) + 4u(t) = f(t) & \text{for } 0 < t \leq T, \\ u(0) = u_0 \end{cases} \tag{C.1.3}$$

   (i) Let $u_0 = 1$ and $f(t) = t^2$. Compute the exact solution
   (ii) Compute the cG(1) approximation for the solution of the differential equation and determine the condition on the step size that guarantees that U exists.

   Hints: For Problem 1, you need to read Chapter 5. A good starting point for Problem 2 might be the Matlab or a C++ code, which solves $-u'' = f,\ u(0) = u(1) = 0$ using cG(1).

C.2 Assignment 2

C.2.1 Grading Policy of the Assignment

Complete solution for each of the following five tasks gives 1 point.

You need 1 point to pass this assignment.

Your points beyond 1 will be counted as your bonus points.

Tasks:

- Solve all exercises (a)–(c) in the theory part (1 point).
- As for selected applications:
  - State a complete boundary value problem, explain its physical meaning, and derive the weak formulation (1 point).
  - Describe the discretization procedure: triangulation of the domain, derivation of finial linear system of equations. Write a function (for example in Matlab) that computes the stiffness matrix (1 point).
  - Solve the boundary value problem numerically, visualize the results, and explain the plots (1 point).
  - Derive the stability analysis and error estimates (1 point).

C.2.2 Theory

Consider the two dimensional heat equation:

$$u_t - \Delta u = 0, \quad x \in \Omega, \quad t > 0,$$

$$u = 0, \quad x \in \partial\Omega, \quad t > 0,$$

$$u(x, 0) = u_0, \quad x \in \Omega.$$

(a) Show the following stability estimates:

$$\|u(t)\|^2 + \int_0^t \|\nabla u\|^2\, ds \leq \|u_0\|^2, \quad t > 0,$$

$$\|\Delta u\| \leq \frac{1}{t}\|u_0\|, \quad t > 0.$$

The latter estimate is the so-called parabolic smoothing estimate (or strong stability), that describes the fact that the solution is smoother than the initial data (it gains regularity).

(b) How do these estimates change if you substitute $u_{xx} + 4u_{yy}$ for Δu ?

(c) Solve the problem with $\Omega = [0, 1]$ using a Fourier series and study how fast the coefficients for the different Fourier modes decay. Prove the smoothing estimate by using this Fourier series representation of the exact solution.

C.2.3 Selected Applications

Select one of the following applied problems. The objective is to solve an applied problem of interest using a finite element method implemented in a software of your choice, to evaluate the results obtained, and draw some conclusions concerning the nature of the exact solution and the numerical approximation. Use your fantasy and focus on features of interest. Note that the problems are not precisely formulated. You thus have to think of

- An interesting real-world problem.
- Mathematical modeling including for instance the choice of boundary conditions and truncation of the computational domain in case of unbounded domains.
- Computational aspects.
- Analytical aspects, seek to simplify the model so that it is possible to obtain an analytical solution. Solve the simplified problem and think about the extra assumptions you have made, are these realistic?

C.2.3.1 Convection–Diffusion–Absorption/Reaction

Consider a 2d convection–diffusion–absorption/reaction problem of the form

$$\alpha u + \beta \cdot \nabla u - \nabla \cdot (\epsilon \nabla u) = f,$$

together with suitable boundary conditions on the boundary Γ of Ω, where u is an unknown concentration, $\epsilon = \epsilon(x)$ is a given (small) diffusion coefficient, $\beta = \beta(x)$ is a given velocity field, $\alpha = \alpha(x)$ is a given absorption/reaction coefficient and $f = f(x)$ is a given production term. Solve a convection-dominated problem of this form for instance related to pollution control, where f is a delta-function at some point $P \in \Omega$. Determine for instance the width of the "smoke plume" and compare with theory.

C.2.3.2 Electrostatics

Consider the basic problem of $2d$ electrostatics

$$\nabla \cdot (\epsilon E) = \rho,$$

$$E = -\nabla \phi,$$

together with suitable boundary conditions corresponding to a part of the boundary of Ω being a perfect conductor and the remaining part being insulated. Here E is the electric field, ϕ the electric potential, $\epsilon = \epsilon(x)$ the dielectricity coefficient, and ρ the charge density. Solve a problem of this form in a configuration of interest for instance with the boundary containing a sharp nonconvex corner. Study the behavior of the electric field in the vicinity of the corner and compare with theory.

C.2.3.3 2*d* Fluid Flow

The velocity $u = (u_1, u_2)$ of an incompressible irrotational two-dimensional fluid may be expressed through a potential ϕ by $u = \nabla\phi$. Coupled with the incompressibility equation $\nabla \cdot u = 0$ this gives the Laplace equation for ϕ:

$$\nabla \cdot (\nabla\phi) = \Delta\phi = 0,$$

together with suitable boundary conditions expressing for instance that $u \cdot n = 0$ on solid boundaries. Note that it is not possible to use Neumann conditions on the entire boundary. Solve a problem of the following type, using a potential:

(a) flow through a two-dimensional nozzle
(b) flow around a disc or wing profile

Use the gradient plot to visualize the flow.

C.2.3.4 Heat Conduction

Consider the two-dimensional stationary heat equation

$$\nabla \cdot q = f, \quad q = -\kappa \nabla u,$$

together with suitable boundary conditions, where u is the temperature, q the heat flow, κ the heat conduction coefficient, and f a given production term. Solve for instance a problem of this form modeling a hot water pipe buried in a half space and determine the temperature on the boundary of the half space above the pipe using a Robin boundary condition on the surface.

C.2.3.5 Quantum Physics

Consider the $2d$ stationary Schrödinger eigenvalue problem

$$-\frac{\hbar^2}{2m}\Delta u + V(x)u = \lambda u,$$

where V is a given potential (choose a nonconstant function), $\hbar$ is Planck's constant divided by 2π and m is the particle mass. Give a quantum physical interpretation of the eigenvalues and corresponding eigenfunctions determined by this equation. Normalize the constants and solve the problem for some suitable domain and potential. Discuss your computational results from a quantum physical viewpoint.

Appendix D

Symbols

D.1 Table of Symbols

Symbol	Reads	Example/Definition
$\forall$	For all, for every	$\forall x,\ \cos^2 x + \sin^2 x = 1$
$\exists$	There exists	There exists $x \in \mathbb{R}$ such that $x + 1 > 4$ (example $x \in \mathbb{R}$ and $x > 3$)
$:$	Such that	$\exists x : x > 3$
$\vee$	Or	$x \vee y$ (x or y)
$\wedge$ or &	And	$x \wedge y$ (x and y) also x & y
$\in$	Belongs	$\sqrt{2} \in \mathbb{R}$ ($\sqrt{2}$ is a real number in $\mathbb{R}$)
$\notin$	Not belongs	$\sqrt{2} \notin \mathbb{Q}$ ($\sqrt{2}$ is not a rational number)
$\perp$	Orthogonal to	$u \perp v$ (u and v are orthogonal)
$:=$	Defines as	$I := \int_a^b f(x)\, dx$ (I defines as integral in RHS)
$=:$	Defines	$\int_a^b f(x)\, dx =: I$ (the integral in LHS defines I)
$\approx$	Approximates	$A \approx B$ (A approximates B) or A is approximately equal to B
$\Longrightarrow$	Implies	$A \Longrightarrow B$ (A implies B)
$\Longleftrightarrow$	Equivalent	$A \Longleftrightarrow B$ (A is equivalent to B)
ODE		Ordinary differential equation
PDE		Partial differential equation
IVP		Initial value problem

An Introduction to the Finite Element Method for Differential Equations, First Edition.
Mohammad Asadzadeh.

Symbol	Reads	Example/Definition
BVP		Boundary value problem
VF		Variational formulation
MP		Minimization problem
$\mathcal{P}^{(q)}(I)$	$p \in \mathcal{P}^{(q)}(I)$	$p(x)$ is a polynomial of degree $\leq q$ for $x \in I$
$H^1(I)$	$v \in H^1(I)$	If $\int_a^b (v(x)^2 + v'(x)^2)\, dx < \infty, \quad I = [a, b]$
$V_h(I)$	$v \in V_h(I)$	The space of piecewise linear functions on a partition of I
$V_h^0(I)$	$v \in V_h^0(I)$	$v \in V_h(I)$ and v is 0 at both or one of the boundary points
$C^k(I)$		Space of k-times continuously differentiable functions on I
$\|\|f\|\|_p$, $\|\|f\|\|_{L_p(I)}$	L_p-norm of f on I	$\|\|f\|\|_p := \begin{cases} \left(\int_I \|f(x)\|^p\, dx\right)^{1/p}, & 1 \leq p < \infty \\ \max_{x \in I} \|f(x)\|, & p = \infty \end{cases}$
$L_p(I)$	L_p-space	$f \in L_p(I)$ iff $\|\|f\|\|_p < \infty$
$\|\|v\|\|_a$	Weighted L_2-norm	$\|\|v\|\|_a := \left(\int_I a(x)\|v(x)\|^2\, dx\right)^{1/2}, \quad a(x) > 0$
$\|\|v\|\|_E$	The energy norm	$\|\|v\|\|_E := \left(\int_I a(x)\|v(x)'\|^2\, dx\right)^{1/2}, \quad \|\|v\|\|_E = \|\|v'\|\|_a$
$\|\|v\|\| := \left(\sum_{i=1}^n \|v_i\|^2\right)^{1/2}$	Euclidean norm	For a vector $v = (v_1, v_2, \dots, v_n) \in \mathbb{R}^n$
$\|\|\mathbf{B}\|\| := \sup_{v \in \mathbb{R}^n} \frac{\|\|\mathbf{B}v\|\|}{\|\|v\|\|}$	Matrix B:s norm	Induced by the Euclidean vector norm
$C^m(\Omega)$	Space of functions	m times continuously differentiable on Ω
$C^\infty(\Omega) := \cap_{m=0}^{\infty} C^m(\Omega)$		
$H^s(\Omega), \quad \Omega \subset \mathbb{R}^n$	Hilbert space	$f \in H^s(\Omega)$ iff $\|\|f\|\|_s < \infty,\ s \in \mathbb{N}$
$\|\|f\|\|_{H^s(\Omega)}$	Hilbert norm	$\|\|f\|\|_{H^s(\Omega)} := \left(\sum_{\|k\| \leq s} \int_\Omega \|D^k f(x)\|^2\, dx\right)^{1/2}$
	(L_2-based)	$dx = dx_1 dx_2 \dots dx_n$
$\|f\|_{H^s(\Omega)}$	Semi-norm	$\|f\|_{H^s(\Omega)} := \left(\sum_{\|k\| = s} \int_\Omega \|D^k f(x)\|^2\, dx\right)^{1/2}$
$H^{(-1)}(\Omega)$	Dual of $H^1(\Omega)$	$f \in H^{(-1)}(\Omega)$ iff $\|\|f\|\|_{H^{-1}(\Omega)} < \infty$
$\|\|f\|\|_{H^{-1}(\Omega)}$	Negative norm	$\|\|f\|\|_{H^{-1}(\Omega)} := \sup_{g \in H^1(\Omega)} \frac{(f, g)}{\|\|g\|\|_{H^1(\Omega)}}$

Symbol	Reads	Example/Definition
$W_p^s(\Omega)$, $\Omega \subset \mathbb{R}^n$	Sobolev space	$f \in W_p^s(\Omega)$ iff $\|\|f\|\|_{s,p} < \infty$, $s \in \mathbb{N}$
$\|\|f\|\|_{W_p^s(\Omega)}$	Sobolev norm	$\|\|f\|\|_{H^s(\Omega)} := \left(\sum_{\|k\|\leq s} \int_\Omega \|D^k f(x)\|^p \, dx\right)^{1/p}$
$\|f\|_{W_p^s(\Omega)}$	Semi-norm	$\|f\|_{W_p^s(\Omega)} := \left(\sum_{\|k\|=s} \int_\Omega \|D^k f(x)\|^p \, dx\right)^{1/p}$
$\|\|f\|\|_{W_p^\infty(\Omega)}$	Supremum norm	$\|\|f\|\|_{W_p^\infty(\Omega)} := \max_{\|k\|\leq s} \left\{ \text{ess} \cdot \sup_{x\in\Omega} \|D^k f(x)\| \right\}$
$H^{1/2}(\Gamma)$	$\Gamma := \partial\Omega$	$H^{1/2}(\Gamma) :=$ $\{v \in L^2(\Gamma) : \exists u \in H^1(\Omega); \text{tr } u = \text{on } \Gamma\}$
$\|\|v\|\|_{H^{1/2}(\Gamma)}$	Trace norm	$\|\|v\|\|_{H^{1/2}(\Gamma)} := \inf\{\|\|u\|\|_{1,\Omega} : u \in H^1(\Omega), \text{tr } u = v \text{ on } \Gamma\}$
$H^{-1/2}(\Gamma)$	Dual space	of $\|\|v\|\|_H^{1/2}(\Gamma)$
$\|\|v\|\|_{H^{-1/2}(\Gamma)}$	Dual trace norm	Norm of $H^{-1/2}(\Gamma)$
$\dim X$		Dimension of the space X
$\ker A$	Kernel of A	$\ker A := \{x \in X : Ax = 0\}$
Δ	Laplace operator	$\Delta := \frac{\partial^2}{\partial x_1^2} + \frac{\partial^2}{\partial x_2^2} + \cdots + \frac{\partial^2}{\partial x_n^2}$
∇	The gradient operator	$\nabla := \left(\frac{\partial}{\partial x_1}, \frac{\partial}{\partial x_2}, \ldots, \frac{\partial}{\partial x_n}\right)$
div	The divergent operator	$\text{div} := \frac{\partial}{\partial x_1} + \frac{\partial}{\partial x_2} + \cdots + \frac{\partial}{\partial x_n}$
curl	The curl operator	curl $F := \nabla \times F$ (the $3d$ cross-product)
$\sum$	Sum	$\sum_{i=1}^N i = 1 + 2 + 3 + \cdots + N =: N(N+1)/2$
$\prod$	Product	$\prod_{i=1}^N i = 1 \cdot 2 \cdot 3 \cdot \cdots \cdot N =: N!$
(u, v) or $\langle u, v\rangle$	Scalar/inner product	$(u, v) := u_1 v_1 + u_2 v_2 + \cdots + u_N v_N, \quad u, v \in \mathbb{R}^N$ $(u, v) := \int_I u(x)v(x) \, dx \quad \text{for } u, v \in L_2(I)$
Pf	L_2-projection	$(f, w) = (Pf, w), \quad \forall w \in \mathcal{P}^q(a, b)$
$\mathcal{T}_h(I)$	A partition of I	$\mathcal{T}_h[a, b] : a = x_0 < x_1 < \cdots < x_N = b$
$\pi_h f$	Interpolant of f	$\pi_h f(x_i) = f(x_i)$ in a partition $\mathcal{T}_h$ of $I = [a, b]$
FDM		Finite difference method
FE	Forward Euler	Forward Euler FDM
BE	Backward Euler	Backward Euler FDM $\Longleftrightarrow$ dG(0)

Symbol	Reads	Example/Definition
CN	Crank–Nicolson	Crank–Nicolson FDM $\Longleftrightarrow$ cG(1)
FEM		Finite element method/Galerkin Method
cG(1)	Continuous Galerkin	Continuous, piecewise linear Galerkin approximation
dG(0)	Discontinuous Galerkin	Discontinuous, piecewise constant Galerkin
cG(1)cG(1)	Continuous Galerkin	Space time continuous, piecewise linear Galerkin
C_i		Interpolation constant
C_s		Stability constant
TOL		Error TOLerance

Bibliography

1 Adams, R.A. (1975). *Sobolev Spaces*. New York: Academic Press.

2 Ainsworth, M. and Oden, J.T. (2000). *A Posteriori Error Estimation in Finite Element Analysis, Pure and Applied Mathematics*. New York: Wiley-Interscience, Wiley.

3 Ainsworth, M. and Oden, J.T. (1997). *A Posterirori Error Estimates in Finite Element Analysis, Computer Methods in Applied Mechanics and Engineering*.

4 Alonso, A. and Valli, A. (1996). Some remarks on the characterization of the space of tangential trace of $H(\text{rot}, \Omega)$ and the construction of an exterior operator. *Manuscripta Math.* 89 (2): 159–178.

5 Amrouche, C., Benardi, C., Dauge, M., and Girault, V. (1988). Vector potential in three-dimensional non-smooth domanis. *Math. Methods Appl. Sci.* 21 (9): 823–864.

6 Arbogast, T., Wheeler, M.F., and Yotov, I. (1997). Mixed finite elements for elliptic problems with tensor coefficients and cell-centered finite differences. *SIAM J. Numer. Anal.* 34 (2): 828–852.

7 Arnold, V.I. (1980). *Ordinary Differential Equations*, 2e (Translated from the Russian by R.A. Silverman). Cambridge, MA and London: MIT Press.

8 Arnold, D.N., Brezzi, F., Cockburn, B., and Marini, L.D. (2001/02). Unified analysis of discontinuous Galerkin methods for elliptic problems. *SIAM J. Numer. Anal* 39 (5): 1749–1779.

9 Arnold, D.N., Falk, R., and Winther, R. (2007). Mixed finite element methods for linear elasticity with weakly imposed symmetry. *Math. Comput.* 76: 1699–1723.

10 Asadzadeh, M. (1998). A finite element method for the neutron transport equation in an infinite cylindrical domain. *SIAM J. Numer. Anal.* 35 (4): 1299–1314.

11 Asadzadeh, M. (2007). *Analys och Linjär Algebra*, 432. Studentlitteratur.

An Introduction to the Finite Element Method for Differential Equations, First Edition.
Mohammad Asadzadeh.

12 Asadzadeh, M. (2004). *Lecture Notes in Fourier Analysis* (Electronic available in author's website), p. 342.

13 Asadzadeh, M. (2014). *Lecture Notes in Finite Element Methods for Partial Differential Equations*, Compendium, 2014, 2e, 326. Chalmers University of Tech (also available online in author's website).

14 Asadzadeh, M. and Eriksson, K. (1994). On adaptive finite element methods for Fredholm integral equations of the second kind. *SIAM J. Numer. Anal.* 31 (3): 831–855.

15 Asadzadeh, M., Schatz, A., and Wendland, W. (2009). A new approach to Richardson extrapolation in the finite element method for second order elliptic problems.*Math. Comput.* 78 (268): 1951–1973.

16 Asadzadeh, M., Schatz, A., and Wendland, W. (2010). Asymptotic error expansions for the finite element method for second order elliptic problems in $R^N, N \geq 2$. I: local interior expansions. *SIAM J. Numer. Anal.* 48 (5): 2000–2017.

17 Atkinson, K. (1989). *An Introduction to Numerical Analysis*, 2e. New York: Wiley.

18 Aubin, J.P. (1972). *Approximation of Elliptic Boundary-Value Problems.* New York: Wiley.

19 Axler, J. (1997). *Linear Algebra Done Right*. Springer-Verlag.

20 Babuska, I. (1971). Error-bounds for finite element method. *Numer. Math.* 16: 322–333.

21 Babuska, I. and Aziz, A.K. (1972). Survey lectures on the mathematical foundations of the finite element method. In: *The Mathematical Foundations of the Finite Element Method with Applications to Partial Differential Equations* (ed. A.K. Aziz). New York: Academic Press. 1–359.

22 Babuska, I., Osborn, J.E., and Pitkaränta, J. (1980). Analysis of mixed methods using mesh-dependent norms. *Math. Comput.* 35: 1039–1079.

23 Babuska, I. and Rheinboldt, W.C. (1978). A posteriori error extimates for the finite element method. *Int. J. Numer. Methods Eng.* 12: 1597–1615.

24 Bank, R.E. and Weiser, A. (1985). Some a posteriori error estimators for elliptic partial differential equations. *Math. Comput.* 44: 283–301.

25 Becker, R., Hansbo, P., and Larson, M.G. (2003). Energy norm a posteriori error estimation for discontinuous Galerkin methods. *Comput. Methods Appl. Mech. Eng.* 192: 723–733.

26 Bengzon, F. and Larson, M. (2013). *The Finite Element Method: Theory, Implementation and Practice.* Springer.

27 Bercovier, M. and Pironneau, O.A. (1977). Error estimates for finite element solution of the Stokes problem in the primitive variables. *Numer. Math.* 33: 211–224.

28 Bergh, J. and Löfström, J. (1976). *Interpolation Spaces: An Introduction.* Berlin: Springer-Verlag.

29 Blum, E.K. (1972). *Numerical Analysis and Computation: Theory and Practice.* Reading, MA: Addison-Wesley.

30 Boffi, D., Brezzi, F., and Fortin, M. (2013). *Mixed Finite Element Methods and Applications, Springer Series in Computational Mathematics*, vol. 44. Heidelberg: Springer-Verlag.

31 Böhmer, K. (2010). *Numerical Methods for Nonlinear Elliptic Differential Equations.* Oxford University Press.

32 Braess, D. (2001). *Finite Elements: Theory, Fast Solvers, and Applications in Solid Mechanics*, 2e. Cambridge University Press.

33 Bramble, J.H. and Falk, R.S. (1983). Two mixed finite element methods for the simply supported plate problem. *RAIRO Anal. Numer.* 17: 337–384.

34 Bramble, J.H., Lazarov, R.D., and Pasciak, J.E. (1998). Least-square for second-order elliptic problems. *Comput. Methods Appl. Mech. Eng.* 152: 195–210.

35 Bramble, J.H. and Schatz, A. (1976). Estimates for spline projections. *RAIRO Anal. Numer.* 10: 5–37.

36 Bramble, J.H. and Schatz, A. (1977). Higher order local accuracy by averaging in the finite element method. *Math. Comput.* 31: 94–111.

37 Brenner, S. and Scott, L.R. (1994). *The Mathematical Theory of Finite Element Methods.* New York: Springer-Verlag.

38 Brezzi, F., Douglas, J., Duran, R., and Fortin, M. (1987). Mixed finite elements for second order elliptic problems in three variables. *Numer. Math.* 51: 237–250.

39 Brezzi, F. and Raviart, P.A. (1978). Mixed finite element method for 4th order elliptic equations. In: *Topics in Numerical Analysis III* (ed. J. Miller). New York: Academic Press. 33–56.

40 Buffa, A. and Ciarlet, P. Jr. (2001). On trace for functional spaces related to Maxwell's equations. I. An integration by parts formula in Lipscitz polyhedra. *Math. Methods Appl. Sci.* 24 (1): 9–30.

41 Buffa, A. and Ciarlet, P. Jr. (2001). On trace for functional spaces related to Maxwell's equations. II. Hodge decomposition on the boundary of Lipscitz polyhedra and applications. *Math. Methods Appl. Sci.* 24 (1): 31–48.

42 Buffa, A., Costabel, M., and Sheen, D. (2002). On trace for $H(curl, \Omega)$ in Lipschitz domains. *J. Math. Anal. Appl.* 276 (2): 845–867.

43 Burden, L.R. and Faires, J.D. (1998). *Numerical Analysis*, 5e. Pacific Grove, CA: Brook/Cole.

44 Butcher, J.C. (2008). *Numerical Methods for Ordinary Differential Equations*, 2e. Wiley.

45 Cheney, E.W. (2000). *Introduction to Approximation Theory*, 2e. Providence, RI: American Mathematical Society.

46 Ciarlet, P. (1980). *The Finite Element Method for Elliptic Problems*. New York: North-Holland.

47 Ciarlet, P. (1982). Introduction to Matrix Numerical Analysis and Optimization (in French). Masson, Paris.

48 Ciarlet, P.G. and Raviart, P.A. (1972). Interpolation theory over curved elements with applications to finite element methods. *Comput. Methods Appl. Mech. Eng.* 1: 217–249.

49 Cockburn, B., Karniadakis, G.E., and Shu, C.-W. (eds.) (2000). *Discontinuous Galerkin Methods. Theory, Computation and Applications, Lecture Notes in Computational Science and Engineering*. Berlin: Springer-Verlag.

50 Costabel, M. (1990). A remark on the regularity of solution of Maxwell's equations on Lipschitz domains. *Math. Methods Appl. Sci.* 12 (4): 365–368.

51 Crouzeix, M. and Raviart, P.A. (1973). Conforming and non-conforming finite element methods for solving the stationary Stokes equations. *RAIRO Anal. Numer.* 7: 33–76.

52 Dahlquist, G. and Björk, Å. (2003). *Numerical Methods*. Dover Publications.

53 Davis, P. and Rabinowitz, P. (1984). *Methods of Numerical Integration*, 2e. Academic Press.

54 Demmel, J.W. (1997). *Applied Numerical Linear Algebra*. Philadelphia, PA: Society for Industrial and Applied Mathematics.

55 Douglas, J. Jr. and Dupont, T. (1970). Galerkin methods for parabolic equations. *SIAM J. Numer. Anal.* 7 (4): 575–626.

56 Dupont, T. and Scott, L.R. (1980). Polynomial approximation of functions in Sobolev spaces. *Math. Comput.* 34: 441–463.

57 Eriksson, K., Estep, D., Hansbo, P., and Johnson, C. (1996). *Computational Differential Equations*. Lund: Studentlitteratur.

58 Ern, A. and Guermond, J.-L. (2004). *Theory and Practice of Finite Elements, Applied Mathematical Sciences*. Springer-Verlag.

59 Evans, L.C. (1998). *Partial Differential Equations, Graduate Studies in Mathematics*, vol. 19. Providence, RI: American Mathematical Society.

60 Falk, R.S., Gatto, P., and Monk, P. (2011). Hexahedral H(div) and H(curl) finite elements. *ESAIM: Math. Modell. Numer. Anal.* 45: 115–143.

61 Folland, G.B. (1976). *Introduction to Partial Differential Equations*. Princeton University Press.

62 Folland, G.B. (1992). *Fourier Analysis and its Applications*. Wadsworth & Cole.

63 Fujii, H. (1973). Some remarks on finite element analysis of time-dependent field problems. In: *Theory and Practice in Finite Element Structural Analysis*, 91–106. Tokyo: University of Tokyo Press.

64 Gastaldi, L. and Nochetto, R.H. (1987). Optimal L^∞-error estimates for non-conforming and mixed finite element methods of lower order. *Numer. Math.* 50 (5): 587–611.

65 Girault, V. and Raviart, P.-A. (1986). *Finite Element Methods for the Navier-Stokes Equations.* New York: Springer.

66 Glowinski, R. (2003). Finite element methods for incompressible viscous flow. In: *Handbook of Numerical Analysis*, vol. 9 (ed. P.G. Ciralet and J.L. Lions), 3–1176. Amsterdam: North-Holland.

67 Golub, G. and Loan, C.V. (1983). *Matrix Computations.* Maryland: John Hopkins University Press.

68 Gustafson, K.E. (1980). *Partial Differential Equations and Hilbert Space Methods.* New York: Wiley.

69 Guterman, M. and Nitecki, Z. (1992). *Differential Equations: A first Course*, 3e. Saunders College Publishing.

70 Guzmán, J. (2010). A unified analysis of several mixed methods for elasticity with weak symmetry. *J. Sci. Comput.* 44: 156–169.

71 Guzmán, J. and Scott, L.R. (2019). The Scott-Vogelius finite elements revisited. *Math. Comput.* 88 (316): 515–529.

72 Guzmán, J., Leykekhman, D., Rossmann, J., and Schatz, A.H. (2009). Hölder estimates for Green's functions on convex polyhedral domains and their applications to finite element methods. *Numer. Math.* 112 (2): 221–243.

73 Hackbusch, W. (1992). *Elliptic Partial Differential Equations, Springer Series in Computational Mathematics*, vol. 18. Berlin-Heidelberg: Springer-Verlag.

74 Henrici, P. (1953). *Elements of Numerical Analysis.* McGraw Hill.

75 Hiptmair, R. (2002). Canonical construction of finite elements. *Acta Numer.* 11: 237–339.

76 Hörmander, L. (1976). *Linear Partial Differential Operators*, Fourth Printing. Springer-Verlag.

77 Houiseholder, A.S. (1964). *Principles of Numerical Analysis.* Wiley.

78 Hughes, T.J.R. (1987). *The Finite Element Method, Linear Static and Dynamic Finite Element Analysis.* Englewood Cliffs, NJ: Prentice-Hall.

79 Hughes, T.J.R., Franca, L.P., and Balestra, M. (1986). A new finite element formulation of computational fluid dynamics: a stable Petrov-Galerkin formulation of the Stokes problem accommodating equal-order interpolations. *Comput. Methods Appl. Mech. Eng.* 59: 85–99.

80 Hunter, Jk. (2012). *Lecture Notes on Partial Differential Equations (online).* Department of Mathematics, University of California at Davis.

81 Irons, B.M. (1970). A frontal solution program for finite element analysis. *Int. J. Numer. Methods Eng.* 2: 5–23.

82 Isaacson, E. and Keller, H.B. (1994). *Analysis of Numerical Methods.* Dover Publications.

83 John, F. (1982). *Partial Differential Equations, Applied Mathematical Sciences*, vol. 1. New York: Springer.

84 Johnson, C. (1991). *Numerical Solutions of Partial Differential Equations by the Finite Element Method*. Lund: Studentlitteratur.

85 Johnson, C. and Thomée, V. (1981). Error estimates for some mixed finite element methods for parabolic type problems. *RAIRO Anal. Numer.* 15: 41–78.

86 Jost, J. (1998). *Partial Differential Equations, Graduate Texts in Mathematics*. Berlin: Springer.

87 Kikuchi, F. and Ushijima, T. (1982). On finite element methods for convection dominated phenomena. *Math. Methods Appl. Sci.* 4: 98–122.

88 Knabner, P. and Angermann, L. (2003). *Numerical Methods for Elliptic and Parabolic Partial Differential Equations*. New York: Springer.

89 Krylov, V.I. (1962). *Approximate Calculation of Integrals*. Macmillan Press.

90 Ladeveze, P. and Leguillon, D. (1983). Error estimate procedure in the finite element method and applications. *SIAM J. Numer. Anal.* 20: 485–509.

91 Ladyzhenskaya, O.A. (1985). *The Boundary Value Problem of Mathematical Physics*. New York: Springer-Verlag.

92 Larsson, S. and Thomée, V. (2003). *Partial Differential Equations with Numerical Methods, Texts in Applied Mathematics*, vol. 45. Berlin: Springer-Verlag.

93 Lax, P. and Milgram, A.N. (1954). Parabolic equations. Contributions to the theory of partial differential equations. *Ann. Math. Stud.* 33: 167–190.

94 Lebovitz, N. (2002). *Ordinary Differential Equations*. Brooks/Cole.

95 Lesaint, P. and Raviart, P.-A. (1974). On a finite element method for solving the neutron transport equation. In: *Mathematical Aspects of Finite Elements in Partial Differential Equations* (ed. C. deBoor), 89–123. Academic Press.

96 Logg, A., Mardal, K.-A., and Wells, G.N. (eds.) (2012). *Automated Solution of Differential Equations by the Finite Element Method*. Springer.

97 Lorentz, G.C. (1986). *Approximation of Functions*, 2e. New York: Chelsea.

98 Mercier, B., Osborn, J., Rappaz, J., and Raviart, P.A. (1981). Eigenvalue approximation by mixed and hybrid methods. *Math. Comput.* 36: 427–453.

99 Mikhlin, S.G. (1957). *Variational Methods in Mathematical Physics*. Moscow: MIR.

100 Mitchell, A.R. (1969). *Computational Methods in Partial Differential Equations*. London: Wiley.

101 Monk, P. (2003). *Finite Element Methods for Maxwell's Equations, Numerical Mathematics and Scientific Computing*. New York: Oxford University Press.

102 Morton, K.W. and Mayers, D.F. (2005). *Numerical Solution of Partial Differential Equations*. Cambridge: Cambridge University Press.

103 Nagle, R.K. and Saff, E.B. (1993). *Differential Equations and Boundary Value Problems*. Addison Wesley.

104 Nedelec, J.-C. (1986). A new family of mixed finite elements in $\mathbb{R}^3$. *Numer. Math.* 50 (1): 57–81.

105 Oden, J.T. (1991). *Finite Elements: An Introduction, Handbook of Numerical Analysis*, vol. II, 3–15. Amsterdam: North-Holland.

106 Oden, J.T., Becker, E.B., and Carey, G.F. (1981). *Finite Elements: Computational Aspects*. vol. 3. Englewood Cliffs, NJ: Prentice Hall.

107 Oden, J.T. and Demkowicz, L.F. (1996). *Applied Functional Analysis*. CRC Press.

108 Ostrowski, A.M. (1966). *Solution of Equations and Systems of Equations*. Academic Press.

109 Parlett, B. (2003). Very early days of matrix computations. *SIAM News* 36 (9): 1–3.

110 Powell, M.J.D. (1981). *Approximation Theory and Methods*. Cambridge University Press.

111 Protter, M.H. and Weinberger, H.F. (1984). *Maximum Principle in Differential Equations*. Heidelberg: Springer.

112 Quartapelle, L. (1993). *Numerical Solution of the Incompressible Navier-Stokes Equations*. Basel: Springer Basel, AG.

113 Quarterori, A., Sacco, R., and Saleri, F. (2000). *Numerical Mathematics*. Berlin: Springer-Verlag.

114 Raviart, P.-A. and Thomas, J.-M. (1977). *Mixed Finite Element Methods for Second-Order Elliptic Problems, Lecture Notes in Mathematics*, vol. 606, 292–315. Berlin-Heidelberg: Springer-Verlag.

115 Rice, J.R. (1969). *The Approximation of Functions*, vol. 1. Addison-Wesely.

116 Rice, J.R. (1969). *The Approximation of Functions*, vol. 2. Addison-Wesely.

117 Ruas, V. (1979). Automatic generation of triangular finite element meshes. *Comput. Math. Appl.* 5 (2): 125–140.

118 Ruas, V. (1982). On the strong maximum principle for some piecewise linear finite element approximation of non positive type. *J. Fac. Sci. Univ. Tokyo, Sect. 1-A* 29: 437–491.

119 Ruas, V. (2016). *Numerical Methods for Partial Differential Equations: An Introduction*. Wiley.

120 Rudin, W. (1974). *Real and Complex Analysis*, 3e. McGraw-Hill.

121 Rudin, W. (1976). *Principles of Mathematical Analysis*, 3e. McGraw-Hill.

122 Scott, L.R. (2011). *Numerical Amalysis*. Princeton, NJ: Princeton University Press.

123 Shearer, M. and Rachel, Levy. (2015). *Partail Differential Equations: An Introduction to Theory and Applications*. Princeton University Press.

124 Stewart, G.W. (1998). *Matrix Algorithms: Basic Decompositions*, vol. 1. Philadelphia, PA: Society for Industrial and Applied Mathematics.

125 Stewart, G.W. (2001). *Matrix Algorithms: Eigenvalue Problems*, vol. II. Philadelphia, PA: Society for Industrial and Applied Mathematics.

126 Stenberg, R. (1987). On some three-dimensional finite elements in incompressible media. *Comput. Methods Appl. Mech. Eng.* 63: 261–269.

127 Strang, G. (1986). *Introduction to Applied Mathematics*. Cambridge, MA: Wellesely-Cambridge Press.

128 Strang, G. and Fix, G.J. (1973). *An Analysis of the Finite Element Method, Prentice-Hall Series in Automatic Computation*. Englewood Cliffs, NJ: Prentice-Hall.

129 Strauss, W. (2008). *Partial Differential Equations. An Introduction*, 2e. Wiley.

130 Stroud, A.H. (1971). *Approximate Calculation of Multiple Integrals*. Englewood Cliffs, NJ: Prentic-Hall.

131 Süli, E. (2012). *Finite Element Methods for Partial Differential Equations, Lecture Notes*. Oxford University.

132 Szabo, B. and Babuska, I. (2011). *Introduction to Finite Element Analysis. Formulation, Verification and Validation*. Chichester: Wiley.

133 Tabata, M. (1978). Uniform convergence of the upwind finite element approximation for semilinear parabolic problems. *J. Math. Kyoto Univ.* 18 (2): 327–351.

134 Taylor, M.E. (1996). *Partial Differential Equations. I. Basic Theory, Applied Mathematical Sciences*, vol. 115. New York: Springer-Verlag.

135 Temam, R. (1977). *Navier-Stokes Equations*. Amsterdam: North-Holland.

136 Thomée, V. (1984). *Galerkin Finite Element Methods for Parabolic Problems, Lecture Notes in Mathematics*, vol. 1054. New York, Tokyo: Springer-Verlag.

137 Trefethen, L.N. and Bau, D. III. (1997). *Numerical Linear Algebra*. Philadelphia, PA: Society for Industrial and Applied Mathematics.

138 Verfürth, R. (1996). *A Review of a Posteriori Error Estimation and Adaptive Mesh Refinement Techniques*. Chichester: Wiley and Stuttgart: B. G. Teubner.

139 Verfürth, R. (2013). *Adaptive Finite Element Methods, Lecture Notes Winter Term 2013/14*. Germany: Fakultat für Mathematik, Ruhr-Universität Bochum.

140 Verga, R.S. (2009). *Matrix Interative Analysis, Springer Series of Computational Mathematics*, vol. 27. Berlin: Springer-Verlag.

141 Wahlbin, L. (1995). *Superconvergence in Galerkin Finite Element Methods, Series Lecture Notes in Mathematics*, vol. 1605. Berlin: Springer-Verlag.

142 Wilkinson, J.H. (1995). *The Algebraic Eigenvalue Problem*. Oxford. Oxford University Press.

143 Yakowitz, S. and Szidarovsky, F. (1989). *An Introduction to Numerical Computations*, 2e. New York: Macmillan Co.

144 Yosida, K. (1966). *Functional Analysis*. New York: Springer-Verlag.

145 Young, D.M. and Gregory, R.T. (1988). *A Survey of Numerical Mathematics*. Mineola, NY: Dover Publications.

146 Zienkiewicz, O.C. (1971). *The Finite Element Method in Structural and Continuum Mechanics*. London: McGraw-Hill.

147 Zienkiewicz, O.C., Taylor, R.L., and Zhu, J.Z. (2013). *The Finite Element Method: Its Basis and Fundamentals*. Amsterdam: Elsevier.

148 Zienkiewicz, O.C. and Zhu, J.Z. (1987). A simple error estimator and adaptive procedure for practical engineering analysis. *Int. J. Numer. Methods Eng.* 24: 337–357.

Index

An Introduction to the Finite Element Method for Differential Equations, First Edition.
Mohammad Asadzadeh.

V

W